INTERNATIONAL
ENERGY AGENCY

Cleaner Coal in China

OECD

INTERNATIONAL ENERGY AGENCY

The International Energy Agency (IEA) is an autonomous body which was established in November 1974 within the framework of the Organisation for Economic Co-operation and Development (OECD) to implement an international energy programme.

It carries out a comprehensive programme of energy co-operation among twenty-eight of the thirty OECD member countries. The basic aims of the IEA are:

■ To maintain and improve systems for coping with oil supply disruptions.

■ To promote rational energy policies in a global context through co-operative relations with non-member countries, industry and international organisations.

■ To operate a permanent information system on international oil markets.

■ To provide data on other aspects of international energy markets.

■ To improve the world's energy supply and demand structure by developing alternative energy sources and increasing the efficiency of energy use.

■ To promote international collaboration on energy technology.

■ To assist in the integration of environmental and energy policies, including relating to climate change.

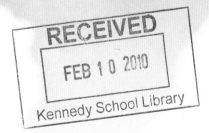

ORGANISATION FOR ECONOMIC CO-OPERATION AND DEVELOPMENT

The OECD is a unique forum where the governments of thirty democracies work together to address the economic, social and environmental challenges of globalisation. The OECD is also at the forefront of efforts to understand and to help governments respond to new developments and concerns, such as corporate governance, the information economy and the challenges of an ageing population. The Organisation provides a setting where governments can compare policy experiences, seek answers to common problems, identify good practice and work to co-ordinate domestic and international policies.

FOREWORD BY THE EXECUTIVE DIRECTOR OF THE IEA

In 1999, the IEA Coal Industry Advisory Board (CIAB) reported on coal in China.[1] Its recommendations focussed on coal sector reform and opportunities for co-operation. Today, much of that report is of merely historical interest, such is the rate of change in China over the last decade. Remarkable progress has been made in a sector that has fuelled China's rapid economic growth, bringing with it a better life for many of China's citizens. Today, coal production in China provides more energy to the world's economy than the whole of Middle Eastern oil production. This report is a timely reminder that the use of coal on such a scale cannot be ignored – it is in everyone's interest to ensure that the environmental concerns associated with coal can be managed, even in these times of economic uncertainty. It contains a wealth of information to guide those with an interest in engaging with China and helping to shape a cleaner future. Its recommendations are pragmatic. They offer opportunities for China to grasp, but only if developed countries show their commitment to clean energy by moving quickly to establish markets for technologies that are currently too expensive and not fully demonstrated. I am thinking in particular of carbon dioxide capture and storage – a critical technology for the world at large.

The IEA has been fortunate to have the support of the National Development and Reform Commission, ably assisted by the China Coal Information Institute, and the UK Foreign and Commonwealth Office in carrying out this work. This allowed us to bring together a team of consultants, some based in China, others in Japan and the UK, to complete the project. Experts from China worked on secondment at the IEA in Paris and made site visits to Germany and the UK. Reciprocal visits were made to many facilities in China, of all ages and sizes. Such exchanges at a working level need to grow, and professional relationships established that can serve the aims and objectives of governments. This initiative has also highlighted some practical difficulties. Reliable information on China remains a much sought-after and valuable commodity. Likewise, China seeks better information on how other countries have implemented effective energy and environmental policies. Language remains a barrier to sharing information – we need more bilingual experts and analysts to collect, assimilate and disseminate information of benefit to Chinese policy makers and their counterparts elsewhere.

In May 2008, the CIAB held a working meeting in Beijing for the first time. Participants made site visits to the Shendong coalfield in Inner Mongolia and heard the IEA Secretariat present a draft of this report to Chinese authorities. They were, without exception, encouraged by what they saw and heard. I wish to echo their sentiments and urge governments and corporations to accelerate their efforts in working with

---

1.   *Coal in the Energy Supply of China*, IEA Coal Industry Advisory Board, OECD/IEA, Paris, 1999.

China. Commercial activity, official government-to-government co-operation, research and development partnerships and personal relationships are all needed to make clean energy a reality. China itself has the opportunity to steal a lead in the development of cleaner coal technologies in response to a growing, global market. In that respect, my organisation is committed to promoting the competitive markets that we know are needed to win the environmental benefits of widespread deployment of cleaner technologies.

The report and key recommendations have been developed by the project team and were subject to peer review by experts inside and outside of China, as well as within the Secretariat. Whilst our member countries have given helpful feedback on the report, it does not necessarily reflect the views or position of IEA member countries, or of any official Chinese body. It is published under my authority as Executive Director as part of the IEA aim to engage more closely with all major energy-consuming countries.

Nobuo Tanaka
Executive Director

FOREWORD BY THE CHIEF ENGINEER OF THE NATIONAL ENERGY ADMINISTRATION OF THE NDRC

Energy is an essential element for human survival and development. Over the history of mankind, each and every significant step in the progress of civilisation has been accompanied by energy innovations and substitutions. The exploitation and utilisation of fossil energy has boosted enormously world economic development and society as a whole. However, large-scale exploitation and utilisation of fossil energy is also one of the major causes of ecological destruction and environmental pollution.

As is well known, coal is China's dominant source of energy. It accounts for about 70 percent of primary energy production and consumption. Furthermore, coal will remain the main energy source in China for a long period of time in the future. The proportion of coal in China's energy mix is far higher than the world average, and coal exploitation and utilisation has become one of the major causes of environmental pollution. Therefore, clean coal technology is a strategic choice for energy development in China.

The central government has adopted clean coal technology as a strategy for adjusting coal industry structure, increasing the commercial value of coal and coal-based products, improving the environment, and realising the sustainable development of the coal industry. In 1995, the State Council announced the "9th Five-Year Plan for Clean Coal Technology and its Development to 2010 in China", and a clean coal strategy was listed as one of four major strategies in the "10th Five-Year Plan for the Coal Industry". In 2005, the State Council issued its "Opinions on Promoting the Sound Development of the Coal Industry" and, in general, the coal industry in China has been improving, with a fast and steady development trend. In 2007, coal output was 2.523 billion tonnes, and coal production and use is following a trend of resource saving, environmental protection, value-added processing and clean utilisation. It is also pointed out in the "11th Five-Year Plan for Coal Industry Development in China" that the coal industry must change its growth vector, accelerate its structural adjustment, and take a safe, reliable, fully harmonised and sustainable path with a high resource utilisation rate and less environmental pollution. The clean development and utilisation of coal is an important part of realising this goal.

In recent years, climate change has become an international hot topic, closely associated with the development and utilisation of fossil energy. It has been noted in the IPCC *Fourth Assessment Report on Climate Change* that anthropogenic emissions of greenhouse gases are the major cause of global warming.[2] Of these gases, carbon dioxide emitted during the course of large-scale fossil fuel use is the major reason for climate change.

---

2. *Climate Change 2007: Synthesis Report*, Fourth Assessment Report of the Intergovernmental Panel on Climate Change, IPCC, Geneva, 2007.

Moreover, according to IEA forecasts, by 2030, the three major fossil fuels (oil, natural gas and coal) will still account for 84 percent of total energy demand, and emissions of carbon dioxide will increase by 57% compared to 2005.[3]

Under such circumstances of domestic and international development, it is necessary to bring new efforts to clean coal technology and its strategic development.

In 2006, Vice Chairman Chen Deming of the National Development and Reform Commission and the Executive Director of the International Energy Agency agreed to co-operate on the research for a *Clean Coal Strategy in China*. The research has been led by the National Energy Administration of the National Development and Reform Commission, and organised and co-ordinated by the China Coal Information Institute. After more than one year's hard work by the research team and experts at home and abroad, we finally see a successful research report. In the report, the development of the coal industry in China has been reviewed and experiences of developed countries in coal industry development have been analysed in terms of coal resources and markets, clean coal technologies and environmental protection, coal industry reconstructing, social welfare and mine safety, and international co-operation. On this basis, recommendations and suggestions have been put forward for the cleaner utilisation of coal in China.

The successful completion of the research reflects the great attention given by the International Energy Agency to the coal industry in China. It is worth noting that the national conditions and development situation of China's coal industry are different from those in the other main coal-producing countries, thus, some research results and suggestions are positive while some still need to be discussed. In any case, we should seek common points while reserving differences, continuously expand communication and promote a common understanding with the aim of guaranteeing sustainable supplies of global energy and promoting continuous improvement in the global environment and sustainable development of the global economy.

I herein express my sincere thanks to all the people participating in this research. Meanwhile, I wish the coal industry in China a cleaner and sustainable future.

Wu Yin
Chief Engineer
National Energy Administration

---

3.   *World Energy Outlook 2007: China and India Insights*, OECD/IEA, Paris, 2007.

ACKNOWLEDGEMENTS

This publication reflects the efforts of many people from a number of organisations. It would not have been possible without the official support of the National Development and Reform Commission (NDRC) in China. At a meeting on 19 January 2007 between Claude Mandil, then Executive Director of the International Energy Agency (IEA), and Chen Deming, then Vice Chairman of NDRC, the seeds were sown for a joint project on cleaner coal technologies. This book is the tangible output of a collaboration between the IEA and organisations in China that is certain to continue. Wu Yin, formerly Director General of the Coal Division, and now Chief Engineer, of the National Energy Administration, delegated Yan Tianke, Zhang Guo and Wei Pengyuan of the Coal Division to oversee this collaborative effort.

The project was designed and managed at the IEA by Dr. Jonathan Sinton, China Programme Manager and Brian Ricketts of the Energy Diversification Division with guidance and support from IEA directors. Ambassador William Ramsay, former Deputy Executive Director responsible for relations with non-member countries, was instrumental in driving the project through to conclusion. Yo Osumi, Head of Asia Pacific and Latin America Division and Ian Cronshaw, Head of Energy Diversification Division inspired and motivated the project team. Brian Ricketts carried overall editorial responsibility.

The IEA's lead collaborator was the China Coal Information Institute (CCII). Under the leadership of its President, Prof. Huang Shengchu, staff from CCII gave tremendous assistance on every aspect of the project. Liu Wenge and Sun Xin managed this collaboration, devoting more of their time and effort than anyone demanded. Han Jiaye, Lan Xiaomei, Wu Jinyan, Chen Weichao, Zhao Yingchun and Zhang Bingchuan contributed to the drafting of Chapters 2 through to 6 and Annex II. CCII was also responsible for translating the publication into Chinese. Dr. Yu Zhufeng, formerly Deputy Director at the China Coal Research Institute (CCRI), was responsible for much of Chapter 5 and Section 8.1, with contributions from Ren Shihua. Of special note, secondees from CCII and CCRI worked at the IEA in Paris during November and December 2007 to complete their tasks: Dr. Li Hongjun (CCII), Sun Jian (CCII) and Wu Lixin (CCRI). Dr. Yang Fuqiang, Vice President of the Energy Foundation, smoothed progress in many ways. He permitted Hou Yanli, a researcher at the Energy Foundation, to work on secondment at the IEA in Paris where she completed Chapter 4. Hu Yuhong, Deputy Director General of the China National Coal Association (CNCA) drafted parts of Chapter 3, notably the section on coal resources and reserves. Prof. Nobuhiro Horii of Kyushu University, Japan, and a recognised expert on China's coal sector, drafted Sections 3.6 and 3.8. He also made contributions to Chapters 4 and 6. Two consultants from the UK gathered information on the coal sectors in IEA member countries and on international collaboration with China: Simon Walker drafted Chapter 7 and Annex V while Dr. Andrew Minchener drafted Sections 8.2 to 8.5, Chapter 9 and Annex IV.

The IEA Coal Industry Advisory Board (CIAB) has been especially supportive during the project under the chairmanship of Steven Leer (Arch Coal), ably supported by Deck Slone and Brian Heath. The CIAB rescheduled one of its regular meetings so that members could attend a project review workshop in Beijing in May 2008. Special thanks are due to Dr. Carl Zipper (Virginia Polytechnic Institute and State University) and to Bill Koppe (Anglo Coal) who made substantive inputs.

The publication was subject to rigorous internal review. Feedback and input from Dr. Sankar Bhattacharya, Dr. Pieter Boot, Dr. Dolf Gielen, Didier Houssin, Tom Kerr, Andrea Nour, Ulrik Stridbæk, Nancy Turck and Dr. Yang Ming led to significant improvements to the content and the way it is presented. Reliable statistics form the bedrock of IEA publications. Michel Francoeur, Paul Tepes, Julian Smith, Jung Woo Lee and Roberta Quadrelli, led by Jean-Yves Garnier, compiled much of the data used in the analysis.

Work was guided by two IEA committees: the Standing Group on Long-Term Co-operation and Policy Analysis and the Standing Group on Global Energy Dialogue. Committee members and IEA Energy Advisors from member countries provided ideas and assistance that helped to improve the conclusions, recommendations and key messages. Those representing Australia, the European Commission, Germany, Poland, the UK and the US gave particular help with Chapter 9 and Annex V.

To inform the IEA Secretariat's thinking, two workshops were held in Beijing and various missions made to coal-related sites in China's provinces, the UK and Germany – many of which illustrate the report. In Shanxi, Xu Yiqing, Director of the Coal Division, Shanxi Development and Reform Commission welcomed the project team in July 2007, along with Wang Chongmei of the Shanxi Coal Bureau, Prof. Wang Yang of the Institute of Coal Chemistry, Chinese Academy of Sciences, Wang Hongyu of the Jincheng Coal Mining Group and other colleagues. Shen Jinming, Vice General Manager of Shanxi Coking Group Co. Ltd., accompanied the team to coal mines, wash plants and power stations around Taiyuan. Song Qiyue, Vice President of Taiyuan Coal Gasification Group Corp. Ltd. kindly interrupted his weekend to lead the team around his company's gasification plant. Li Cunzhu, General Manager of Datong Mine Group Co. Ltd. hosted the project team at industrial sites in northern Shanxi. Later, during September 2007, Huang Qin, Vice Director of the Henan Coal Industry Bureau facilitated site visits to Yima gasification plant and Zhengzhou Coal Group's Jianye coal mine. Zhang Weidong, Executive Director of Henan Coal Gas Group Co. Ltd. made every effort to answer the many questions asked at Yima. The project team enjoyed endless hospitality and thanks are due to the many staff, too numerous to mention, involved in making preparations for their foreign visitors.

In November 2007, Philip Lawrence, Chief Executive of the UK Coal Authority welcomed a delegation of Chinese experts with whom Ian Wilson and John Delaney shared their practical and administrative experiences. At Drax Power plc, Chris McGlen and David Loveday were able to explain how a large coal-fired plant is managed within what has become a complex regulatory environment. Few underground mines remain in the UK, but at Thoresby colliery, John Brough and Andy Leitch of UK Coal plc showed the Chinese delegation how modern standards were being met with old equipment. UK Coal was also proud to show its land restoration, managed by Derek Harrison, at Waverley opencast coal mine. The co-operation of Jon Lloyd, Chief Executive and assistance of Tim Marples and Martin Mee are gratefully acknowledged.

In Germany, Dr. Franz-Josef Wodopia, Executive Director of the German Hard Coal Association (GVSt) facilitated site visits which included CentrO Oberhausen, a former mine site converted into a large shopping centre, and the former Zollverein mine and coking plant in Essen. Dr. Detlef Riedel, Reinhard Rohde and Roland Lübke each assisted with the organisation while, separately, Bernd Bogalla made valuable contributions to Chapter 9.

Dr. Zhang Yuzhuo, Chairman of China Shenhua Coal Liquefaction Corp. Ltd., Vice President of Shenhua Group Corp. Ltd. and member of the CIAB, kindly offered to host an underground visit to Shangwan coal mine, a surface visit to Daliuta mine and a tour of the Shenhua CTL plant under construction near Erdos. Wang An, General Manager, Shenhua Shendong Coal Branch, Cui Minli, Vice President of China Shenhua Coal to Liquids and Chemical Co. Ltd. and their staff generously shared their knowledge at these sites in Inner Mongolia during May 2008. Zhang Zhilong, Manager of Shenhua Group's International Co-operation Department, organised this weekend visit for three members of the project team, plus ten from the CIAB. These site visits were, without doubt, a highlight of the project and gave representatives from Australia, Russia, South Africa, Turkey, the UK and the US first-hand experience of the impressive developments that are taking place in China's coal sector.

Thanks also go to the following international experts for their feedback and advice to this book: Prof. Philip Andrews-Speed (University of Dundee); Jim Brock (CERA), who sadly passed away in 2008 and is remembered with affection; Dr. Xavier Chen, Tee Kiam Poon, Dr. Stephen Wittrig and Dr. Bruce Yung (BP China); Dr. David Creedy (Sindicatum Carbon Capital); Phillip Dobbs (Anglo Coal); Dr. Mark Dougan (Barlow Jonker); Dr. Kelly Sims Gallagher and Zhao Lifeng (Harvard University); He Ping (formerly UNDP, now Energy Foundation); Takaji Kigasawa (NEDO); Dr. Ma Linwei (Tsinghua University); Dr. Eric Martinot (Institute for Sustainable Energy Policies, Tokyo); Russ Phillips (Peabody); Masaki Takahashi and Dr. Zhao Jianping (World Bank); and Dr. Xu Hailong and Dr. Zhongxin Chen (Shell).

Rebecca Gaghen and her team in the IEA Communication and Information Office ensured that a high quality was maintained throughout the process of review, production and promotion: the patience and professionalism of Muriel Custodio, Jane Barbière, Bertrand Sadin and Sylvie Stephan were especially appreciated. Much of the work relied on efficient and reliable communication. Jim Murphy and his small but dedicated team ensured that the best IT systems were always available. Amanda Watters, Caroline Gill, Alette Wernberg, Michelle Ewart, Virginie Bahnik and Julie Calvert provided administrative support to the project. Thanks are also due to Nancy Turck and Andrea Nour for overseeing all contractual relations.

Financial support from the Strategic Programme Fund of the UK Foreign and Commonwealth Office allowed a much broader collaboration with Chinese organisations than would otherwise have been possible.[1] John Fox, First Secretary

---

1.   www.fco.gov.uk/en/about-the-fco/what-we-do/funding-programmes/strat-progr-fund

and Richard Ridout, Second Secretary at the British Embassy in Beijing, managed this invaluable support, assisted by Leanne Wang, Sherry Ma, Xu Hao and James Godber.

Any questions or comments about this publication should be directed to the Asia Pacific contact at the IEA (dalsa@iea.org).

PHOTOGRAPH CREDITS

© The Coal Authority 2008 (Figure 7.12) / Nobuhiro Horii (Figure 5.10) / Brian Ricketts (Figures 5.8, 5.11, 5.12 and 7.15) / © Thomas Robbin 2006 | staedtefoto. de (Figure 7.10) / © Steve Roe 2005 (Figure 7.14) / © Jacques de Selliers 2008 (Figure 5.6) / Jonathan Sinton (Figure 5.7) / Simon Walker (Figure 5.9) / © David Wild 2005 (Figure 7.11).

TABLE OF CONTENTS

ANNEXES

The annexes are freely available on-line from the IEA (www.iea.org).

I.   Data on coal resources, reserves, production, mine worker numbers
and safety, and transport in China

II.   Major national laws and regulations pertaining to the coal sector
in China, with examples of local laws from Shanxi, Inner Mongolia
and Henan

National legislation currently in force

Local legislation currently in force

I. EXECUTIVE SUMMARY

This report presents an overview of coal in China, examines coal-related policies and issues, and recommends ways the country – both on its own and in co-operation with others – might improve the sustainability of coal use.

Coal meets just over one quarter of the world's demand for primary energy. In 2007, 2.5 billion tonnes of coal were mined across China, almost one half of global hard coal production. Coal is the nation's most important fuel, accounting for 63% of total primary energy supply (including all biomass fuels), much greater than the global average.[1] However, coal production and use bring heavy social and environmental burdens. Within China, there is agreement that urgent challenges are posed by the ever-larger volumes of coal that the economy requires to meet national development goals. Internationally, the implications for the regional and global environment, for world coal trade and for China's comparative economic position have attracted growing attention. China's challenges are shared in many other countries, since coal use is anticipated to grow worldwide for many years. All of us, directly or indirectly, have a strong stake in a "cleaner coal" future.

To help address the challenges, the IEA makes ten key recommendation, together with suggestions on how these might be implemented in China. Each recommendation is important – all the issues must be tackled and none ignored. The challenges created by coal use in China are no longer just a national issue – they transcend boundaries. Finding solutions in our increasingly globalised world demands much greater international engagement. The most powerful form of co-operation is international trade and this forms a central theme to the recommendations. All governments need to make sure that trade, linked to clean energy, grows quickly.

This report provides policy makers with the information needed to appreciate the scale of the challenges faced and the role of international co-operation and collaboration in solving them. Providing insight for those outside of China is only one objective of the report; another is to share the experiences of developing coal-related policy in IEA member countries with policy makers in China. These experiences have been distilled into a single chapter, which, of course, cannot do justice to the efforts made over many years to improve the way coal is mined and used. Wherever possible, links to more detailed sources of information are suggested. By simultaneously publishing this report in English and Chinese, the concepts and ideas will at least be accessible to the widest possible audience in China and beyond.

---

1. Official Chinese sources report that coal provided nearly 70% of China's primary energy in 2007, a higher figure than IEA statistics which include biomass use and treat primary electricity sources differently.

CHINA'S SURGING COAL USE

In the past quarter of a century, China has created wealth for many of its people, lifted many out of poverty, and helped drive and sustain global economic growth. Coal has underpinned China's massive and unprecedented growth in output, fuelling an economic miracle that has helped to improve the standard of living in many countries. Since 1997, annual coal output has increased by 1.1 billion tonnes, more than the United States produced in 2007, and led to approximately 2.2 billion tonnes of additional annual carbon dioxide (CO_2) emissions. The recent *annual growth* of China's coal production has been over 200 Mt, or not much less than Russia's *total annual* production. Projects to sink new mines, build washeries, establish worker communities, add capacity on dedicated rail lines and expand ports all serve the largest expansion of coal-fired power generation capacity in history. The imperative to ramp up output quickly has seen the coal industry undergo many structural changes, and it continues to evolve in response to challenges ranging from inefficient resource exploitation to safety issues at thousands of small mines.

China's reliance on indigenous coal, notably for over 80% of its electricity generation, brings benefits in terms of energy security. The conversion of coal to chemicals, liquid fuels and synthetic natural gas can and does allow an even greater reliance on its indigenous coal resources than seen in most other countries. Recently, the Chinese government has encouraged its coal mining companies to invest in coal mines outside of China to secure coal supplies, while discouraging majority foreign ownership of mines in China for strategic reasons. Most developed nations have found that energy resource ownership does not equate to energy security, since many factors determine the destination of energy supplies. In any event, efficient domestic energy production and import diversity enhance energy security, regardless of ownership. The important point for all governments is to promote greater competition and transparent markets in a way that enhances energy security for all.

From an international coal trade perspective, there is an appetite for better information and data on China. Small imbalances in China's huge internal coal market – which, in 2007, was three times larger than total world seaborne coal trade[2] – can have a significant impact on global coal flows. While the country has large coal reserves and is likely to remain largely self-sufficient, it is often more economic for coastal customers to import coal. Their demand is likely to become an important component of international trade. In recent years, overall coal demand in China has soared faster than indigenous supply, leading to higher prices and government actions aimed at fulfilling domestic demand, particularly in the power sector to avoid electricity shortages. These actions affect the balance of coal trade. China's rapid growth of exports, much welcomed by coal users around the world, made it a major player from 2000. However, exports peaked in 2003 at 94 million tonnes and have fallen markedly since then, while imports have risen rapidly, adding to the combination of

---

2. Exports of hard coal from all countries in 2007 totalled 917 million tonnes, of which 834 million tonnes was seaborne.

circumstances that led to unprecedented hikes in traded coal prices during 2007 and the first half of 2008.

Managing the exploitation of China's coal resources

China's coal resources are vast, over 5 500 billion tonnes, and its proven reserves of 189 billion tonnes would last for over 70 years at the current rate of production. Although China is not about to run out of coal, it does face a number of challenges: average mining depth is increasing, adding to costs; resource recovery rates are low; many mines are located in environmentally sensitive areas with limited water resources; the number of mining fatalities is falling, but remains unacceptably high; coal transport routes are relatively long and congested; restructuring to eliminate small mines will lead to rising job losses; and large-scale, timely investment in new mines and transport infrastructure will be needed to meet the forecast growth in demand.

Coal exploitation would be improved by a variety of changes, including fairer and more transparent resource allocation, perhaps through auctioning. Non-discriminatory mine permitting would promote greater competition, and would open the door to international participation – leading to more rapid penetration of the most-efficient mining practices and technologies from around the world. Conditions imposed during permitting could set standards for land restoration and treatment of subsidence damage, and a bond system, as widely used in other countries, could help to ensure such remediation is properly carried out. Operating mines need to be regularly inspected by independent pollution control officers. Beyond that, meeting the highest environmental standards should be seen as a key business objective for all mining companies – the fee-based pilot scheme in Shanxi is a step in this direction.

Recommendation

Environmental charges on coal mining have been introduced, but more should be done to directly link them to levels of pollution (i.e. the widely accepted "polluter-pays principle"). Funding for environmental protection agencies should be guaranteed separately and not be linked to revenues from environmental charges.

Industry restructuring

China has taken some very effective steps towards improving the economic and technical efficiency of coal mining, such as ordering large numbers of small unsafe mines and small inefficient power plants to be shut down. No other country has had to effect such a wide-scale industrial restructuring and it should be no surprise that the local authorities in China lack the resources to ensure that it is carried out as intended by the state government in Beijing. China can learn from experiences in other countries to establish competitive markets with many players, from small to large, properly regulated to achieve economic and environmental protection goals. The provision of training and social welfare assistance during restructuring will be as necessary in China as it has been in other countries during periods of change. The report draws on experience in Europe, Australia and the United States to highlight those good practices that China should embrace.

Recommendation

Coal-industry restructuring should be founded on a belief in the power of properly regulated markets to deliver economically efficient mines, operated by competing companies of varying sizes, from small to large.

Coal mining safety

Safety should always be the first priority in mining. In addition to the human costs, mining accidents result in productivity losses and economic costs associated with treating injuries and compensating dependents. A viable mining industry avoids these through improved safety. In China, there is a pressing need to strengthen the resources and capabilities of the mines inspectorate to ensure current safety regulations are enforced. Perhaps just as important is the need to enhance training of underground workers, who should be given greater responsibility for ensuring their own safety and that of their fellow workers. These have proven to be essential elements in bringing down accident rates elsewhere.

Recommendation

A properly resourced, national mines inspectorate is central to ensuring mine worker safety. China needs to strengthen its own inspectorate, and complement this by training and empowering coal miners to take greater responsibility for their own safety.

Competitive markets

A stable coal supply is fundamental to achieving other goals. Over the last decade, coal shortages, volatile prices, poor product quality, transport bottlenecks, financial losses and other, near-term issues have all, at times, distracted leaders, government officials and enterprise management from giving their full attention to longer-term, sustainability issues. A properly functioning coal market, with effective supply and demand responses, has clear and immediate benefits that give the space and freedom to address the more difficult problems associated with coal use.

Removing all forms of subsidy (including from the coalbed methane industry) would allow the coal industry to grow on a more commercial footing – consistent with moves towards cost-reflective pricing. This should extend to the power sector, where a timetable could be set for incorporating the full costs of fuel for power generation into wholesale and retail electricity rates. This last point is one of the hardest problems of energy regulation in any country and is often subject to political interference, but can be eased by protecting low-income customers through the simultaneous roll-out of targeted assistance programmes.

Recommendation

Market-based, energy and resource pricing should be used as the primary means of balancing supply and demand in China, so that resources are exploited, transported and used efficiently and effectively, including those that are imported and exported.

STEPS TOWARDS SUSTAINABLE COAL USE

Coal use brings environmental challenges. China emits more sulphur dioxide (SO_2) than any other nation and coal use also adds significantly to dust and NOx emissions. China is the largest emitter of CO_2, although its cumulative contribution to the atmospheric stock of CO_2 and its per-capita emissions remain well below those of the world's industrialised nations. Further growth in coal use – potentially to 2.8 billion tonnes in 2010 and 3.2 billion tonnes in 2020 under a low-energy intensity scenario, and substantially more under business-as-usual scenarios – makes it more urgent than ever to develop a strategy that marries the clear economic benefits of coal use with China's sustainable development goals.

Any comparative advantages built on resource wastage and environmental degradation are not sustainable, and the Chinese government is making efforts to eliminate unsustainable practices. Successive Five-Year Plans and recent energy and environmental policies provide a framework for sustainable development. The 11th Five-Year Plan (2006-10) sets a target to reduce energy use per unit of GDP by 20% by 2010 compared to 2005, and calls for a 10% reduction in key pollutant emissions. Data for 2006 suggests that good progress is being made in the case of particulate emissions, but that greater effort will be needed to reverse the rising trend of SO_2 emissions and to further reduce energy intensity which remains above its 2002 level. In April 2006, Premier Wen Jiabao announced three new policy directions: to place environmental protection and economic development on an equal footing; to make environmental protection an integral part of economic development, not simply an afterthought; and to integrate environmental protection into all administrative activity. More recently, in October 2007, President Hu Jintao emphasised the pressing need for resource conservation and environmental protection – principles of sustainable development through which China would, "make new contributions to protecting the global climate". The government is promoting vigorous development of renewables, natural gas and nuclear power; but, even under the most optimistic scenarios, it would take decades for them to push coal from its dominant position in China. Cleaner coal technologies are therefore critical. In June 2007, China unveiled its *National Action Plan on Climate Change*, which includes goals to develop clean coal technologies, from more efficient coal mining equipment to CO_2 capture and storage (CCS). The country has also strengthened its international engagement on technologies and policies to improve the way coal is exploited.

Promoting cleaner coal technologies

Experience worldwide shows that deployment of clean coal technologies must encompass the entire coal supply chain, and that parallel progress is needed in technical and non-technical areas for coal to remain an acceptable component in a country's energy mix. A modern coal-fired power plant cannot be considered in isolation from the coal mines, transport infrastructure and coal markets that supply it. The setting that allows clean coal technologies to be deployed effectively at a power plant is complex and includes: the grids and power markets that receive its output; the regulatory apparatus that approves its construction, and oversees its operation and eventual decommissioning; the banks and investors that join the EPC (engineering-procurement-construction) contractor and utility company to build, manage and maintain it; the neighbouring residents who work at it and live with it; and the increasingly global technical community that designs, manufactures and services it.

Making a nation's coal-based energy system cleaner is not just about improving access to better technology. China already hosts facilities that feature some of the largest-scale and most-advanced equipment in the world from fully automated longwall mining equipment and modern coal washeries, to ultra-supercritical power plants, with 1 000 MW units, and industrial coal gasifiers. An inability to produce critical components is often the main barrier to manufacturing such systems in China, but many of the major equipment suppliers operate in China, so imported components are commercially available. Adaptation of imported technology is often desirable to reduce costs and meet local market needs. Chinese companies have been successful here, for example in adapting flue gas desulphurisation (FGD) systems to their own needs. Technology transfer could deliver more, and China should consider relaxing any remaining barriers to participation of foreign companies in key energy industries, since joint ventures and foreign direct investments are an effective means to technology diffusion on commercial terms. It is the movement of people that allows effective technology transfer, and not simply the transfer of information, such as contained in engineering drawings.

Recommendation

The government should further encourage joint ventures and foreign direct investments in the energy sector to promote technology transfer, both into and out of China.

Developing new cleaner coal technologies

For technologies where R&D is needed, China is in a similar position to other nations. In the area of direct coal liquefaction, China is a pioneer, and is also making significant progress in demonstrating some of the components and processes needed for CO_2 capture and storage. But overall, greater efforts are needed globally in R&D; spending simply does not reflect the challenges faced by the energy industry as a whole. China has shown a willingness to participate in international partnerships and joint ventures in many fields to research, develop and demonstrate new technologies. In the case of cleaner coal, such active participation can speed progress towards those technologies that are most appropriate for commercial markets within China and elsewhere. If a technology is not commercially viable, then it will not be deployed.

Recommendation

International and national partnerships, supported by governments, industry and academia, can stimulate the development of new technologies before their commercialisation.

Deploying well-proven technologies and practices

Sometimes the newest and biggest is not the best for a particular application. There are tremendous opportunities to make improvements using techniques and equipment already widely available in China. For instance, more rational mining would raise the recovery rate of coal resources – a major priority in China – and can be achieved with modern management practices that maximise economic rent under a well-regulated system of resource allocation. Matching fuel quality to users' specifications is another area where coherent policy and effective market regulation are far more important than acquiring new technologies. Simple housekeeping measures during transport, at power plants and at other end users would raise efficiency and reduce unnecessary emissions, especially of dust.

Recommendation

Even as it pursues innovative new technical and policy solutions, China should quickly adopt well-proven technologies, management practices and policies that deliver immediate and sustainable improvements along the entire coal supply chain, from mine to end user.

Importance of effective regulations

In China and elsewhere, the constraint to energy demand growth is not the resource base, but humanity's ability to use fossil fuels without creating unacceptable environmental impacts. Thus, this report returns repeatedly to the need for stronger implementation of well-designed environmental regulations, without which there is no reason to install and operate cleaner systems. As with mine regulation, the key is well-trained, adequately funded and independent regulatory bodies. In most countries, greater public information and involvement of citizens in approvals and monitoring processes have been essential to making environmental policies work.

Recommendation

Greater accountability and transparency that allow reliable delegation to lower levels of Chinese government are prerequisites to the proper functioning of existing environmental laws and hence the successful deployment of clean coal technologies.

Other countries can contribute to progress through bilateral and multilateral collaboration at all levels. For example, individuals with regulatory experience in other countries should share their first-hand experience with their Chinese counterparts who face similar issues. Official secondments of staff to foreign government departments and regulatory agencies should become an established element of career development in China and elsewhere.

Creating an international price for pollution

China has recently made great strides in deploying FGD equipment; the next challenge is to ensure that these and other pollution control systems are operated in a way that achieves China's emission reduction goals and reduces transboundary air pollution in northeast Asia. With rising coal use, China's SO_2 emissions have followed a rising trend since 1999 to reach 26 million tonnes in 2006, a trend that confounds the rapid increase in FGD capacity to around 50% of total installed thermal capacity. National legislation on atmospheric pollution prevention and control, and detailed regulations that include emission standards for power plants have not reduced emissions, largely because of inadequate enforcement by provincial authorities. Without stronger financial incentives than recent actions have provided, this situation may persist. An alternative would be to make it more profitable to generate electricity and heat, and to produce coke, cement and other products at clean and efficient plants rather than at inefficient plants with poor pollution control. Various means are available to signal the higher value of cleaner production: taxes on emissions, feed-in tariffs, emissions trading and pollution charges. Such market mechanisms have been effective in the US at reducing SO_2 emissions and can be used in combination with appropriate emission standards to give flexibility and the incentive that is missing in China. Fungible emission reduction credits could bring value to pollution control projects, especially if they can be traded between countries.

Recommendation

Market-based mechanisms, such as sulphur and carbon trading, should be central to China's pollution abatement strategy and the key incentive to develop cleaner coal technologies for domestic and international markets.

China's role in a cleaner future

An even greater challenge will be to deploy systems for CO_2 capture and storage – which is a critical technology for coal's long-term future, but which has not yet been demonstrated at a commercial scale at any coal-fired power plant anywhere. Such demonstrations are 5 to 10 years away and China is already participating in R&D initiatives that aim to accelerate progress. Now is the time to look ahead and envision how to encourage deployment. One way would be through international carbon trading systems. Negotiations over the next couple of years will shape a long-term international carbon market, so China needs to move swiftly and with determination so that its domestic actions are compatible with a global effort, and that international flows of funds can be harnessed to build CCS facilities in China. In the near term, the "CO_2 capture-ready" concept needs to be better defined so that the unprecedented number of new, coal-fired power plants built each year in China and other developing countries, and the likely replacement of many ageing coal-fired plants elsewhere, do not lock in CO_2 emissions for decades to come.

More broadly, China has an unprecedented opportunity to become a major player in the global market for cleaner, more efficient coal technologies. It has already developed some unique technologies that other countries should sensibly adopt, and will certainly create more. It should work with other governments to create a global market for clean energy technologies, and allow its manufacturing industry to respond with commercially relevant products, for local markets and for export. The IEA believes that such commercial activities, some in partnership with foreign companies, some in competition with them, will have a far greater impact than piecemeal co-operation on individual, government-supported projects, important though these are during the early stages of product R&D.

Recommendation

China should co-operate with other nations to establish common technical standards for coal-fired plants and their sub-systems, and so allow the wider deployment of more affordable clean coal technologies, both in China and elsewhere.

INTERNATIONAL ENGAGEMENT ON CLEANER COAL

Globally, two market imperfections currently limit the uptake of cleaner coal technologies: it costs less to pollute than to control pollution and barriers, such as high development costs, slow technological change. Accelerating deployment will require changes at the national and international levels. Commercial deployment of cleaner coal technologies requires investment certainty through stable policies that recognise the costs and risks of long-term capital investment in pollution control, ultra-supercritical, IGCC (integrated gasification combined cycle) and CCS technologies. Hence, the three priorities for international engagement with China are:

■ negotiations leading to successful international accords that create national, regional and global markets for clean, low-carbon technologies;
■ government-industry partnerships to develop and demonstrate low-carbon, cleaner coal technologies; and
■ technology transfer and deployment of cleaner coal technologies through commercial arrangements that respond to the market demand created in China and elsewhere.

China will need to decide for itself how to proceed, but its actions, more than those of any other country, will shape the global approach to the cleaner use of coal that is urgently needed to avoid the worst effects of climate change.

II. INTRODUCTION

BACKGROUND

China's remarkable economic expansion and the ensuing global economic benefits are predicated on the largest escalation in coal use ever seen anywhere. The challenges to economic, social and environmental sustainability of obtaining and using coal on this scale are enormous. Concerned stakeholders in China want to know how experience in other countries can help them face these challenges. Those in other countries, many with a new interest in China's coal sector, want to know what is happening, and what the opportunities are for them to make a difference. This project was conceived to help people inside and outside China reach a common understanding of where China's coal supply and transformation sectors stand today, what is needed to make them cleaner, and how international collaboration can contribute to that objective.

This is not a new field. China has engaged in many co-operative bilateral, multilateral and commercial activities related to coal over the past two decades. The mounting impacts of coal use, however, bring new urgency. Moreover, despite rising international interest to help improve coal use in China, the understanding of key market, technological and regulatory issues remains shallow. Coal industry players in China must contend with a range of competing priorities, from output and economic performance, to worker safety and local pollution control. It is often difficult for them to see how international co-operation might contribute to their businesses. Stakeholders with an international perspective may be focused more on regional and global environmental security or the dynamics of international coal trade (Figure 2.1).

Figure 2.1 .... Regional and global environmental security, stable international coal trade and familiar coal-sector issues

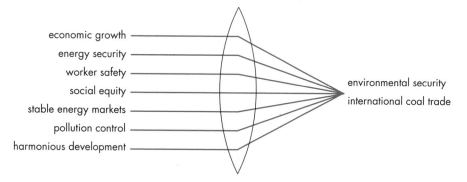

Note: "Harmonious development" is a chief goal of the *Scientific Concept of Development*, introduced by President Hu Jintao. A harmonious society would redress the inequalities that have resulted from the recent pattern of Chinese growth. This has been tremendously successful in raising total economic output, but less so in distributing benefits equally across regions and groups.

We need to move beyond simplistic, often outmoded prescriptions. Calls for "energy price reform" are, alone, insufficient to aid in the complex regulation process needed to ensure that prices do fully reflect economic and external costs. Requests to "transfer better technology" seem dated when much of the best equipment that the world has to offer can now be found in China's mines and power plants. This report is an attempt to learn from the successes and failures of the past, to apply those lessons to the rapidly evolving conditions we find today in China, and to provide a reference point to aid all who care about cleaner coal in China to work together more effectively.

DRIVERS FOR CLEANER COAL

The incentives for making coal use cleaner are much the same worldwide, though the particular circumstances of each country, and the influence of individual actors, often produce different orders of priority. Box 2.1 summarises the most pressing challenges facing China's coal sector. Each is addressed in this report, conclusions are drawn, many based on experience in other countries, and recommendations are made on how government policy and international engagement might respond to these challenges and others. To meet China's rising coal demand whilst avoiding local, regional and global environmental damage is, in essence, the challenge of sustainable development that faces the world today.

Box 2.1 .... Immediate challenges facing China's coal sector

Coal resources

■ **Scarcity of resources.** China has huge coal resources and reserves (Chapter 3). On a per-capita basis, its resources are more than twice the world average, although significantly below Russia's. Similarly, its per-capita reserves are above the world average, but much less than both the United States' and Russia's. China's proven reserves of the best quality coking coal, used for steelmaking, are scarce.

■ **Difficult geological conditions.** With increasing mining depth, it becomes more difficult to exploit coal resources. Today, the average depth of existing large- and medium-scale coal mines in China is over 400 m. In 2010, the average depth will have increased to around 500 m and those deeper than 600 m will account for 35% of all mines, with fourteen shafts reaching a depth of 1 000 m or more.

■ **Serious resource wastage.** Recovery of coal resources at Chinese mines averages 30-35%, when >50% might be expected.[1] This low rate is mainly due to poor and outdated mining technologies, but also because recovery rates at many small mines can be as low as 10-15%.

---

1. Under ideal conditions, an efficient underground longwall mining operation might recover 80% or more of in-situ reserves. Some coal has to be left behind to support the roof between the blocks of coal being mined, although the roof is allowed to collapse behind the working face itself. In room and pillar mining, the roof is left completely supported by pillars of coal, allowing up to 60% of the coal to be recovered. In 2006, the average recovery rate at underground coal mines producing more than 10 000 short tons (9 072 metric tonnes) per year in the United States was 58.6% (EIA, 2007). Recovery rates at surface mines are universally high – 80.3% in the case of the United States in 2006 (ibid.) – since entire coal seams can be recovered, even multiple seams.

Environmental protection and energy conservation

■ **Mining in sensitive areas.** Among the coal reserves suitable for exploitation, nearly 90% are located in arid and semi-arid regions, where the ecological environment is fragile.

■ **Energy conservation targets.** Implementation of environmental protection and energy-saving policies will tend to damped coal demand. The National Development and Reform Commission is supervising a plan to conserve the energy equivalent of 340 Mtce over the 11th Five-Year Plan period (2006-10) through a 20% reduction in the energy intensity of key industries. It is implementing ten major energy-saving projects, and enhancing investment support by central and local governments. In addition, the Plan sets a target to reduce emissions of major pollutants by 10% (Chapter 6).

Coal supply

■ **Expanding production capacity.** During the 10th Five-Year Plan period (2001-05), coal mine construction expanded, both in number and scale, and productivity increased. With the new mines planned, production capacity will continue to grow. Investment in fixed assets of about RMB 300 billion will result in 800 Mtpa of additional production capacity by 2011.

■ **Coal import/export policy.** At the end of the 1990s, the Chinese government encouraged licensed companies to export coal and so absorb surplus production. Then, since 2002, the government changed this policy, eventually eliminating tax rebates and other benefits, before introducing tariffs in 2006 and reducing quotas to limit coal and coke exports. These measures were designed to increase domestic supply and avoid shortages (Chapter 6).

■ **Cost pressures.** A number of new policy measures will increase mining costs, which have been typically in the order of RMB 100-200/t. For example, in Shanxi, coal resource taxes have been increased from RMB 0.9/t to RMB 2.5-8.0/t, in addition to a royalty of RMB 6.0/t. Resource guarantee fees have been increased from 1% of sales to 3-6% and various other fees, related to the external cost of coal mining, can add a further RMB 50/t (Chapter 3). The new measures also stipulate that coal mines should improve the health allowances of underground workers, establish work injury insurance schemes, rebuild shantytowns in areas prone to subsidence and repair other subsidence damage.

Coal industry structure and mining safety

■ **Industrial rationalisation.** In 2006, the domestic market share of the top-four Chinese coal enterprises was 18%, a rise from the 12% market share in 2000, but low compared, for example, to the 47% market share of the top-four US coal companies. Although China has a number of world-class, highly productive coal mines, the production efficiency at other mines is very variable such that the average is quite low. In 2005, the average efficiency of Shenhua Group, one of China's top performing coal enterprises, was nearly 120 t/man-shift, while the average achieved at key state-owned coal mines was only 4 t/man-shift, and even lower at the thousands of small mines.

■ **Mining safety.** Several risks threaten work safety: flooding, mine gas, coal dust, rock and gas outbursts, fire and explosion, and natural disasters. The equipment at many state-owned coal mines is outdated: around one third of the equipment is beyond its design life, creating risks for operators. Most township and private mines still use primitive mining methods – the death toll resulting from accidents at these mines accounted for 74% of total coal mining fatalities in 2005.

Coal transport

■ **Supply distant from demand.** The distribution of coal and water resources leads to many challenges in meeting demand. Coal demand in most parts of China, notably in coastal areas, continues to rise, but supply will remain concentrated in western and northern regions. Water shortages in those regions limit the opportunities for large-scale coal processing and conversion, hence the pressures on West-East and North-South coal transport corridors will increase.

■ **Transport bottlenecks.** With coal production concentrated in Shanxi, Shaanxi and Inner Mongolia, coal transport from these regions attracts much attention. In 2007, the capacity of the Daqin line (Datong to Qinhuangdao port) was increased by 50 Mtpa to 300 Mtpa. Qinhuangdao port, with a coal-handling capacity of 220 Mtpa – the world's largest, will guarantee East China's coal supply. Coal supply capacity from the mid-west will only be limited by the number of barges able to traverse the Three Gorges locks on the Yangtze River. However, despite new railways and line reconstruction work, rail capacity cannot yet fully meet demand.

Sources: China National Coal Association; China Coal Information Institute; Shanxi Fenwei Energy Consulting (2007-08); CASS (2007); and BGR (2007).

For IEA member countries, concern about climate change is perhaps now the strongest driver for new clean coal technologies. In the latter part of the 20th century, reducing sulphur dioxide emissions was a key objective and, before that, limiting particulate emissions. Now, the prospect of increasingly stringent limits on emissions of carbon dioxide raises questions about the long-term role of coal and other fossil fuels in the wider energy system. On the coal supply side, land use, solid waste disposal and water quality issues demand ever-more rigorous coal mining and processing practices, while greenhouse gas trading has joined worker safety as an incentive to ensure that coal mine methane is collected and utilised.

As in most other countries, profit has become an overriding objective of firms that mine, transport and use coal in China. Most technology improvements result from a blend of socially and politically inspired imperatives, coupled with the business desire to gain competitive advantage and maximise revenues. Greater efficiency has often delivered these goals simultaneously. For example, developing new mining methods to recover a greater fraction of coal resources or to convert a higher proportion of coal feedstock into useable coke can have both financial and environmental returns.

STAKEHOLDERS

Broadly speaking, there are five stakeholder groups considered in this study: industry, labour, government, experts and ordinary citizens. The industrial group, typically represented by managers and owners, covers upstream and downstream sectors, and includes industry associations, as well as coal transporters and traders. Important differences exist between the sectors, and between segments within a sector. For example, small rural mines are a world away from China's large coal groups that have international standing. Labour is considered as a separate stakeholder group since, historically and currently, the interests of ordinary workers and management have not always coincided. The government grouping is not uniform since viewpoints may differ at national, regional and local levels. Coal-producing regions often have different interests than those that use coal or suffer the consequences of coal use, and local administrations may emphasise different goals from those of their national government. Experts, whether serving one of the other groups or independent, play an important role in developing and deploying new technologies and policies, and in shaping ideas about what is desirable and possible. The interests of ordinary citizens are nominally looked after by government, but non-governmental organisations and private individuals can also exert an influence on the decision-making process in China.

None of these stakeholder groups can be omitted from an assessment that aims to truly reflect the current situation. Representatives of these groups, therefore, have been consulted in the course of preparing this report. Any practical resolution of the complex issues facing the coal sector requires a balanced response to their various interests.

SCOPE AND STRUCTURE OF THE REPORT

The terms "clean coal" and "cleaner coal" are broad. Box 2.2 lists many of the technologies that the IEA regards as important for today's state-of-the-art systems and for future systems. Much past work on clean coal technologies and policies has focused on power generation. While the power sector is the largest coal-using sector in China and worldwide, we believe that coal technologies must be seen in wider context. To take but one example, a stable supply of in-specification fuel is needed to achieve the best performance from power generation equipment. It would be unwise to assume that the geological, technical and institutional issues involved in providing a compliant fuel supply will simply resolve themselves.

Box 2.2............... Clean coal technologies

The IEA Clean Coal Centre provides a public database with descriptions of all the clean coal technologies available today.[1] Separately, the Centre also gives free access to Coal Online, a major resource for scientists and engineers.[2] This has comprehensive

1. *Clean Coal Technologies* database, www.iea-coal.org.uk/site/ieacoal/clean-coal-technologies.
2. *Coal Online*, www.coalonline.org/site/coalonline/content/home.

and detailed information on coal and coal utilisation technologies, drawn from hundreds of the Centre's published reports and other sources. China already has experience – sometimes considerable experience – with most state-of-the-art clean coal technologies. The Coal Industry Clean Coal Engineering Research Center (CCERC), supported by the Ministry of Science and Technology, offers a range of services to those involved in the development and commercialisation of clean coal technologies, including the China Clean Coal Technology database.[3] Technologies for future systems and certain underpinning technologies would benefit from joint development efforts that bring in knowledge and experience from outside China.

State-of-the-art systems

- efficient coal mining equipment – longwalls and continuous miners
- heavy-duty gas turbines
- integrated gasification combined cycle (IGCC)
- supercritical (SC) and ultra-supercritical (USC) pulverised fuel units
- large-scale SC circulating fluidised-bed combustion units
- combined heat and power (cogeneration), and polygeneration
- coal mine methane and coalbed methane exploitation
- particulate emissions controls (ESPs, fabric filters, scrubbers, cyclones)
- SO_2 emission controls (scrubbers, sorbents)
- NOx emission controls (burners, air/fuel staging, flue gas recirculation, SCR/SNCR)

Future systems

- coal drying
- >620°C USC units with a target of 700°C
- high-temperature, high-pressure particulate removal
- combined SO_2 and NOx removal
- polygeneration with hydrogen production
- CO_2 capture and storage (CCS)
- coal liquefaction with CCS

Underpinning technologies

- manufacture and fabrication of advanced materials, including nickel-based alloys
- material coatings for steam turbine and gas turbine systems
- coal handling, especially dry-feed systems for IGCC
- process modelling and simulation

Source: IEA Clean Coal Centre reports.

---

3. *China Clean Coal Technology* database, www.cct.org.cn/cct.

Table 2.1 shows the chain of coal supply and use, with technical issues listed to the left and institutional responses listed to the right. This study deals with the first four links of the coal supply chain; final end-use in sectors such as iron and steel is the

Table 2.1 ....International co-operation can contribute to the institutional responses required to deal with technical issues encountered in China's coal supply chain

Technical issues	COAL SUPPLY CHAIN	Institutional responses
Resources & reserves Quality		Exploration programme Access to geological data
Property rights		Licensing Management of rights of interacting parties Tendering / auctioning of rights Ownership title
Industry structure	COAL MINING & COAL MINE METHANE (CMM) COAL BED METHANE (CBM) PRODUCTION	Structure & responsibilities Restructuring & rationalisation Small mines closure programme Policing of illegal mining
Mine design Mechanisation Working practices Extraction efficiency		Mining & equipment standards H&S legislation, regulation & inspection Training
Subsidence Spoil disposal Coal mine methane control & use Water treatment Land reclamation techniques		Surface damage repair & compensation Environmental impact assessment Environmental pollution control legislation, regulation & enforcement Insurance & bonds
Cost structure Supply forecast		Private investment / FDI policy
Coal quality standards	PREPARATION	Minimum quality requirements in coal supply contracting
Water demand		Abstraction licensing
Tailings disposal		Waste disposal legislation & regulation
Capacity & bottlenecks Access to remote reserves	TRANSPORT	Rail transport policy Infrastructure ownership Rolling stock ownership Regulation of freight operators
Comparative costs Use of trucks		Freight rate transparency Road transport policy
Industry structure	TRANSFORMATION • electricity • heat • coke • chemicals • coal-to-liquids (CTL)	Power sector reform
Conversion efficiencies		Efficiency standards
Emissions monitoring Emissions control equipment		Air quality standards Acidification & global climate policies Emissions legislation & regulation
Water demand Ash disposal CO_2 capture & storage		
Energy efficiency Alternative energy sources Alternative energy technologies	CONSUMERS industrial residential	Energy strategy
Demand forecast		Pricing (inc. taxes & carbon trading) Demand-side measures (inc. efficiency standards & alternatives) Targets & incentives
Supply-demand balance	IMPORTS / EXPORTS	Export licensing / quotas / tax policy

Source: IEA.

subject of other ongoing projects and studies.[1] In addition to the themes introduced in the table, this study considers existing coal-related programmes and the expected evolution of Chinese administrative structures and socio-economic conditions when assessing policies and measures.

The first two chapters of this report are an executive summary and this introduction. Chapter 3 is a review of the current situation of coal supply, transport and utilisation in China today, including technical, policy and regulatory aspects. Chapter 4 surveys forecasts of where China's coal system is headed and of the country's energy system as a whole. The potential role of cleaner coal technologies in that future is explored in more depth in Chapters 5 and 6, with the former evaluating technologies that will be needed, and the latter exploring the legal and regulatory changes that would be needed to deploy these new technologies. Lessons from the experience of IEA countries are described in Chapter 7, and their significance for China highlighted. As this study is certainly not the first to consider clean coal in China, the findings of other recent studies are reviewed in Chapter 8, and the key recommendations from them assessed. Chapter 9 examines recent and current international collaborations on cleaner coal in China, with a view towards improving the effectiveness of such joint activities. Findings and recommendations are drawn together in Chapter 10. With these recommendations, the aim has been to strike a balance between the immediate and long-term objectives for China's coal supply and transformation sectors. Achieving those objectives with well-designed policies can deliver the overarching objective of national and global environmental security. Supporting materials are collected into five appendices and are available on-line.[2]

A single report cannot exhaustively cover such a vast subject as coal in China, but the IEA hopes that it will provide a readable reference for many interested parties. In order to meet the goal of effective international collaboration, this report is published in English and Chinese.

REFERENCES

BGR (Bundesanstalt für Geowissenschaften und Rohstoffe – Federal Institute for Geosciences and Natural Resources) (2007), *Reserves, Resources and Availability of Energy Resources – Annual Report 2006*, BGR, Hannover, Germany, www.bgr.bund.de.

CASS (中国社会科学院 – Chinese Academy of Social Sciences) (2007), 能源蓝皮书 2007中国能源发展报告 *(The Energy Development Report of China 2007 or "Blue Book of Energy")*, Social Sciences Publishing House, Beijing.

EIA (2007), *Annual Coal Report 2006*. DOE/EIA-0584(2006), EIA, US Department of Energy, Washington, DC.

Shanxi Fenwei Energy Consulting (2007-08), *China Coal Weekly*, various issues, Shanxi Fenwei Energy Consulting Co. Ltd., Taiyuan, Shanxi, China, http://en.sxcoal.com.

---

1. These include initiatives such as the UNDP/GEF *End-Use Energy Efficiency Programme*, the European Commission-sponsored *Energy and Environment Programme*, and efficiency policy programmes of the Energy Foundation's *China Sustainable Energy Program*.
2. www.iea.org/textbase/publications

III. COAL IN CHINA TODAY

This scene-setting chapter presents a detailed overview of coal and coal supply in China. Resources and reserves are considered first to determine if any obstacles exist to the future exploitation of China's large endowment of coal. To better understand the coal sector today, a brief account is provided of how the industry has evolved – under the planned economy and, latterly, under a socialist market economy – to become one of China's most important industrial sectors, with some world-class enterprises. Alongside the achievements, there is also a need to report on some less attractive aspects of the industry: worker safety; environmental protection; and social welfare. These are being addressed by the Chinese government and further progress can be anticipated. Bringing ever-more coal to market presents some significant transport challenges which are described. Coal demand is now dominated by the power sector, but other sectors are also significant coal users, such as iron and steel, cement and chemicals. These are each examined in relation to coal supply and China's coal import-export balance. The history of China's evolving coal market is explored, starting with an explanation of how coal was allocated under the centrally planned economic system, and the challenges of moving away from such micro-management by the state. This leads onto coal pricing which, outside of the power sector, is now determined by market forces. The laws and regulations relevant to mining and using coal are summarised, with a description of who administers these and some of the issues faced. Safety is a major one, and the chapter closes with an assessment of China's programme to close the small mines where most accidents occur.

COAL RESOURCES AND RESERVES

Alongside Russia and the United States (US), China holds some of the world's largest coal resources in coal-bearing regions encompassing 6% of the country's 9.6 million km^2. The China Geological Survey has reported that China's inferred coal resources total 5 555 billion tonnes (Gt) (MLR, 1999). Most resources are in the west and north (Figure 3.1); Shanxi, Shaanxi and Inner Mongolia together account for 65% of the nation's proven coal reserves, while just 13% lie in the southern part of the country, mainly in Guizhou and Yunnan. Over 90% of identified coal reserves are in less-developed, arid areas that are environmentally vulnerable.

In 2003, the Ministry of Land and Resources (MLR), in accordance with international norms for coal resources reporting (UNECE, 1997), stated that China's total coal reserves stood at 1 021 Gt across the country's 6 111 mining districts, comprising 334 Gt of "basic reserves" and 687 Gt of "prognostic reserves" (Table 3.1).[1] "Proven reserves" were reported to be 189 Gt, suggesting a reserve-to-production ratio of over

---

1. "Basic reserves" are defined as those resources that can be potentially exploited under current techno-economic conditions. "Prognostic reserves" include those amounts that are not economic to recover or for which economic significance is uncertain because data is insufficient. "Proven reserves" are the economically recoverable fraction of basic reserves.

Figure 3.1 .... Location of major coal resources in China

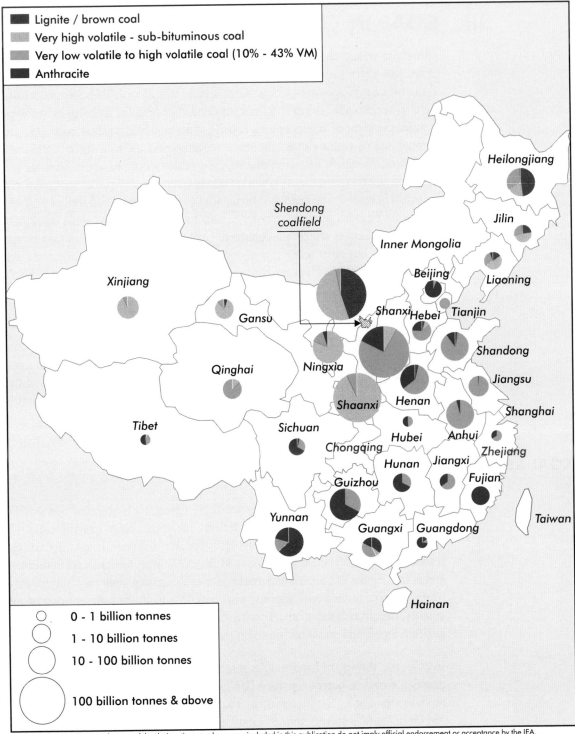

Legend:
- Lignite / brown coal
- Very high volatile - sub-bituminous coal
- Very low volatile to high volatile coal (10% - 43% VM)
- Anthracite

Size legend:
- 0 - 1 billion tonnes
- 1 - 10 billion tonnes
- 10 - 100 billion tonnes
- 100 billion tonnes & above

The boundaries and names shown and the designations used on maps included in this publication do not imply official endorsement or acceptance by the IEA.

Source: Beijing HL Consulting (2006).

Table 3.1 ....Coal resources and reserves in China at the end of 2003
(billion tonnes)

| Planning area | No. of mining districts | Resources | | | Proven reserves |
		Basic reserves	Prognostic reserves	Total reserves	
Beijing, Tianjin, Hebei	274	9.8	8.1	17.9	4.2
Liaoning, Jilin, Heilongjiang	641	15.9	15.5	31.4	6.8
Jiangsu, Anhui, Shandong, Henan	781	37.0	41.6	78.6	17.4
Zhejiang, Fujian, Jiangxi, Hubei, Hunan, Guangdong, Guangxi, Hainan	1 618	4.7	4.5	9.1	2.4
Shanxi, Inner Mongolia, Shaanxi, Ningxia	1 208	213.4	472.6	686.0	126.7
Guizhou, Yunnan, Chongqing, Sichuan	1 051	36.8	49.5	86.2	24.4
Tibet, Gansu, Qinghai, Xinjiang	538	16.7	95.1	111.7	7.3
Total	**6 111**	**334.2**	**686.9**	**1 021.1**	**189.3**

Source: MLR (2003).

70 years. China has a broad range of coal ranks, from lignite to the best quality coking coal, but per-capita resources and reserves, particularly of coking coal, are below some of the world's other major coal-producing nations. There are 276 Gt of coking coal reserves, 27% of total coal reserves.

Most of China's coal reserves lie deep underground; the average mining depth is currently 400 m and is expected to be 500 m by 2010. Of the total reserves, 36% lie within 300 m of the surface, 45% between 300 m and 600 m, and the remainder between 600 m and 1000 m. Only 4% of reserves are suitable for opencast mining. Serious hazards threaten underground coal mining operations: complicated hydrological conditions, with the attendant risk of flooding; methane releases that pose an explosive risk, if not properly ventilated; and geothermal heat that makes for hot and arduous working conditions.

The sole right to manage coal resources rests with MLR under the Mineral Resources Law 1996. Rules promulgated in 1998 provide the legal basis for the transfer of mining rights and the establishment of tradable mining rights (MLR, 1998). Thus, China's approach to coal resource management changed from one of quantity management to that of mining rights management. In 2006, the State Council approved a pilot programme, jointly proposed by MLR, the Ministry of Finance and the National Development and Reform Commission (NDRC), to award coal exploration and mining rights for cash payments based on the coal resource's open market value, as determined through tender or auction, with few exceptions (State Council, 2006).

Around 100 Gt of proven coal reserves could be exploited from existing mines. According to MLR, at the end of 2003, 55.7 Gt of these reserves lay with China's 736 key state-owned coal mines, mainly in Shanxi, Shaanxi and Inner Mongolia. Among the key state-owned mines, 194 mines, with a combined production capacity of 440 million tonnes per year (Mtpa) had excellent reserves, accounting for 61% of total coal reserves at key mines. A further 222 mines (aggregate capacity of 193 Mtpa) had good reserves, 27% of the total. The remaining 320 mines (102 Mtpa) had poor reserves, together amounting to 13% of the reserves at key mines. At the end of 2003, the nine largest local state-owned coal mines, each with a production capacity above 1.2 Mtpa, had a combined design capacity of nearly 20 Mtpa. The remaining mining reserves at these mines were 1.7 Gt. A further 88 mines with production capacities of 0.45-1.2 Mtpa (46 Mtpa in total) had 4.3 Gt of reserves. Smaller mines, numbering 1 497 (aggregate design capacity of 205 Mtpa) had remaining reserves of 14.2 Gt. Around 16 000 township and village enterprise (TVE) mines[2] and private mines account for a large proportion of China's coal resources, with varied potential for exploitation. At present, TVE and private coalmines are estimated to have 22.4 Gt of proven reserves.

There are less than 70 Gt of unexploited reserves, of which 9% or 6.2 Gt are based on "detailed" survey data, 11 Gt on "fine" data from general exploration, and the remainder on "normal" prospecting data. Within the detailed reserves figure, there are 1.2 Gt having excellent geological conditions for mining, 1.9 Gt with good conditions, and 3.0 Gt with poor geology. Reserves in the first two categories are suitable for mining by large- and medium-scale operations. Within the fine reserves, 4.2 Gt have excellent conditions, 5.1 Gt good conditions, and 1.4 Gt poor conditions. Exploration of these unexploited reserves has been limited, with few feasibility studies. Only a small number of surveys have been carried out to produce the "detailed" data needed prior to sinking shafts to exploit the reserves. Nevertheless, among the reserves identified by prospecting data are many with promising prospects for exploitation.

The MLR has identified a widening gap between the availability of prospected coal reserves and those needed for developing new mines to meet China's growing coal demand (MLR, 2003). There is a general trend towards deeper, more distant mines in the west. Many factors limit future coal mining potential which have an impact on coal supply economics. Chinese authorities recognise the need to improve coal prospecting efforts to yield better data for mine planning. Some policies are aimed at consolidating coalfields that have been fragmented into mining blocks too small for efficient mine design. A more commercial approach is emerging, with tradable rights that should bring focus to prospecting activities, but this is still in its infancy. Greater efforts are also being made to mitigate the environmental impacts of mining in regions suffering from a serious lack of water, desertification and vulnerable ecology. For example, the coal-rich provinces and autonomous regions of Shanxi, Shaanxi, Inner Mongolia and Xinjiang hold 74% of China's total coal resources, but only 1.6% of the nation's water resources. Addressing all these issues would allow China's coal supply to continue to grow as rapidly in the future as it has over the past three decades.

COAL SUPPLY

Before the economic reforms introduced in 1979, coal dominated China's commercial energy structure even more so than today, but short supplies were a bottleneck to the rapid economic growth planners envisioned. Far-reaching regulatory changes were

---

2. Also known as "township mines".

enacted, enabling coal production to expand very quickly (Figure 3.2), and its growth far outstripped that of oil and natural gas until the mid-1990s. In the latter half of the 1990s, official figures show that coal production and use dropped significantly. Demand picked up again after 2002, and, more than any other energy source, coal has fuelled the current decade's phenomenal expansion. China became the world's largest coal producer in the early 1980s. Output in 2006 was 2 320 Mt (and an estimated 2 549 Mt in 2007), far beyond the second-largest producer, the US, with 1 068 Mt (IEA, 2008a).

Figure 3.2 .... Primary energy consumption by source and share of coal in total, 1953-2006

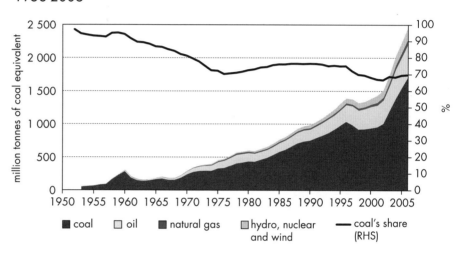

Note: Data excludes biomass. If estimates for biomass fuel use are included, then the share of coal is smaller. For instance, in 2005, biomass supplied 13% of total primary energy in China, and coal supplied 63% (IEA, 2007). Source: NBS (various years).

The surge in coal output in the 1980s came from new small local mines, mainly TVEs owned by local governments. In 1979, key state-owned coal mines accounted for 56% of total production, but fell to 37% in 1995, while, over the same period, the share of coal from TVE mines went from 17% to 46% (Figure 3.3). In the mid- to late-1990s, restructuring of state-owned enterprises allowed them to shed excess labour and to invest in expansion projects that enabled the post-2000 surge in output. Additional reforms encouraged the sale of TVE coal mines to private individuals, in part to separate managerial interests from the regulatory functions of local governments, so that most small mines are now in private hands.

The principles used by policymakers in developing the coal industry during the economic reform period since 1979 were summarised in three slogans. "Walking on two legs" referred to development of both key state-owned mines and local mines, the latter comprising state-owned, TVE and private mines. "Going forward in parallel with large, medium and small" signified that development should not favour only large-scale mines. Both policies aimed to move away from the planned economy towards markets with multiple actors. In this, the nation was largely successful; the ownership structure of coal mining enterprises has diversified, production scale has grown and production efficiency has gradually improved. Once exclusively under

Figure 3.3 .... Coal production by mine ownership, 1960-2006

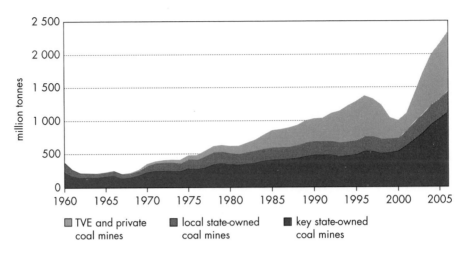

Note: Aggregate coal production data was subsequently revised: the fall in output around year 2000 was not as sharp as shown here (*cf.* Figure 3.2).
Sources: CCII (various years), CNCA (2006) and SIC (2008).

public ownership, China's coal mines now include those where the state is the owner or major shareholder, at the national, provincial and lower levels, and those that are collectives, co-operatives, private or joint ventures with foreign enterprises (Table 3.2). The last slogan, "let stagnant water run quickly", was a call for unexploited coal resources to become new supply streams. This call, too, was answered, but at a cost to long-term recovery rates.

Table 3.2 .... Coal mining enterprises by ownership, 2005

Ownership	Number of enterprises	Share of national total, %	Design capacity in 2005, Mtpa	Share of national capacity, %
State-owned or state-held shares	4 185	16.9	1 340.5	59.2
Collective and co-operative	11 695	47.1	642.9	28.4
Private	8 919	35.9	278.8	12.3
Foreign joint venture*	14	0.1	2.1	0.1
Total	**24 813**	**100.0**	**2 264.3**	**100.0**

*Notably Shanxi Asian American Daning Energy Co. Ltd., a joint venture company 56% owned by Asian American Coal Inc. to develop and operate the Daning Anthracite Mine that produced 2.3 Mt of saleable coal in 2007.
Source: China National Coal Association.

Deregulation in the 1980s enabled TVE and private coal mines to develop rapidly, exploiting the vitality of a sector that had been restricted under the planned economy. In 1985, entities other than the state were allowed to invest in and run coal mines and resource allocation rules were relaxed. Other reforms concerned project approvals, management, taxation, and mining rights (Ye and Zhang, 1998; Horii and Gu, 2001). Fostering the non-state sector circumvented difficulties in reforming state-owned mines, which had to maintain high levels of employment and social services. From 1979 to 1995, output of key state-owned mines rose by 125 million tonnes (Mt), just 19% of the total increase, compared to the additional 487 Mt from TVE mines which had developed quickly in response to rising demand (Annex I). After the 1980s, key state-owned mines suffered serious deficits as huge subsidies to cover the difference between their production costs and the low "plan" prices that end users paid for their allocations were withdrawn. Subsidies had given key state-owned mines little incentive to reduce their costs, while straining the government's ability to invest in new mines. The new TVE and private mines, which were not subject to the planning system's price controls, allowed non-state investment to flow in and to relieve shortages.

In 1993 – the first year in which output from TVE and private mines exceeded that from key state-owned mines – the government liberalised some coal prices, expecting this to reduce growing losses at key state-owned mines. Low "in-plan" prices were the main reason behind these losses, so further price liberalisation and exposure to market competition was expected to force greater efficiency. Other measures to reform the high-cost structure of key state-owned mines were intended to allow phasing out within three years of subsidies, which, nevertheless, were nearly RMB 6 billion in 1995, compared to RMB 350 million in 1985. Despite reforms, the market share of key state-owned mines continued to fall until 1995. These coal mining enterprises suffered enormous losses as they worked deeper and more difficult coal seams at ageing mines, employed huge numbers of surplus workers, faced high costs for social services and failed to improve their low productivities.

In 1998, the final step of institutional reform to affect the key state-owned coal mines was to introduce market-based trading for all coal mines. Most key state-owned coal mines were handed over to provincial governments who took responsibility for all social costs, which had become unbearable. By creating a level playing field, upon which all players would compete, it was hoped that the key state-owned mines would either survive, by becoming more efficient, or disappear from the market.

Meanwhile, the adverse effects of relying on small non-state coal mines was becoming apparent: disorderly exploitation of resources, deteriorating environmental performance and an unacceptable number of coal mining accidents. By 1998, the economy had slowed down following the Asian financial crisis of 1997, so demand was sluggish and the Chinese coal market became oversupplied. In response, the government adopted policies to regulate TVE and private coal mines to address not only the rising external costs of their operations, but also to aid the reform process at key state-owned mines. In 1996, production volume from TVE and private coal mines reached a peak, and coal prices fell to new lows. In that year, the government introduced a permit system for mining operations and, in 1997, it began to control output of TVE and private coal mines, closing many of them. Statistics show that in 1996 and 1997 around 14 700 illegal coal mines were shut down. In 1998 and 1999, a further 31 000 were closed, removing 253 Mtpa of production or 19% of China's coal output in 1997. The market, however, did not improve because many small mines simply raised output to compensate for low sales prices. Still, in 2000, the government declared the policy to close small mines had been successful, and announced that a further 18 900 mines with a combined annual capacity of 120 Mtpa would be closed. Throughout this period, some closed mines reopened surreptitiously, and many analysts believe a certain amount of coal output went unreported.

These closures, combined with a fall in investment at state-owned mines during the 1990s, foreshadowed coal shortages. In the past, investments at key state-owned coal mines mainly came from the government, but these declined from the early 1990s, giving way to bank financing and other sources of capital (Figure I.A, Annex I). At first, financing came as government-backed loans from banks such as the State Development Bank. By 1997, commercial banks were the main source, but investment fell dramatically from 1998 to 2000, consistent with the sluggish coal demand seen in the period immediately after the Asian financial crisis. At that time, coal-sector reforms focused mainly on mine closures with few measures to support capacity expansion, hence few new large coal mines were opened.

The appearance of a tight supply situation in 2002 forced many mines to operate at full capacity and accident fatalities rose to a peak of almost 7 000 in that year (Section 3.3). The boom in coal demand led to a resurgence of investment at key state-owned mines, through retained earnings, bank loans, and, later, public sales of shares. TVE and private coal mines also responded quickly to the tight market, demonstrating their extraordinary ability to meet changes in the market. From 2000 to 2006, their annual output rose by 232% or 623 Mtpa, compared to 109% or 584 Mtpa for key state-owned mines. Production from TVE and private coal mines in 2006 was at an historic high of 892 Mt, 38% of national output (Figure 3.3 and Table I.D, Annex I).

Most TVE and private coal mines are still quite small. In 1995, when township and private coal mines numbered 72 919, the annual production of the average mine was just 1% of a typical key state-owned coal mine (Figure 3.4). However, in aggregate, the 596 key state-owned mines produced less coal than the TVE and private mines. By 2005, the number of small mines had fallen to 16 276, but output per mine was up by almost an order of magnitude. At the same time, key state-owned mines doubled in size and grew in number to 735; they now account for the largest share of coal output.

Most of China's coal mines are small by world standards, but a growing share of output is coming from larger new mines. In 2005, 39% were classified as large (>1.2 Mtpa), 7% medium (0.45-1.2 Mtpa) and 54% small (<0.45 Mtpa); large and medium mines accounted for 54% of total coal production. Ten coal mining enterprises now each produce over 30 Mtpa, including two over 100 Mtpa (Figure 3.5). In 2006, Shenhua Group alone produced 137 Mt and had total sales of over 170 Mt which included coal purchased from smaller mining companies. Other large enterprises – Datong Coal Mining Group, China National Coal Group and Shanxi Coking Coal Group – each surpassed 50 Mt. The average mine size has risen from below 20 ktpa in 2000 to over 90 ktpa currently. Meanwhile, coal mines have strengthened their co-operation with power and chemical companies; through diversification and expansion, 23 coal enterprises have entered the ranks of China's top-500 companies.

Figure 3.4 .... Coal mine categories: number of mines and average annual production per mine, 1995 and 2005 (total annual production from each category shown by area of circles)

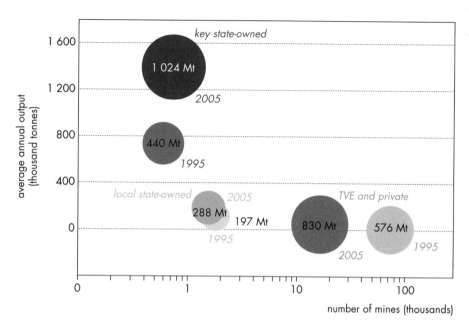

Sources: 1995 data: Ye and Zhang (1998); 2005 data: CNCA (2007).

Figure 3.5 .... Coal output by province and production by major coal group, 2006

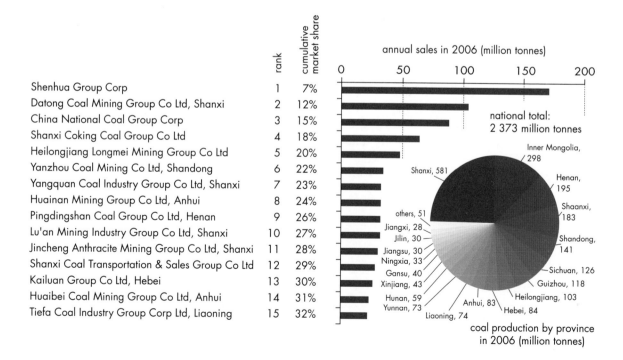

	rank	cumulative market share
Shenhua Group Corp	1	7%
Datong Coal Mining Group Co Ltd, Shanxi	2	12%
China National Coal Group Corp	3	15%
Shanxi Coking Coal Group Co Ltd	4	18%
Heilongjiang Longmei Mining Group Co Ltd	5	20%
Yanzhou Coal Mining Co Ltd, Shandong	6	22%
Yangquan Coal Industry Group Co Ltd, Shanxi	7	23%
Huainan Mining Group Co Ltd, Anhui	8	24%
Pingdingshan Coal Group Co Ltd, Henan	9	26%
Lu'an Mining Industry Group Co Ltd, Shanxi	10	27%
Jincheng Anthracite Mining Group Co Ltd, Shanxi	11	28%
Shanxi Coal Transportation & Sales Group Co Ltd	12	29%
Kailuan Group Co Ltd, Hebei	13	30%
Huaibei Coal Mining Group Co Ltd, Anhui	14	31%
Tiefa Coal Industry Group Corp Ltd, Liaoning	15	32%

Sources: company annual reports for 2006 and NBS (2008).

Still, the hallmark of China's coal mines (as in other industries) remains variety; the average output from a key state-owned mine is almost 30 times that from a TVE and private mine. At one extreme, Shenhua Group's Bulianta underground mine produced over 20 Mt in 2006, making it the world's largest, while the majority of TVE and private mines produced under 50 ktpa. Larger mines mainly use the longwall mining method, while smaller mines employ room-and-pillar or shortwall methods using continuous miners, with blast mining still common using explosives. Parallel to this, the average degree of mechanisation (*i.e.* proportion of coal produced entirely by mechanical means) at key state-owned mines was nearly 83%, with some reaching 100%, like most Shenhua mines, while the national average is 45%. Some 58% of small mines continue to use blasting and manual extraction methods that are not only dangerous, but leave a great deal of coal behind. While some mines exhibit world-class performance, China's average recovery rate is 30-35%, and can be as low as 10-15% at small mines where only the thickest of multiple seams are exploited (CASS, 2007).

While the early part of this decade saw a resurgence in output from small mines in response to strong market demand, by 2005, closure or rectification of small mines had resumed, with 5 290 small coal mines closed during that year. In recent years, MLR has set strict prospecting and mining procedures for large- and medium-size coal mines. Of the categories of unregulated coal mines scheduled for closure, TVE coal mines make up a large proportion.[3] Although the number of TVE and private coal mines has fallen, the segment is buoyant and remains important as the swing producer. As for the future, the tight supply of coal in the market should improve as industry consolidation and the massive investments in key state-owned coal mines bear fruit.

Under the 11th Five-Year Plan (2006-10), 13 large coal bases are planned: Shendong in Inner Mongolia and Shaanxi; North Shaanxi; West Shaanxi; Jincheng in North Shanxi; Central Shanxi; East Shanxi; Luxi in West Shandong; Huainan and Huaibei in Anhui; Jizhong in Central Hebei; West Henan; the northeast region of East Inner Mongolia; Yunan and Guizhou; and East Ningxia. These areas have rich reserves of a broad range of coal types, generally of high quality, good mining conditions and the ability to support large-scale mining operations and ancillary facilities, like washeries and CTL plants. Together they have resources of 853 Gt, and in 2005 their collective output was 84% of China's total, with a slightly higher than average proportion coming from large and medium mines. Most larger new mines are to be established within

---

3. Five categories are to be closed immediately, pending remedial work and approvals: mines producing beyond their ventilation capability; mines with no methane drainage system, inadequate gas monitoring facilities and abnormally gassy mines; mines with signs of methane outburst but no precautionary measures; new, upgraded or expanded mines put into production prior to completion of official safety inspection, violating construction procedures, without approval or exceeding terms of approval; and mines that fail to apply for work safety permits within the prescribed time limit. A further four categories must be permanently closed: illegal mines with no licence; mines resuming production after being closed; mines unable to meet work safety standards and hence without work safety permits; and mines refusing government safety supervision and inspection.

these coal bases. By 2010, output from these 13 bases is projected to reach 2.24 Gt, with 60% from large mines and 20% each from medium and small mines, with many mines to be upgraded or consolidated into larger mines.

COAL MINE SAFETY, ENVIRONMENTAL PROTECTION AND MINER WELFARE

Coal mine safety is one of the key issues facing China's energy sector. Mining has always been a dangerous occupation in all countries, but the sheer size of China's industry raises it to the level of a major social concern. Accident fatality rates have been dropping each year since 2000, from over 6 fatalities per Mt, to 2.04 per Mt coal mined in 2006 (Figure 3.6). Several thousand miners still lose their lives each year, but total fatalities have fallen in recent years, a tribute to the effectiveness of recent efforts to improve safety. Besides enhancing safety laws and regulations, the government introduced an accountability system for leaders with decision-making power over mine operations and safety. Local agencies charged with work safety have intensified inspections and enforcement. State and corporate outlays on better safety systems and training have risen, and many of the least safe mines have been closed.

Figure 3.6 .... Coal mining fatalities: annual total and rate per million tonnes of coal mined, 1949-2006

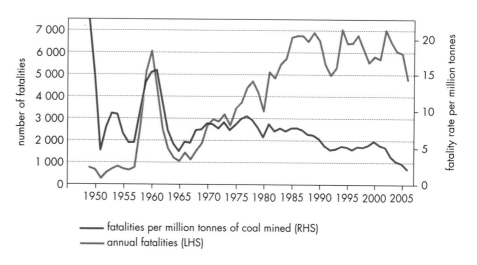

Source: SACMS (2007).

The fatality statistics, however, do not account for mortality from occupational disease, also a major killer of miners. China now has 600 000 former coal miners who suffer from pneumoconiosis. In 2003, key state-owned mines reported 12 000 new cases of pneumoconiosis, 1.5% of the total number of workers. There is no universal reporting system for local state-owned and TVEs mines – the total number at these mines is certainly much larger, possibly five times the number, assuming a pneumoconiosis prevalence rate equal to that at key state-owned mines.

In recent years, financial losses from coal mining accidents amounted to about RMB 1.5 billion per year. Treatment costs for occupational pneumoconiosis are an estimated RMB 1.75 billion per year, yet this problem is receiving much less attention than accidental deaths. From the time of its founding in 2003 to early 2007, the China Coal Miner Pneumoconiosis Treatment Foundation had dispensed nearly RMB 12 million and aided 1 818 pneumoconiosis patients with whole-lung lavage treatment and rehabilitation.

The reasons for the grim coal mine health and safety situation are not easily overcome, as most mining is underground in often difficult geology. In China, coal reserves suitable for opencast mining account for only 4% of the total. At underground mines, complex geology, with dipping coal seams, faulting and roof fracturing, affects one-third of mines. Since underground mining will continue to dominate, including many small-scale mines, further investment in safety systems and in worker training is needed. Levels of general education and professional training are particularly limited among the two million workers in small mines, which account for over 70% of both accidents and mining deaths (Table I.F, Annex I). Even when safety equipment is available at such mines, workers often lack the ability to use it effectively. Health and safety will remain major preoccupations in China for many years.

Similarly, concern about environmental issues is sharpening. In 2005, subsidence from coal mining affected 45 000 km^2, and an even greater area suffered loss of surface waters and soils. Mine water drainage is a problem, although of the 4.5 billion m^3 pumped from mines in 2005, 44% was treated. The biggest coal industry in the world produces spoil equivalent to 10–20% of its raw coal output, and waste material from coal washing adds a further 15–20%. The current total volume of spoil exceeds 5 Gt, occupying 16 000 km^2 of land. Of the 350 Mt of spoil produced in 2005, 43% was reportedly put to productive use (CCII, 2007a). Most mine gas is vented to the atmosphere for safety reasons. In 2005, total emissions of coal mine methane were 15 bcm, of which 2.3 bcm were recovered and 40% of this used as a fuel. Government estimates of damage from coal mining, including wasted resources, environmental pollution, ecological destruction and surface subsidence, total about RMB 30 billion per year.

Before the 1980s, people acting independently attempted to plant crops, graze animals and build property in abandoned mining areas. In the early 1980s, the government launched a subsidence reclamation plan, and in 1988 introduced the first *Land Reclamation Provisions*. Meanwhile, the obligations of industrial enterprises, mine owners and individuals responsible for damage were determined, and local branches of land, water resources and environmental protection administrations became responsible for implementing the provisions. In 1995, the former State Bureau of Land Administration approved trial standards for land reclamation. These rules and later laws established the principle that mining enterprises themselves must restore any environmental damage they cause and reclaim land affected by their mining activities. Failure to comply results in the enterprise being liable for payment of a reclamation fee.

As for restoration and reclamation of the 9 000 km^2 of land that has been destroyed by past coal mining activity, there are a number of potential funding sources. One is financing from central and local governments, including a proportion of resource taxes, land reclamation fees, compensation payments for the occupation of farmland, sewage charges, and soil and water conservation fees. In recent years, the central and provincial governments have made large investments to rectify subsidence damage and for resettlement: RMB 7 billion in Shanxi, RMB 1.2 billion in Huainan, Anhui, and RMB 800 million in Fuxin, Liaoning. The other sources are social investment funds, investors who gain the right to use abandoned land after funding reclamation, charitable donations and international financial assistance.

The consolidation and modernisation of China's coal industry around 13 large coal bases will still leave other legacy issues to address: closure and bankruptcy of depleted mines, redundancy and resettlement, and replacement of the social functions formerly provided by state-owned mines. At the end of 2003, coal mines with a capacity of 65 Mtpa faced closure, bankruptcy or revocation of production licences because of resource depletion. One such facility is the Haizhou opencast mine of the Fuxin Mining Group in northern Liaoning. Once the largest mechanised open pit in Asia, it is also the largest opencast mine to have been closed in China after producing 210 Mt of coal over 50 years. In 2002, the mine was included in a group of mining enterprises receiving national bankruptcy support to resettle employees and reduce other burdens on the enterprises following depletion of their coal reserves. The state budgeted RMB 860 million to close Haizhou, including compensation of RMB 20 000 for each of Haizhou's 10 100 workers and costs of resettlement and re-employment. An industrial heritage site is planned as part of the mine reclamation.

Most mines are in remote areas and have dependent communities, with hospitals, schools, police, fire protection and provision of water, electricity and heat. Some mines even have jurisdiction over rural towns. Reforms have gradually separated enterprises from social functions, but progress is slow and many enterprises still carry large social burdens. In 2005, key state-owned coal enterprises spent over RMB 10 billion on social functions. (CCII, 2007b)

While wages for coal miners and other workers have gradually increased in the past few years, they are still much lower than in other industries. In some places, the average income of coal mine workers is only one fifth that of electricians and others face unpaid wages. In 2004, key state-owned mines owed their employees RMB 2.9 billion in back wages (CCII, 2007b). Living space for employees and their dependents at key state-owned mines is only 8 m^2 per person, just one-third that of the average city dweller. In the northeast and other old mining areas, many miners still live in shantytowns. These legacy challenges are, in principle, similar to those faced in developed countries where coal mines have closed. Their experience is summarised in Chapter 7.

COAL TRANSPORT

China's main coal mines are in the north and northwest, while energy demand is greatest in the eastern and south-eastern coastal areas (Figure 3.1). As a consequence, coal must be transported long distances from west to east and from north to south, which has always been a limit to coal use. Of the three means of transport – rail, water and road – rail is by far the most important. The volume of coal and coal products moved by rail increased from 629 Mt in 1990 to 1 390 Mt in 2006 (1 120 Mt on the national railway) when it accounted for 48% of total rail freight tonnage and 58% of total coal production (Figure 3.7). Water transport, especially coastal shipping from north to south, is the second-largest carrier. From 2000 to 2006, the annual growth rate of coal freight carried by national rail was 8.5%, much lower than the 11.1% production growth rate over the same period. The rapid rise in coal output has meant that, even with large new rail projects, a great deal of incremental output has had to be carried by truck, sometimes over long distances. Nevertheless, in terms of tonne-kilometres (t-km), road transport remains the smallest carrier of coal since it is used mainly for local distribution. The average delivery distance is 550-575 km by rail, 1 900 km by water and about 80 km by road. Rail rates are formulated by NDRC and the Ministry of Railways, and then approved by State Council. The rate structure is complex because, for example, charges are different for each route depending on their technical specification, such as whether the route is electrified. In general, the rate from the largest coal-producing regions to coastal ports is around RMB 100/t, while it is around RMB 150/t to eastern regions of China. The current rate for trucks is about RMB 0.6/t-km, compared to RMB 0.07/t-km for rail and RMB 0.03-0.04/t-km for water, making the transport of coal over long distance by road very uneconomic.

Figure 3.7 .... Coal transport by national rail, inland waterway and coastal vessel, 2001-06

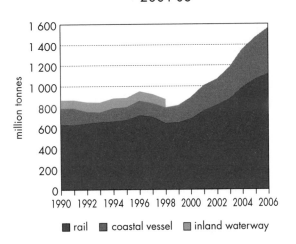 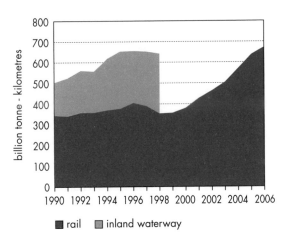

Notes: Rail transport data does not include the volume of coal carried on Shenhua Group's Shuohuang railway or local railways. The National Bureau of Statistics of China stopped reporting tonnage of coal carried by inland waterways in 1999. When last reported in 1987, road transport accounted for 20% of the total volume of coal freight.
Source: NBS (various years).

Coal is transported by rail along three distinct corridors. The northern corridor comprises the Daqin, Fengsha, Jingyuan, Jitong and Shenshuohuang railways. These lines transport mostly steam coal from northern Shanxi, northern Shaanxi and the Shendong region to Beijing, the ports at Qinghuangdao and Tianjin in Hebei, and customers in northeastern and eastern areas of China. The central corridor comprises the Shitai and Hanchang railways, which mainly bring coking coal and anthracite from eastern and central Shanxi to the port at Qingdao and to consumers in the East and

South. The southern corridor is made up of the Taijiao, Houyue, Longhai, Xikang and Ningxitie railways. These transport coking coal, steam coal and anthracite from northern Shaanxi, central Shanxi, Shendong, Huanglong and eastern Ningxia to ports at Rizhao and Lianyungang, and to South and East China.

In 2006, the Ministry of Railways, in accordance with the 11th Five-Year Plan and the Long-Term Railway Network Plan, accelerated expansion of rail networks to relieve bottlenecks that had hindered coal deliveries into growing markets. Planned investment for 2006 was RMB 165 billion for 87 new railway projects, resulting in additional rail transport capacity of 80 Mtpa. According to the Ministry of Railways, demand for the rail transport of coal is expected to be 1 700-1 800 Mtpa by 2010 and 2 000-2 200 Mtpa by 2020. Planned rail capacity falls short of these figures, so bottlenecks are expected to persist, particularly in Shaanxi and the central southern part of Shanxi as coal production becomes further concentrated in these provinces and Inner Mongolia.

In recent years, road transport of coal has played an important role in relieving the pressure on rail capacity, which has risen more slowly than coal output. Shanxi, the country's largest coal producer, shipped out 430 Mt of its total 2005 output of 543 Mt. Of this, 100 Mt was delivered by road, putting tremendous strain on the newly built network of highways. During the 11th Five-Year Plan period, Shanxi will build or renovate 3 000 km of roads, linking its three coal bases and improving the ability to move coal within Shanxi and to other provinces.

Coastal transport of coal from the north to the south is via the ports of Qinhuangdao, Tianjin, Jingtang at Tangshan and Huanghua (Hebei), Qingdao and Rizhao (Shandong) and Lianyungang (Jiangsu). In 2005, there were 42 dedicated berths for coal loading in the north, with total capacity of 343 Mtpa. The 122 dedicated coal unloading berths (including 75 deep-water berths), with total capacity of 270 Mtpa, located in East and South China are important in securing energy supplies to that region. Throughput of coal at seaports and the main ports on the Yangtze River for onward shipment to customers was 408 Mt in 2007, including domestic sales of 345 Mt and the remainder for export (Table I.G, Annex I). The State Council approved in 2006 a National Coastal Ports Plan to strengthen this crucial link in the coal transport system which is expected to reach 600 Mtpa by 2010 and 750 Mtpa by 2020.

COAL DEMAND AND STOCKS

Just as it is the largest producer, China is also the world's largest coal user, with consumption trends roughly following the production trends described above. Over the past two decades, China's coal demand has undergone a remarkable structural change, becoming concentrated in power and heat generation and three industrial sectors (Figure 3.8). In 1985, the power and heat sector accounted for just 23% of total primary coal use, followed by residential (20%), non-metallic minerals production[4] (11%), iron and steel (11%) and chemicals (6%). Since then, all sectors have become much more dependent on electricity, as is typical in developing economies. Direct

---

4.   Mainly cement for the construction industry.

burning of coal by end users has become less important except in certain sectors where substitution is not possible, and where the relative abundance of coal compared to oil and gas has led to continued reliance on coal. By 2005, the share of primary coal use going to power and heat generation was over 57%, while non-metallic minerals used 8%, iron and steel 14% and chemicals 3%. Residential use of coal had fallen to just 4%, and absolute consumption of coal by households had halved due to fuel switching, mainly to natural gas.

Figure 3.8 .... Share of coal consumption by sector in China, 1980-2005

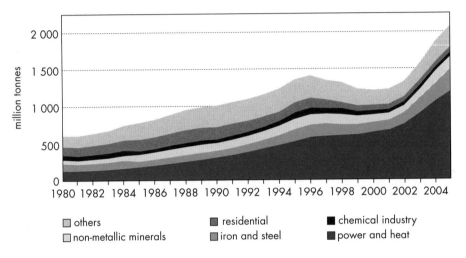

Source: calculated using IEA database of coal and coke production-consumption balances (World Coal Statistics.ivt available at http://data.iea.org).

In 2006, coal production and sales volumes continued to increase steadily, with, supply and demand in close equilibrium, and relatively stable prices. Total coal consumption increased by 9.8% compared with the previous year, a lower growth rate than the 11.3% seen in 2005 (IEA, 2008a). Growth in demand continued to be strong in 2007 at 10.3%, with prices continuing to rise. Looking forward, the power sector will continue to be the main driver of coal demand for years to come, supplemented by demand for steel, building materials and chemicals. These sectors are treated in turn below.

China's power generation capacity has grown at an incredible pace, and continues to be dominated by coal-fired plants. In 2007, total installed capacity grew by more than 15% to reach 718 GW, with total generation growing at a similar rate to 3 264 TWh (CEC, 2008). This included 556 GW of fossil- (mainly coal-) fired capacity, which generated 83% of the nation's power output. Fossil-fired units typically have higher load factors than hydropower and other units, so their share of installed capacity was lower (77%). Of the over 100 GW of new capacity installed in 2007, around 70 GW was coal fired. The rate of construction slowed compared to 2006, when China's total capacity rose by 107 GW, including 93 GW of coal-fired capacity. To illustrate further, China built the equivalent of one typical coal-fired unit of 500 MW about every two and a half days in 2007, compared to one every two days in 2006.

More than any other factors, demand for electricity and the efficiency of coal-fired generation will determine China's future coal demand. Average efficiency has been rising steadily as larger new units come on line (Figure 3.9). The mean size of new coal-fired units is about 500 MW, while increasingly larger units of up to 1 000 MW are being commissioned that are close to state-of-the-art in their class. The rapid pace of construction, combined with closures of small, less-efficient units (reportedly over 14 GW in 2007), means that China's average power plant efficiency is rapidly catching up with that of developed countries. It is estimated that the average efficiency of China's coal-fired fleet was 32% (LHV, gross output) in 2005 and is expected to approach 40% by 2030 as more large supercritical units come on line and older subcritical units are phased out (IEA, 2007). Some power utilities and coal mining enterprises have agreed to jointly build mine-mouth power plants, a vertical integration that holds promise of building and operating larger-scale, more-efficient plants. Chapter 5 has more details on the economics and technologies of coal-fired generation in China.

Figure 3.9 .... Average efficiencies of coal-fired power plants in China by age and unit size

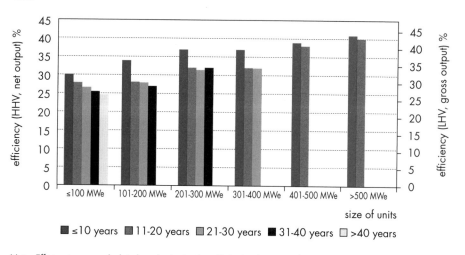

Note: Efficiencies are calculated on the basis of coal's higher heating value (HHV) and the net electrical output from a generation unit. The right-hand scale, showing efficiency on a lower heating value (LHV), gross output basis, is indicative since the conversion depends on coal quality and unit size.
Source: IEA (forthcoming).

Iron and steel output has expanded rapidly since 2000, driven by a boom in construction and infrastructure development, as well as strong demand for consumer durables, with a knock-on demand for coal. According to the National Bureau of Statistics of China, the production of coke, iron and steel in 2006 consumed 439 Mt of steam and coking coal – 18% of China's total coal consumption – having grown at an annual rate of 13.6% since 2000.[5] China's iron and steel output continued to grow strongly in 2006 and 2007, producing respectively 419 Mt and 489 Mt of crude steel (second-ranked Japan produced 120 Mt in 2007). While the country's steel plants are far more

---

5. Some coke is used outside the iron and steel sector, and some is exported, hence Figure 3.8 shows a lower share.

efficient than they were a decade ago, significant opportunities remain to become even more efficient. Nevertheless, even rapid efficiency gains will not prevent coal demand in this industry from rising so long as steel demand keeps going up.

Annual coal consumption in the non-metallic materials sector was 165 Mt in 2006, mainly for cement production. China is the largest producer in the world, making 1 204 Mt in 2006, up 19% over 2005. Manufacturing processes tend to be more energy-intensive than elsewhere, although substantial progress has been made. In the cement industry, for instance, about 70% of output now comes from modern dry-process preheater and precalciner kilns. In the 1990s most cement still came from vertical kilns, a process that has been improved considerably through domestic R&D, but which remains more energy-intensive – and produces lower-quality product – than modern kilns. Steady improvements in efficiency are expected to continue as shaft kilns are replaced by rotary kilns. Similar trends are observed in glass and porcelain manufacturing, but the brick and tile industries have been slower to change.

Demand for coal in the chemical industry comes mainly from the production of chemical products, *e.g.* synthetic ammonia, urea, methanol and dimethyl ether (DME). As demand for chemical fertilisers has grown, so has the chemical industry's demand for coal, since coal is the primary feedstock for ammonia synthesis in China. Output of chemical fertilisers rose 14% in 2006 to 53 Mt and synthetic ammonia production increased 7% to 49 Mt. Methanol production has also risen fast, to nearly 11 Mt in 2006. Recently, in a bid to mitigate rising reliance on imported oil, coal-to-liquids projects – producing chemicals and/or fuels – have received a great deal of attention. For instance, DME has been listed in the national medium- and long-term development plans for science and technology, billed as a clean alternative energy. By 2010, DME production is expected to increase by 4 Mt. The central government, which first encouraged such projects, is now taking a more cautious approach, as they promise not only to raise demand for coal in a tight market, but also to place heavy new burdens on the environment and on scarce water resources in coal-bearing regions.

Quantities of coal consumed in other sectors have been declining, as have coal exports. Urban households have been turning away from coal towards electricity, natural gas and delivered heat, although rural households continue to rely mainly on coal and solid biofuels. Coal and coke demand for households was 76 Mt in 2006, and is anticipated gradually to decline. Similar declines are expected in industrial uses other than those mentioned above, as industrial kilns and boilers become more efficient, or are switched to other fuels.

Influenced by national policies, coal exports have reduced since 2003 while imports have increased. In 2006, China's coal exports were 63.2 Mt, down 12% on 2005, and in 2007, they fell a further 15% to 53.7 Mt (IEA, 2008a). In 2006, coal imports were 38.1 Mt and they rose to 47.6 Mt in 2007. On recent trends, the nation would become a net importer in 2009. China's coal trade represents a significant share of the international market for coal, where it has an influence on prices, but it remains a very small proportion of the total domestic market.

In December 2006, the national coal inventory stood at 144 Mt, nearly 5 Mt higher than at the beginning of the year. Stocks at key state-owned coal enterprises rose by 44% to 42 Mt, while those at power stations decreased by 0.8% to 24 Mt and those at key steel plants fell by 9% to 3 Mt. Coal stored in transhipment ports decreased by 27% to 10 Mt. The growth of producer stocks and fall in consumer stocks reflected the ongoing transport capacity shortages and declining enthusiasm among consumers to hold coal stocks, which, along with a period of severe weather, helped lead to shortages in early 2008. The government is encouraging consumers, particularly power plants, to raise stocks, with the aim of reducing fluctuations in the coal market caused by transportation problems and to help to stabilise coal prices. There is also a public consultation on whether China should build a strategic coal reserve by 2015 (China Daily, 2008).

EVOLUTION OF COAL MARKETS, PRICING AND GOVERNMENT OVERSIGHT

Until the 1980s, China had no coal market to speak of, as coal was allocated through centralised planning. The reforms that allowed small mines to spring up led to local markets for their products. Later, state-owned mines were allowed to join in market activity, selling any "beyond plan" production in these new markets, while still providing most of their output to customers through state-mandated transactions. Throughout the 1990s, state-owned mines participated ever more deeply in markets, becoming more exposed to demand fluctuations and gradually freer to respond. Since the late 1990s, China's coal market has evolved from one where supply outstripped demand to one where demand frequently outpaces supply. Between 2001 and 2004, coal output rose by over 200 Mt each year, making it the fastest growing period in the history of China. Nevertheless, in both 2003 and 2004, strong demand, including coal destined for export, led to coal shortages. Since 2005, the rate of production growth has slowed somewhat, as measures have been taken to close some small mines and to improve safety at others, as well as to limit exports of coal and energy-intensive products.

From a centrally planned economic system to a functioning coal market

Under the centrally planned economic system that China adopted from 1949 to 1979 when economic reform began, all goods were categorised as either allocated goods within the state-owned system ("raw materials and producer goods") or goods transferred to or between non-state entities ("retail commodities and consumer goods"). This distinction was of vital importance to the coal sector, since its production fell under different ownership systems. A lump of coal could either be a "raw material" or a "retail commodity", depending upon the ownership of the producer, and follow entirely different distribution systems.

Coal mines were classified into three groups. *Centralised-allocation* mines were directly managed by the central government; since 1993, these have been called *key state-owned* mines. *Local state-owned* mines were managed by provincial or county-level governments. *Township and village enterprise* (TVE) mines were managed by township or village governments and *private* mines by individuals. Centralised-allocation mines provided "raw material" to other state-owned enterprises under allocation plans drawn up by the State Planning Commission (SPC), the State Administration for Materials (SAM) and the Ministry of Coal Industry (MOCI). Similarly, each provincial Planning Commission, Bureau of Materials and Coal Industry Administration Bureau made plans for local state-owned coal mines. Coal from TVE mines was distributed as a "retail commodity" and bought by local, collectively owned industrial enterprises, by service industries for heating, by households and by agricultural associations. SAM ensured that almost all the coal from centralised-allocation mines was supplied as "raw material" to state-owned enterprises; it was not allowed to flow to enterprises under the management of local governments or to TVEs.

Each year, state-owned coal users applied to their administrative superiors (*e.g.* Ministry of Electric Power Industry and Ministry of Metallurgical Industry) for quantities and qualities of coal for the following year based on production targets. The ministries in turn would submit an aggregated request to SPC. At the same time, MOCI summarised the production targets of centralised-allocation mines and local state-owned mines for the SPC, which then, with SAM, calculated annual coal balance plans and set allocation quotas. Then MOCI would match mines to coal users through its annual "coal-ordering conference" (first held in 1961), where mines, end users and transport enterprises would confirm specific deals.

The distribution of coal as a "retail commodity" was performed through the Coal Industry Equipment Corporation (Coal-Equipment Corp.) under the Ministry of Commerce (MOFCOM). From its branches in each province, it centralised purchases of coal from TVE mines and some local state-owned mines, handled coal sales and facilitated inter-regional redistribution. Even these transactions, however, were subject to MOFCOM allocation plans and MOCI sales plans.

Fulfilling industrial demand was considered the top priority, particularly since coal had been constantly in short supply. Supply to service and residential users was distinctly secondary. Thus, most coal was distributed as "raw material"; 71% of coal was distributed under the centralised allocation system as "raw material" in 1965 and 56% in 1978. From the mid-1970s, however, the two distribution systems failed as China's economic structure changed. Supply from centralised-allocation coal mines could not keep pace with rising demand, thus allowing TVEs to show their vitality.

The non-state enterprises that fuelled growth in the 1980s built many small and inefficient plants in the power, steel, cement and chemical sectors. Throughout the 1990s, however, these smaller facilities were replaced by progressively larger ones, though often still small by world standards. The power industry is a good example of these changes. In 1988, there were only 32 generating units with capacities of 300 MW or more, accounting for 14% of total capacity. By 2003, the number of such units had grown to 339, representing 45% of overall capacity. At the same time, from 1988 to 1998 the number of small units (<100 MW) nearly doubled and their combined capacity grew by 1.8 times, though their share of capacity fell.[6] The number of boilers in the industrial and residential sectors rose by 300 000, adding 840 000 tph of steam

---

6. From 1998 to 2003, there was no power shortage, and even oversupply, so the government was able to shut down the smallest generation units (<50 MW) and discourage the construction of new units. Thereafter, as the economy began to suffer from power shortages, the number of small units grew again.

generating capacity, over the two decades up to 1998. Average capacity remained small, with a typical boiler consuming under 1 000 tpa of coal, but the combined number of boilers was massive and total boiler demand eventually exceeded 300 Mtpa.

During the early 1990s, coal production and consumption saw significant changes due to deregulation and rapid economic growth driven by the non-state sector, yet the key and local state-owned mines continued to supply consumers according to the planning system. As output from TVE and private mines rose, so did the share of "retail commodity" coal, while output from key state-owned mines remained sluggish, and by the 1990s small, often primitive mines were selling nearly as much coal as the modern key state-owned mines. Until 1993, state-owned mines could only sell coal produced above plan quotas into the market. Even though all output from state-owned mines nominally went to the market from 1993, the coal-ordering conferences continued, handling the majority of the state-owned mines' transactions and setting rail transport quotas. When coal demand fell in the late 1990s, the conferences served as a stable distribution channel for the large mines that took part. This practice continued until 2002, when demand growth created a sellers' market.

Since 1993, coal produced by TVE and private mines has been freely sold in the market. Most product in this category, other than from Shanxi TVE mines, was consumed by non-state users within the county in which the coal mines were located (Table 3.3). The market for small mines was restricted by transport limitations and the generally poor quality of their product.[7] Rail and water are the cheapest modes of long-distance transport for bulk commodities like coal, while trucks are more cost-effective for shorter distances. Railway quotas for coal, however, were mostly set through the coal-ordering conferences. Enterprises not parties to the conference had to apply separately for their own railway transport quotas. Since the 1980s, rail capacity had been chronically inadequate, so for TVE and private mines, transport by truck was often the only option available.

Table 3.3 ....Markets for key state-owned coal mines and TVE and private coal mines, 1995 (million tonnes)

	Total production	**Distribution beyond county**	**Distribution beyond province**
Key state-owned coal mines	482	n/a	301 (62.4%)
TVE and private coal mines	579	249 (43.0%)	112 (19.3%)

Source: *National Township Coal Mine Survey Report* quoted in Li, Dou, Huang et al. (1999).

Markets in China have undergone massive changes in the last few years through consolidations and closures. In 2001, three Shanxi coking coal producers combined into the Shanxi Coking Coal Group, which instantly became the nation's largest such enterprise. In 2002, mines in Ningxia formed the Ningxia Coal Group. In 2003, the Datong Coal Mine Group in Shanxi acquired over ten other mines in the area. In 2004, ten coal companies consolidated to form the Shanxi Coal Group, while four key state-owned mines in Heilongjiang formed the Longmei Mining Group. In 2005, key

---

7. Very little output from small mines was washed, just 7% in 1998. Furthermore, with limited production from each coal mine, blending the output from several mines is needed to assure product consistency.

state-owned mines in Sichuan formed the Sichuan Coal Industry Group and three key state-owned mines in Anhui have been similarly consolidated. Government policy has encouraged this consolidation; the 11th Five-Year Plan calls for the creation of five to seven large coal companies, each producing in excess of 100 Mtpa, with a focus on the 13 coal bases (Section 3.2). From 1998 to 2006, the top-10 coal producers have moved from a 14% share of national output to 27% (Figure 3.5). Closure of TVE and private mines has also contributed to a rising market share for key state-owned coal mines.

Coal usage has also consolidated. Since the 1980s, the number of small facilities in the four main consuming sectors have declined, closed for economic reasons and squeezed by campaigns to phase out inefficient and highly polluting equipment, like small furnaces, vertical cement kilns and small chemical fertiliser plants. However, past efforts to close small power generation units (<50 MW) were relaxed due to power shortages. By 2006, as tight supply was alleviated, enforcement became easier, and the threshold was raised to include units up to 100 MW. NDRC announced that over 14 GW of small power plants were closed in 2007.[8] Moreover, rising coal prices coupled with the availability of domestically produced supercritical generation units makes replacing small units with large ones more economically attractive. Reforms in the power sector further disadvantage small units. For example, the 2007 *Administrative Measures on Energy-Saving Dispatching* encourage the dispatching system to prioritise more-efficient units (NDRC, 2007).

Coal pricing

Thanks to the better-functioning market structures in place today, China has made a great deal of progress in establishing market-based coal-pricing mechanisms. After many years of price control under the planned economy, domestic prices have risen in line with international markets (Figure 3.10).

Beginning in 1958, national quality standards and uniform coal prices were set. At first, prices were allowed to float within a 20% range (or more for local state-owned mines) according to market demand and supply. In 1966, the price of "raw material" coal was incorporated into the national uniform pricing system, and prices of "retail commodity" coal essentially followed. Before 1993, there was an official benchmark price list based upon ash content and size of coal (sulphur content was ignored). When inflation raised mining input costs without any corresponding rise in coal prices, mines suffered worsening deficits. Adjustments to benchmark prices in 1958, 1965, 1979 and 1985 were not enough to relieve coal producers' financial difficulties. In the 1980s, a two-track price system emerged by which out-of-plan coal production could be sold at higher prices, easing matters somewhat.

---

8. While some small generation units were shut down, others were installed for cogeneration of heat and power, or for power generation from coal tailings (waste coal).

Figure 3.10 .... Coal prices in China compared with international market prices, 2000-08

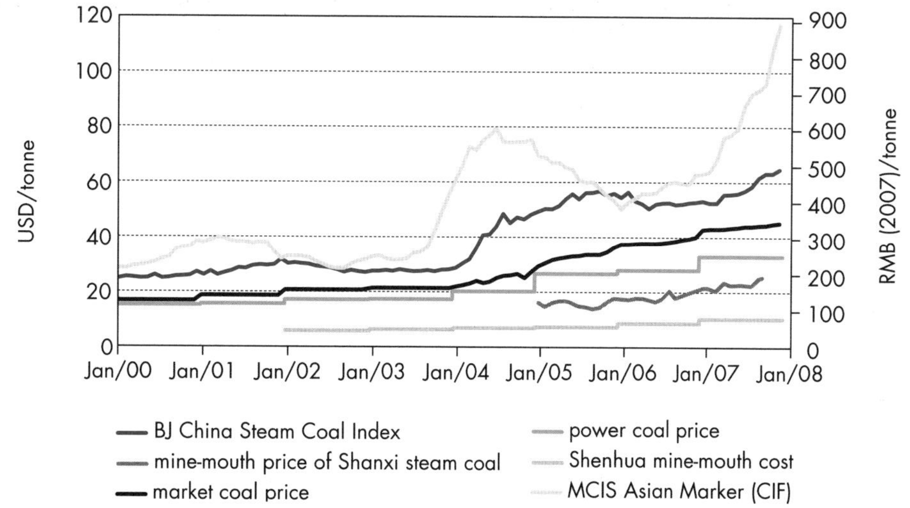

Notes: The BJ (Barlow Jonker) China Steam Coal Index is a weighted average of steam coal export prices (FOB) calculated from Chinese customs returns that includes both spot and contract trades of all coal qualities. The MCIS Asian marker price reflects delivered (CIF) prices to ports in Japan, Korea and Chinese Taipei. For pre-2000 coal prices in China, see Chen (1998) and issues of *China Price*, a monthly journal of the China Price Information Center (www.chinaprice.gov.cn).
Sources: McCloskey Coal Information Services (MCIS), Barlow Jonker (BJ), FACTS (2008), Beijing HL Consulting (2006), Shenhua Group (2005 prospectus and 2006/2007 annual reports), and China Coal Resource (www. sxcoal.com).

In 1993, coal pricing was liberalised, though with significant exceptions. The SPC issued guidance price bands for coal supplied to power plants and negotiators at the national coal-ordering conference simply used these in place of the earlier benchmark prices. Proposals in 1997 and 2002 to eliminate this guidance went unheeded, since power prices remained fixed and generators needed below-market prices for coal to remain solvent. Indeed, the gap between coal prices for power plants and other users widened after 2002, reaching 28% in 2006 (Table 3.4). Negotiations between coal mines and power plants at the national coal-ordering conferences became strained and the number of contracts fell sharply. In 2002 and 2003, respectively, only 30% and 40% of the expected number of contracts were signed.

In June 2004, NDRC tried to eliminate government intervention in setting coal prices, to facilitate direct transactions between mines and users, and to encourage longer-term (two- to three-year) contracts. At the 2004 conference, the guideline price for utility coal was removed, but power prices remained fixed while coal prices soared. The government was forced to mediate between mines and power plants, suggesting a price cap 8% above September 2004 spot prices – too small given market realities. Only 124 Mt of coal were contracted during the conference, 35% of the expected volume, at prices 10-12% above the previous year.[9]

---

9. It was feared that raising retail power prices, as demanded by power utilities, would spur inflation. It was agreed that power plants would absorb 30% of the higher fuel cost and the remainder would be reflected in the wholesale price. Interference was not limited to the power sector. At the 2004 conference, steel, chemical and other industrial customers accepted a 30% coal price hike, but in January 2005 the government suppressed price rises to fertiliser factories in order to protect farmers.

Table 3.4............Price gap between coal for non-utility and utility use, 1997-2007

	"Market coal" for non-utility use, RMB/tonne	"Power coal" for utility use, RMB/tonne	Discount for utility use
1997	167	137	18%
1998	160	133	17%
1999	144	121	16%
2000	140	121	14%
2001	151	124	18%
2002	168	137	18%
2003	174	139	20%
2004	206	163	21%
2005	270	213	21%
2006	302	218	28%
2007	330	246	25%

Source: CCII (2006).

Still, market forces were coming into play as mines and their customers began to negotiate freely, avoiding intermediaries. They began to sign medium- and long-term contracts to safeguard their respective interests, like the three-year contract signed in July 2004 between the large power generator Huaneng Power International and the coal producer Shenhua Group (9 Mt in 2005, 10.5 Mt in 2006 and 11 Mt in 2007). In August 2004, Huaneng entered into a five-year deal with China National Coal Group, taking 5 Mt, rising to 10 Mt, at a benchmark price, adjustable by 5%. Yuedian Group in Guangdong also signed a three-year contract with Shenhua Group.

The supply shortages and price hikes since 2003 triggered more intense pressures between coal producers and the power industry, weakening the role of the coal-ordering conferences and deepening the coal market. Many companies began transacting outside the conference and customers began sending buyers directly to coal mines. A new class of coal broker – with their own yards and trucks to fulfil orders and helping to better match demand with supply while speculating for higher prices – emerged to replace older-style brokers, who had made their living through connections with coal transporters under the former planned economy. In 2007, NDRC ceased hosting coal-ordering conferences. Although coal-ordering conferences are still held to allocate transport quotas, there is expected to be much less pricing intervention from government.

Coal prices in the residential market and the price of commercial coal from key state-owned coal enterprises fluctuate according to the seasons. The recent trend has been for market prices to rise while the price of contract coal for power generation has remained stable. In 2006, the China Coal Trade and Development Association coal index rose to RMB 405.2/t, an annual rise of 4.0%. By contrast, from January to December 2006, the price of contract coal to key state-owned power generators increased 2.5% from RMB 206.5/t to RMB 211.7/t. During the first three quarters of 2006, coal prices at Qinhuangdao Port were largely stable, but in the forth quarter, prices went up steeply, especially the prices of Datong's high quality coal, and have continued rising through to 2008, in line with international prices. Coking coal prices largely reflect international prices. In 2004, the CIF price (cost, insurance and freight) of coking coal imported from Australia was RMB 638/t, while the producer price of prime coking coal from Xishan Coal and Electricity Power Co. Ltd., the major coking coal supplier, was RMB 610/t. Domestic coking coal prices went up again in January 2005, when the price from Baosteel Co. Ltd. rose above export coal prices, and have continued to follow international prices since then.

Establishing markets and reforming prices: the case of Shanxi

In 2007, the government began testing reforms in Shanxi that are expected to fundamentally transform the coal market. The reforms will be extended to the rest of the country if, after two years, the measures are judged successful. One of the two main aims was to establish an open coal market in 2008, in which all transactions are market-based. An electronic trading platform will increase price transparency and eliminate the distortions of the past. Suppliers and customers should be better matched, leading to a better balance between prices, contract terms and conditions, delivery method and distance, and coal quality. More parties are expected to obtain rail transport quotas, not through non-transparent quota allocations, but based upon market demand and willingness to pay. More generally, it is hoped that a more-efficient market will enhance Shanxi's economic development.

The second aim is to reflect all external costs, including those linked to resource use, the environment and worker safety, into coal pricing. The ever-growing external costs could be curbed by encouraging coal mining enterprises to take steps to reduce their external impacts, rather than to profit from them. In the past, a resource tax of RMB 0.3-5.0/t was levied, with a collection rate of below RMB 2.5-3.5/t (Market Avenue, 2008). In the future, the rate may rise to RMB 2.5-8.0 – still a small fraction of mining costs which were in the RMB 100-200/t range during 2007. When granting mining rights, royalties of RMB 1 000/km^2 are paid based upon the reserve area, but, again, payments are negligible compared to expected earnings (Mao *et al.*, 2008). No provisions were made in the past to rectify environmental damage, which is often severe, *e.g.* subsidence affects 4% of Shanxi's land area. On the other hand, a very significant safety surcharge of RMB 15/t has been collected since 2005 to establish a fund for initiatives that have begun to have an impact.

Since January 2007, under a pilot scheme, the provincial government in Shanxi has collected surcharges reflecting resource and environmental costs. The coal resource tax has been increased to RMB 2.5-8.0/t, and the resource compensation fee has also been increased from 1% to 3-6%. In an earlier reform from 2002, the royalty per tonne of coal has been determined through public auction. To begin with, successful bids were around RMB 2/t of the coal reserve, which did not reflect its true market value, but this has now increased to RMB 6/t. In 2007, a Coal Sustainable Development Fund surcharge has been collected at nominal rates of RMB 14/t for steam coal, RMB 18/t for anthracite, and RMB 20/t for coking coal, multiplied by factors intended to reflect typical resource recovery rates at small, medium and large mines. Mines under 450 ktpa pay twice the nominal rate, those producing 450-900 ktpa pay 1.5 times and larger mines pay the nominal rate. Revised rates are proposed (Table 3.5).

Table 3.5 .... Coal production surcharges under pilot system in Shanxi

Surcharge[1]	Before 2007	Temporary rates for 2007	After 2007
Coal resource tax	RMB 0.9/t		RMB 2.5-8.0/t
Resource compensation fee	1% of sales		3-6% of sales
Resource royalty[2]	RMB 2/t of reserves		RMB 6/t of reserves
Coal Sustainable Development Fund[3]	RMB 20-45/t		
steam coal	–	RMB 14/t	RMB 5-15/t
anthracite	–	RMB 18/t	RMB 10-20/t
coking coal	–	RMB 20/t	RMB 15-20/t
Environmental recovery deposit	–		RMB 10/t
Coal Mining Industry Transition Fund	–		RMB 5/t
Production Safety Fund[4]	RMB 15/t		RMB 15/t

Notes:
1. Other fees previously collected included a production allowance (RMB 10-15/t), a special maintenance fee (RMB 5-7.5/t) and water resource fees (RMB 2-3.5/t).
2. Determined by auction, so values are indicative only.
3. The Coal Sustainable Development Fund is shown as the nominal surcharges that apply to mines producing >900 ktpa – smaller mines pay higher surcharges – it replaces the similar Energy Base Construction Fund, which was collected at the rate shown before 2007.
4. The Production Safety Fund has been collected since 2005.
Source: Shanxi Fenwei (2007).

Progress has been made also with environmental fees; a fee of around RMB 10/t is reported to be under consideration. For historic environmental damage, statements issued by the State Council clearly indicate that the state would finance restoration efforts. In addition, since 2007, a RMB 5/t Coal Mining Industry Transition Fund levy has been charged, designed to ensure that when coal mines shut down due to depletion, workers can be assisted to move successfully into other employment (Shanxi Fenwei, 2007).

MAJOR CURRENT COAL-RELATED LEGISLATION AND REGULATIONS

Legislation and regulations

The Chinese government has established the rule of law as fundamental to the country's long-term transformation from a policy-oriented model under a planned economy to a law-oriented model suitable for a market economy. The legal and regulatory system for the coal industry has been reformed along these lines, but remains incomplete, at a stage between initial establishment and refinement through practical application. Laws and regulations currently in force deal with five core aspects of the coal industry: resource administration, safety supervision, environmental protection, industry administration and energy conservation. Annex III presents the long titles and other details of all the laws introduced below.

Resource administration

In March 1986, the Standing Committee of the 6th National People's Congress enacted the Mineral Resources Law (revised in August 1996). In March 1994, the State Council issued *Rules for Implementation of the Mineral Resources Law*.

Safety supervision

In November 1992, the Standing Committee of the 7th National People's Congress enacted the Mine Safety Law. This was followed by the State Council's *Coal Mine Inspection Ordinance*, issued in November 2000. In June 2002, the Standing Committee of the 9th National People's Congress enacted the Work Safety Law.

Environmental protection

In December 1989, the Standing Committee of the 7th National People's Congress issued the Environmental Protection Law, replacing an earlier law dating from 1979. Then, in June 1991, the Standing Committee issued the Water and Soil Conservation Law. In October 1995, the Standing Committee of the 8th National People's Congress enacted the Solid Waste Pollution Prevention and Control Law (revised in December 2004).

Industry administration

In August 1996, the Standing Committee of the 8th National People's Congress enacted the Coal Law. As a comprehensive basic law for the coal industry, the Coal Law established legal systems for regulating coal mining and coal trading, promoting proper coal industry development and coal utilisation, and protecting coal resources.

Energy conservation

In November 1997, the Standing Committee of the 8th National People's Congress issued the Energy Conservation Law (revised in October 2007).

These have been important milestones, but further work is needed to deal with current and emerging challenges (Box 2.1). The system for coal resource prospecting and development could be revised to be more responsive to rising coal demand. Improvements to the way mining rights are allocated could ensure that enterprises develop coal resources rationally, without serious wastage. Changes to resource taxes and charges have been discussed that would better reflect the true value of coal resources. Coal mining is widely regarded as too unsafe and environmental damage from coal mining persists. This situation has attracted the concern of citizens and, in response, the government is improving the legal, regulatory and enforcement systems with clearer lines of authority among the many agencies involved.

Coal mining licences

In a development touching on all licensing issues, the Standing Committee of the 10th National People's Congress enacted the Administrative Licence Law in August 2003. According to Article 12 of this law, coal mining is a business that involves public security, macroeconomic control, environmental protection and the development and utilisation of limited natural resources, so is therefore subject to administrative licences. Government supervision and administration of the coal industry is mainly performed through licensing of market entry, work safety, trading and processing, and other activities in the sector. The current system has established eleven distinct procedural requirements, as follows.

Project approval

According to the *Decision on Reform of the Investment System* (State Council, 2004), coal mining and development projects in the state-planned mining areas should be examined and approved by the investment administration departments of the State Council, while other coal development projects should be examined and approved by local governments. In the case of coal liquefaction, projects with a production capacity of over 500 ktpa should be approved by the investment administration departments of the State Council and smaller projects by local governments.

Permission to establish a coal mining enterprise

Article 19 of the Coal Law 1996 stipulates that those establishing a coal mining enterprise must apply to the relevant coal industry administration departments, depending on the location and size of the proposed enterprise. Examination and approval should be carried out in accordance with the Coal Law and the authority stipulated by the State Council. The department in charge of geology and mineral resources should review the proposed mining scope and resource utilisation scheme, and submit its opinions during the approval procedure.

Mineral licensing

According to Article 16 of the Mineral Resources Law 1996, and Article 3 of the *Rules on the Administration of Coal Exploitation Licences*, the department in charge of geology and mineral resources is responsible for approving and issuing mineral licences.

Work safety qualification certification of mine managers

According to Article 20 of the Work Safety Law 2002, Article 27 of the Mine Safety Law 1992, and Article 4 of the *Rules on the Supervision and Inspection of Coal Mine Work Safety Training*, the safety qualification certificates of mine managers are to be issued by production safety supervision administration departments or the coal mine safety supervision departments.

Qualification certification of mine managers

According to Article 23 of the Coal Law 1996, qualification certificates of mine managers should be approved and issued by provincial coal industry administration departments or work safety supervision departments.

Work safety permitting

According to Article 16 of the Work Safety Law 2002, the *Regulations on Work Safety Permitting* and the *Implementation Rules for Work Safety Permitting in Coal Mine Enterprises*, coal mine work safety permits should be examined and issued by work safety supervision administration departments or coal mine safety inspection departments.

Coal mining licensing

According to Article 24 of the Coal Law 1996 and the *Rules on the Administration of Coal Mining Licences*, a coal mine must obtain a coal mining licence before commencing operation. At present, coal mining licences are mostly issued by provincial development and reform commissions or coal industry administration departments (*e.g.* Coal Industry Administration Bureaus, Coal Industry Bureaus, Coal Industry Offices or Economic and Trade Commissions).

Business licensing

According to the Company Law 2005 and the *Regulations on the Administration of Enterprises Licences*, business licences of coal enterprises should be examined and issued by provincial industrial and commercial administration departments.

Coal trade licensing

According to Article 48 of the Coal Law 1996 and the *Rules on the Administration of Coal Trade*, coal trade licences should be issued by provincial reform and development commissions or other coal industry administration departments with equivalent power.

Environmental impact assessments

In accordance with the Environmental Protection Law 1989 and the Environmental Impact Assessment Law 2002, environmental impact assessment reports should be submitted for coal mine construction projects and should be examined and approved by environmental protection departments.

Water and soil conservation schemes

According to Article 19 of the Water and Soil Conservation Law 1991, when launching a coal enterprise, a water and soil conservation scheme approved by the water and soil administration department must be included in the environmental impact report.

Agencies responsible for licensing differ from place to place; not every province has a coal industry bureau and other agencies assume responsibility. A larger problem is variation in practice for issuing qualification certificates to mine managers. Key state-owned enterprises are administered by NDRC, but the situation at the local level is more complicated. Enterprises in some provinces are administered by coal industry bureaus, others by work safety supervision bureaus, still others by provincial economic and trade commissions, and some by provincial reform and development commissions. Issuance of coal mining licences is restricted to the provincial level; county and city agencies are only responsible for initial or partial examination of applications. Time limits are stipulated by law for examination and approval of licence applications; ordinarily, decisions should be made within 20 days. Only after obtaining the necessary administrative licences, can a coal enterprise carry out coal mining and development activities (Figure 3.11).

Figure 3.11 .... Coal mine licensing procedures

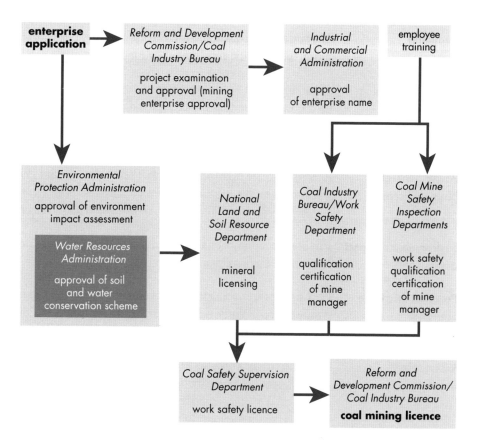

Source: CCII.

Coking sector licensing

Approvals for coking plants are not as extensive as for coal mines, but follow a similar logic designed to properly administer and regulate a sector that has the potential to be highly polluting.

Project approval

Coke plant construction projects must accord with NDRC's *Access Conditions of the Coking Industry* before being granted market access. After that has been demonstrated, projects must obtain approvals for land use and environmental protection.

Environmental impact assessments

Environmental impact assessments must be approved for all projects by provincial environmental protection administrative departments before other agencies can issue other administrative licences.

Pollution discharge licensing

Coking plants must obtain flue gas and waste water discharge licences in accordance with the Atmospheric Pollution Prevention and Control Law 2000, the Water Pollution Prevention and Control Law 2008 and other related laws and regulations.

Export licensing

If an enterprise wishes to export coke, it must obtain an export licence in accordance with the Foreign Trade Law 2004, the *Regulations on the Administration of the Import and Export of Goods* and the *Qualification Standards and Application Procedures for Coke Export Enterprises*.

Coal-fired power generation licensing

The power sector has administrative procedures similar to those applying in the coking sector, with specific regulations covering, for example, the installation of FGD.

Project approval

According to the *Decision on Reform of the Investment System* (State Council, 2004), coal-fired power generation projects should be verified and approved by the investment administrative departments of the State Council, *i.e.* NDRC or equivalent agency of the local government.

Power generation business licensing

According to the *Regulations on Electric Power Supervision* and *Provisions on the Administration of Electric Power Business Licences,* separate business licences are needed for engaging in electricity generation, in electricity transmission and in electricity distribution.

Environmental impact assessments

Environmental impact assessment must be conducted and reported to provincial environmental protection departments for approval.

Pollution discharge licensing

Coal-fired power generators must obtain discharge licences in accordance with the Atmospheric Pollution Prevention and Control Law 2000, the Water Pollution Prevention and Control Law 2008 and other relevant laws and regulations.

Work safety licensing

According to the *Ordinance on Work Safety Licensing* and the requirements of the *Circular on Two Power Plant Tailing Warehouse Accidents* issued by the State Administration of Work Safety (SAWS) and the State Electricity Regulatory Commission (SERC), coal-fired power generation enterprises which handle bottom ash should obtain work safety licences from work safety supervision departments in accordance with the *Rules on the Implementation of Work Safety Certification at Non-Coal Mine Enterprises*.

COAL INDUSTRY ADMINISTRATION: CUSTOM AND PRACTICE

Under central planning, enforcement of rules at state-owned enterprises was easier than after post-1979 reforms, although TVE and private firms had never fallen within the centrally supervised system. Formerly, large groups of coal mines were overseen by several Bureaus of Mines; these nominally managed mines, but in practice simply strove to achieve production targets. The Bureaus were responsible for enforcing laws and regulations formulated by SPC and MOCI. Aside from a brief period of local government administration (1970-75), the industry had long been administered directly by the central government. From 1975 to 1988, coal mines under centralised allocation (later key state-owned mines) came under the control of MOCI. Following the introduction of the contract responsibility system (under which managers signed contracts giving them personal responsibility for fulfilling certain objectives) and other measures, enterprises became more independent.

When the Ministry of Energy (MOE) was established in 1988, all centralised-allocation mines were integrated into the China Coal Corporation, with the exception of those northeastern mines consolidated into the Northeast and Inner Mongolia Coal Industry Corporation. Local coal mine development corporations were established to administer the TVE and private mines. The China Coal Industry Coal Mine Management Association was established to guide these corporations and to enforce policy. The Ministry of Energy did not last long, however; it was abolished in 1993 along with the coal corporations. Coal mines fell again under direct MOCI administration, but with changes from the past. Centralised coal allocation was ended, the key state-owned mines renamed and the Bureaus of Mines were transformed into independent group companies. As the government and enterprise functions were gradually separated, the government sought ways to improve policy and enforcement in the liberalising sector.

Prior to 1998, macroeconomic management, including policy making and the setting of production plans, was handled by the SPC (later renamed the State Development Planning Commission [SDPC] and, later still, NDRC), while MOCI was responsible for coal-specific policies and regulations. In 1998, MOCI was downgraded to become the State Coal Industry Bureau (SCIB) under the State Economic and Trade Commission (SETC), and authority over key state-owned mines was devolved to local governments. A new industrial administrative system was created, with one branch covering key state-owned mines, and the other dealing with local state-owned and non-state mines. At the centre of the system sat SCIB and SETC, with provincial Coal Industry Administration Bureaus under SCIB responsible for implementation and monitoring. The coal group companies and a few remaining Bureaus of Mines were treated similarly. The loss of direct administrative control over the coal sector shifted attention to the design and enforcement of laws and regulations. The central government formulated policies, basic laws and regulations and issued implementation rules in the form of circulars and notices, while provincial governments designed rules in line with local circumstances for enforcement and monitoring by cities and counties.

Many provincial governments set up Coal Departments, which often shared staff with the Coal Industry Administration Bureaus (not an uncommon arrangement in Chinese government). These provincial Coal Departments and county-level Coal Industry

Bureaus administered local state-owned coal mines within their respective jurisdictions. No administrative agency was established for TVE and private mines (except in Shanxi, which has many such mines). Instead, TVE administrative agencies (often within local branches of the Ministry of Agriculture) oversaw these mines, like other sectors. Since the 1990s, TVE and private mine operators signed agreements with township and village governments allowing them to manage their businesses, and avoided intervention so long as they paid fees to the these local governments. This hands-off approach continued until the mid-1990s, when supervision was enhanced.

In 2001, SCIB was abolished and most of its functions were handed over to SAWS. In 2003, SETC was also abolished and most of its coal industry administrative functions were handed over to NDRC. The current organisational structure at the central level is as follows:

National Development and Reform Commission (NDRC)

In 2003, SETC was merged with SDPC to create NDRC, which inherited most functions of the former MOCI. The Coal Division of NDRC's Energy Bureau (and, since March 2008, the Coal Department of the National Energy Administration) prepares national plans for the coal industry, sets policy and approves new mines. Separately, the Economic Operations Bureau of NDRC has a Coal Division, responsible for day-to-day management of the industry. It formulates annual production plans, guides coal transport plans and issues production licences. It is also responsible for CDM projects under the Kyoto Protocol.

Ministry of Land and Resources (MLR)

MLR is responsible for coal resource management. The Geological Prospecting Division examines, approves and issues prospecting licences. The Mineral Development and Administration Division issues mining licences.

State-Owned Assets Supervision and Administration Commission (SASAC)

Key state-owned coal mines are owned, but not administered by SASAC, which is nevertheless concerned with operational efficiency. SASAC is authorised to close or declare bankrupt those mines that are not viable, including those with exhausted coal reserves.

State Administration of Work Safety (SAWS)

The State Administration of Coal Mine Safety (SACMS), under SAWS, oversees safety at coal mines. Recent activity has focused on dealing with hazardous mine gas, part of a greater effort to improve mine safety.

State Environmental Protection Administration (SEPA)

SEPA is responsible for approving environmental impact assessments for new mines, submitted during the feasibility stage, and monitors coal mining operations for any breaches of pollution control regulations. In March 2008, SEPA acquired full ministerial status to become the Ministry of Environmental Protection (MEP).

State Administration for Industry and Commerce (SAIC)

This Administration issues and updates business licences and conducts field inspections.

Ministry of Commerce (MOFCOM)

The Foreign Trade Division of MOFCOM issues coal import and export licences, after NDRC has granted approval.

Ministry of Science and Technology (MOST)

The Transport and Energy Section of the High Technology Industrialisation Division under MOST carries out coal-related R&D projects, which continue to play an important role in developing clean coal technologies.

III. COAL IN CHINA TODAY - **67**

China National Coal Association (CNCA) This association inherited some administrative functions from the former MOCI. It is a non-profit organisation funded by membership fees from coal mining enterprises. It serves as a bridge between enterprises and government, and in practice also complements NDRC, which has only a small professional staff in the Coal Department of the National Energy Administration. Some work is outsourced to CNCA, *e.g.* preparation of drafts of industrial policies. CNCA collects coal industry statistics.

The local administrative system has also changed substantially. Some provincial Coal Industry Bureaus remained intact, *e.g.* Shanxi, Henan and Heilongjiang. Other provinces transferred administrative functions to local trade and economic commissions (many of which survived, even after SETC was abolished), *e.g.* Zhejiang, Hebei and Anhui, or to local DRCs.

COAL MINING SAFETY AND CLOSURE OF SMALL COAL MINES

Occasional campaigns arise to counter specific problems, *e.g.* initiatives to close small coal mines (1998 to 2000, and since 2004), and to improve work safety at coal mines. Task forces are often established to take charge of implementation. For small mine closures, an inter-agency national leadership group, headed by a minister, was set up and closure targets agreed. An implementation office was established, headed by an official of SCIB and drawing other members from other relevant agencies to better co-ordinate inter-agency enforcement, and to speed decision making. Analogous local leadership groups were established, typically headed by provincial governors. They prepared interim provisions and drew up lists of coal mines for closure.

The problems in enforcing policies and regulations in the coal industry are not unique to the coal sector, nor to China. No matter how rational the regulation, enforcement often meets local opposition. Although powerful task forces led the campaign to close mines, many continued production illegally after official closure. Like other state-owned enterprises, coal mines are jointly administered by a local branch of a central government agency and the local government itself, but the latter typically exerts greater influence. Generally, the lower the branch of government, the more closely involved it is with mine owners and managers, sometimes leading to corruption. In 2003, newspapers reported that many local officials owned stock in TVE and private mines; in acting as both players and umpires, they overtly supported policy but covertly opposed it. Even without such collusion, local governments might still balk at policies that would depress economic activity and local fiscal revenue.

In contrast, the campaign to improve mine safety has gone more smoothly. Outcomes were monitored by families of accident victims; mine managers and local officials, aware of the power of such direct monitoring, actively enforced policy. This holds lessons for future improvements that capitalise on grassroots feedback. For example, complaint hotlines direct to the central government could provide early warnings of problems. Beyond this, better enforcement would seem to require enhanced institutional and budgetary autonomy of watchdog agencies like the local Coal Industry Bureaus, where local governments currently control budgets and staffing. If the offices were given more autonomy, it would be necessary for central government to ensure proper accountability through monitoring and periodic inspection.

REFERENCES

Beijing HL Consulting (2006), *China Coal Report*, Beijing HL Consulting, Beijing, 20 June.

CASS (中国社会科学院 – Chinese Academy of Social Sciences) (2007), 能源蓝皮书 2007 中国能源发展报告 (*The Energy Development Report of China 2007 or "Blue Book of Energy"*), Social Sciences Publishing House, Beijing.

CEC (中国电力企业联合会 – China Electricity Council) (2008), 中电联发布 2007 年电力工业统计数据 (*National Power Industry Statistics 2007*), CEC, Beijing, www.cec.org.cn/html/news/2008/8/13/20088131524443838.html.

Chen Zhiping eds. (1998), *Prices in China over Five Decades (1949-1998)*, China Price Publishing House, Beijing.

China Daily (2008), "Coal Reserve Plan a Bid to Tame Prices", *China Daily*, 2 September, www.chinadaily.com.cn/china/2008-09/02/content_6987995.htm, accessed 2 September 2008.

CCII (China Coal Information Institute) (2006), *China Coal Development Report*, CCII, Beijing.

CCII (2007a), *China Coal Outlook 2006*, CCII, Beijing, March.

CCII (2007b), *Collected Coal Industry Data*, CCII, Beijing.

CCII (various years), *China Coal Industry Yearbook* (various years), China Coal Industry Publishing House, Beijing.

CNCA (中国煤炭工业协会 – China National Coal Association) (2006), 中国煤炭工业统计资料汇编 1949-2004 (*China Coal Industry Statistical Compendium 1949-2004*), 煤炭工业出版社 (China Coal Industry Publishing House), Beijing.

CNCA (2007), personal communication, CNCA, Beijing, July.

FACTS (2008), "China's Coal Market: Review in 2007 and Forecast for 2008", *China Energy Series*, No. 45, FACTS Global Energy, Honolulu, Hawaii, June.

Horii, Nobuhiro and Gu Shuhua, eds. (2001), *Transformation of China's Energy Industries in Market Transition and its Prospects*, Institute of Developing Economies – Japan External Trade Organisation (IDE-JETRO), Chiba, Japan.

IEA (International Energy Agency) (2007), *World Energy Outlook 2007 – China and India Insights*, OECD/IEA, Paris.

IEA (2008a), *Coal Information 2008*, OECD/IEA, Paris.

IEA (forthcoming), *Prospects for Replacement and Upgrading of Older Coal-Fired Power Plants*, OECD/IEA, Paris.

Li Xilin, Dou Qingfeng, Huang Shengchu *et al.* (1999), *World Coal Industry Development Report*, China Coal Industry Publishing House, Beijing.

Market Avenue (2008), "Operation of China's Coal Industry 2007-2008", *China Energy Review*, Market Avenue, Beijing, February, www.marketavenue.cn/upload/review/chinamarketreview_13.htm.

Mao Yushi, Sheng Hong and Yang Fuqiang (2008), *The True Cost of Coal*, Greenpeace / The Energy Foundation / WWF, Beijing, September, http://act.greenpeace.org.cn/coal/report/TCOC-Final-EN.pdf.

MLR (Ministry of Land and Resources) (1998), *Administration by Block of Mineral Resource Allocation, Exploration and Surveying, Mining Rights and Transfer of Mineral Exploration and Mining Rights,* MLR, Beijing.

MLR (2003), *National Coal Resources and Reserves Circular,* MLR, Beijing.

NBS (National Bureau of Statistics of China) (2008), *China Energy Statistical Yearbook 2007,* China Statistics Press, Beijing.

NBS (various years), *China Statistical Yearbook* (various years), China Statistics Press, Beijing.

NDRC (国家发展改革委 – National Development and Reform Commission) (2007), 《节能发电调度办法实施细则(试行)》 ("Administrative Measures on Energy-Saving Dispatching [for trial implementation]"), 国办发[2007]53号 (Guo Banfa [2007] No. 53, NDRC / State Environmental Protection Administration / State Electricity Regulatory Commission / National Energy Leading Group, Beijing

SACMS (State Administration of Coal Mine Safety) (2007), *China Coal Mining Safety Statistics,* SACMS, Beijing.

Shanxi Fenwei (2007), *China Coal Weekly,* Shanxi Fenwei Energy Consulting Co. Ltd., Taiyuan, Shanxi, China, 6 April, www.sxcoal.com.

SIC (国家信息中心 – State Information Center) (2008), 2008 年中国煤炭行业年度报告 (*China Coal Industry Annual Report 2008*), SIC, Beijing, 2008 年04 月 (April 2008), http://ar.cei.gov.cn/web/Column.asp?CId3=334.

State Council (国务院) (2004), 《国务院关于投资体制改革的决定》 ("Decision on Reform of the Investment System"), 国发[2004]20号 (Guo Fa [2004] No. 20), Office of the State Council, Beijing, 16 July.

State Council (国务院) (2006), 《国务院关于同意深化煤炭资源有偿使用制度改革试点实施方案的批复》 ("Pilot Program on Deepening Reform of the System of Obtaining Coal Mining Rights for Value"), 国函[2006]102 号 (Guo Han [2006], No. 102), State Council, Beijing, 30 September.

UNECE (1997), *United Nations International Framework Classification for Reserves/Resources – Solid Fuels and Mineral Commodities* (联合国化石能源和矿产资源分类框架), ENERGY/WP.1/R.70, United Nations Economic Commission for Europe Committee on Sustainable Energy, 7th Session, Geneva, November (latest version dated 2004 translated by Yundong Hu, www.unece.org/ie/se/reserves.html).

Ye Qing and Zhang Baoming eds. (1998), *Chinese Town-and-Village Coal Mines,* China Coal Industry Publishing House, Beijing.

IV. COAL AND ITS ALTERNATIVES IN CHINA'S ENERGY FUTURE

China is the world's largest coal producer, accounting for 39% of global production in 2007 and, at 46%, an even greater share of hard coal production, *i.e.* excluding brown coal production (IEA, 2008). As China's dominant energy source, coal accounts for around 70% of total national primary energy production and 63% of total national primary energy consumption.[1] For a considerable period of time to come, China's energy structure will be dominated by coal. The World Bank has calculated that if China's elasticity of energy use to gross domestic product (GDP) remained close to unity (*i.e.* the average value observed between 2000 and 2005), then China's total energy needs by 2020 could double to more than 5 billion tonnes of coal equivalent (tce), with coal accounting for 60% or more (Berrah, 2007). In contrast to such a scenario, most forecasts assume China's energy efficiency will improve and alternative energy sources will be deployed. This chapter explores this thesis, not only by comparing energy supply and demand projections from national and international bodies, but also by considering the influence that energy efficiency and alternative energy sources may have on future coal demand. Even when optimistic assumptions are made about efficiency and alternatives, the absolute growth in coal use over the next one or two decades may well be even more significant than the remarkable growth seen in China over the last decade. The conclusions support the need for cleaner and more-efficient coal technologies to limit the impacts of this continued coal use.

COAL SUPPLY AND DEMAND PROJECTIONS

Prediction of coal demand and supply by NDRC

There are few official projections of China's future coal production and consumption beyond those linked to the current Five-Year Plan. However, coal is seen as crucial to China's future energy supply security. The 11th Five-Year Plan for Coal Industry Development, issued by the National Development and Reform Commission (NDRC, 2007a), states that China's total coal production will be limited to 2.6 Gt (2.0 Gtce) in 2010. This would then match coal demand, taking into account various factors, such as structural adjustments to the economy, technological progress and energy conservation. Coal consumption in the power and steel sectors is expected to continue to increase steeply, while coal-based chemical production is also likely to show strong growth. Consumption in the construction materials industry is forecast to remain largely unchanged.

---

1. Calculated using data from IEA database (wed_bal.ivt available from http://data.iea.org). Chinese statistical sources quote higher shares for coal because, unlike the IEA, estimates of non-commercial, combustible renewables and waste are not included in total energy production and supply.

Coal demand projections by other Chinese organisations

In general, energy supply and demand projections from Chinese organisations are closely aligned with those from NDRC reported above. For example, the China Coal Transport and Marketing Association expects China's coal demand to reach 2.6 Gt (2.0 Gtce) in 2010 – the same as NDRC and very similar to the IEA Alternative Policy Scenario. The China National Coal Association looks out to 2020 when it predicts domestic coal consumption could reach 2.9 Gt per year (2.2 Gtce), this being significantly below projections from the IEA or EIA because of lower growth rate assumptions. Finally, the China Coal Industry Development Research Center has also provided an unpublished projection of coal demand to 2020 (Table 6.2). This is shown in Figure 4.1 to have certain characteristics of the IEA Alternative Policy Scenario – initially high coal demand growth followed by a period of much lower growth. While none of the projections made by Chinese organisations forecast a continuation of the recent high growth in China's coal demand, the possibility cannot be dismissed given China's lack of alternative energy sources and energy efficiency challenges, topics which are considered further in Section 4.2.

China Coal Transport and Marketing Association's prediction of coal demand

During the 11th Five-Year Plan period (2006-10), new developments will affect all major coal-consuming sectors; the China Coal Transport and Marketing Association (CCTMA) expects that some will stimulate coal demand, while others will dampen demand (CCTMA, 2007). First, there will be changes in the size and structure of investments as fixed asset investment slows, particularly in downstream sectors such as the power and metallurgical sectors where surplus capacity exists. At the same time, structural changes in industry mean that energy consumption per unit of output will continue to shrink. Import-export policy changes will tend to reduce coal demand as the government takes measures to end trade in some energy-intensive products by adjusting export quotas, taxes and VAT refunds for crude oil, coking coal, coke, pulp and certain other products. Energy-conservation policies will also play a role, especially as high energy prices make efficiency more strategically important. These changes will affect demand for coal by the major coal-consuming sectors, as explained by the CCTMA:

■ **Heat and power sector.** Power generation accounts for about 90% of total coal consumption in this sector, while heating accounts for about 10%. Coal-fired power plants produce about 80% of China's electricity supply. Although specific coal consumption per unit of output for power generation and heating is decreasing, the CCTMA expects coal demand to grow.

■ **Iron and steel sector.** Coal consumed by the iron and steel industry is roughly equivalent to the tonnage of iron produced and this is expected to grow, although it is predicted that coal consumption per tonne of iron produced will fall. Coking coal accounts for about 75% of total consumption in this sector, the remainder being steam coal – proportions that will remain largely unchanged in the future. Investments, iron and steel exports, and domestic consumption will affect the total quantity of coal consumption.

■ **Construction materials sector.** Coal use for cement production amounts to about 50% of total coal consumption in this sector, wall materials and lime production accounts for about 20%, while the production of glass, ceramics and other non-metallic materials consumes about 15%. Coal consumption in the construction materials sector is not predicted to change, despite reductions in the consumption of coal per unit of cement, wall materials and lime produced.

■ **Chemical sector.** The fertiliser industry, basic chemical production and other sub-sectors consume about 60%, 20% and 20% of the total coal consumed by the chemical sector, respectively. Due to technical advances and other factors, specific coal consumption is decreasing in this sector, but absolute growth is forecast.

CCTMA predicts that coal consumption in other sectors will continue its downward trend. These sectors include households, agriculture, construction, communications, catering, mining and other manufacturing. Improved living standards in cities and stringent environmental protection standards should see residential users turn away from coal to cleaner sources of energy.

During the 11th Five-Year Plan period, it is expected that the recent trend of falling coal exports and rising imports will continue in response to state policies on coal exports and imports (Figure 6.1).

Overall, coal demand in China is expected to continue rising during the 11th Five-Year Plan period. China's domestic coal demand rose 10.4% to 2 392 Mt in 2006 which, with coal exports of 63 Mt, coal imports of 38 Mt and stock changes, was met by production of 2 373 Mt, 7.6% more than in the previous year (NBS, 2008). It is predicted by the China Coal Transport and Marketing Association that domestic coal demand during China's 11th Five-Year Plan period will grow at about 3-5% per year, about 4 percentage points lower than in the 10th Five-Year Plan period (2001-05). This can be translated into a domestic coal demand of about 2 590 Mt in 2010. Adding net coal exports of 80 Mt means that total domestic coal production would then need to be about 2 670 Mt.

China National Coal Association's prediction of coal demand

China's coal demand is concentrated in four industrial sectors: the electric power industry, the iron and steel industry, the construction materials industry and the chemical industry; together, these accounted for 84% of total national coal consumption in 2004 (NBS, 2008). The China National Coal Association has considered these sectors in turn to arrive at demand projections for 2010 and 2020, as summarised below (Pan, 2005).

China's electricity consumption during the period 1953-2003 increased at an average annual rate of 11.5% with a consumption elasticity coefficient of 1.49. The coefficient fell slightly over the 2000-04 period, averaging 1.45 and indicating some decoupling of economic growth and power consumption – a trend that is likely to continue. Assuming an annual GDP growth rate of 8% and an electricity consumption elasticity coefficient of 1.00, Pan calculates that total national power production would need to be 3 470 TWh in 2010. Even with the priority given to hydropower, natural gas-fired power generation and nuclear power, Pan assumes that coal-fired power generation will still account for almost 80% of the national total in 2010, at 2 736 TWh. According to the targets proposed in the state's long- and medium-term energy conservation plan, specific coal consumption should drop to 360 gce/kWh in 2010, so Pan estimates consumption for power generation will be 985 Mtce or 1.31 Gt of raw coal.

Coal accounts for around 70% of total energy use in the Chinese iron and steel industry, mainly in the form of coking coal, pulverised coal for blast furnace injection and steam coal for process heating. The 2004 national pig iron output was 252 Mt for a total coal consumption of 254 Mt. According to industry predictions, pig iron output in the next few years will increase strongly, reaching 350 Mt in 2010 and 380 Mt in 2020. Taking into account measures for energy conservation, the China National Coal Association expects that, by 2010, China's total coal use in the iron and steel industry will reach 290 Mtce and remain around this level until 2020.

In 2004, China's construction materials industry consumed about 200 Mtce of coal. Due to the cement sector's high energy consumption, about 50% more than international best practice, there is great potential for energy conservation. According to targets in the state's medium- and long-term development plan, the sector's specific energy consumption should fall to 148 kgce/t of cement produced by 2010 and 129 kgce/t by 2020. Based on the industry's own predictions, China's cement production will be 1 300 Mt in 2010 and 1 400 Mt in 2020, respectively consuming 192 Mtce and 181 Mtce.

Between 1995 and 2004, production of the major coal-consuming products increased rapidly in the chemical sector, but total coal consumption showed little increase, reaching 102 Mt in 2004. This was mainly due to structural changes to the energy input mix and a comparatively rapid improvement in energy efficiency. Based on the state's energy conservation plan, specific energy consumption in the large-scale synthetic ammonia sector should fall from 1 372 kgce/t in 2000 to 1 140 kgce/t in 2010 and 1 000 kgce/t by 2020. Accordingly, Pan predicts that the ammonia fertiliser industry will require a total of 120 Mtce of coal in 2010.

Coal consumption in other sectors, outside of the four major sectors considered above, amounted to 368 Mt in 2004. With increasing coal consumption in the power sector, the share in total coal consumption of these other sectors has gradually decreased. It is predicted by the China National Coal Association that the volume of coal consumed by the other sectors will continue to decline to 320 Mt in 2010 and 240 Mt in 2020.

In his study for the China National Coal Association, Pan aggregates the above sector forecasts to predict that China's total domestic coal consumption in 2010 will reach 2 430 Mt and 2 820 Mt in 2020. Adding his estimate of 70-80 Mtpa net coal exports, the total demand for Chinese coal would be 2 500 Mt in 2010 and 2 900 Mt in 2020.

Coal demand and supply scenarios from the IEA *World Energy Outlook 2007*

In the IEA *World Energy Outlook* Reference Scenario, which presumes no change to existing policy, China's economy is assumed to grow at an average of 6.0 percent per year through to 2030,[2] largely fuelled by coal (IEA, 2007a). Coal consumption grows

---

2.   7.7% per annum from 2005 to 2015 and 4.9% per annum from 2015 to 2030.

at a forecast average annual rate of 3.2% to reach 3 427 Mtce in 2030 (Figure 4.1). After surging since the start of the decade to reach 1 740 Mtce in 2006, Chinese coal production is projected to increase further to 3 334 Mtce in 2030. Output of steam coal, which currently accounts for 85% of production in energy terms, is projected to increase faster than that of coking coal; by 2030, power generation accounts for 62% of total coal consumption, up from around 55% today. Shanxi province is expected to continue to dominate coal production, with output from Inner Mongolia, Shaanxi, Ningxia and Guizhou also growing significantly. The projected expansion of coal output in the Reference Scenario hinges on the continued restructuring and modernisation of China's coal-mining industry and massive investment in transport infrastructure to move coal to market. Similarly in the power sector where coal-fired capacity is anticipated to reach 1 259 GW, generating 6 586 TWh in 2030.

Under the Alternative Policy Scenario considered by the IEA, China is assumed to follow a less energy-intensive growth path, adopting the energy efficiency policies currently under consideration by the Chinese government and taking further measures to stem rising coal demand such that, by 2030, coal demand is forecast to be 2 631 Mtce, some 23.2% less than under the Reference Scenario (Figure 4.1).

Figure 4.1 .... Forecasts by national and international bodies of coal consumption in China to 2030

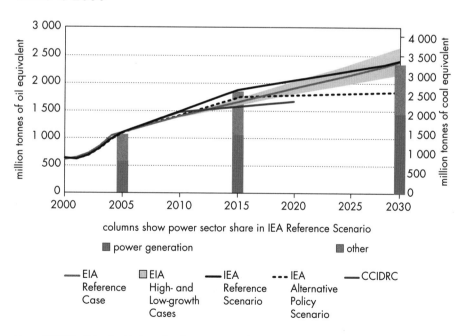

Notes: CCIDRC: China Coal Industry Development Research Center; EIA: US Energy Information Administration; IEA: International Energy Agency.
Sources: CCIDRC (2007); EIA (2007); and IEA (2007a).

In the Reference Scenario, China would become a major coal importer such that imports total more than 300 Mt in 2030, mainly of steam coal. With its high-quality coking coal resources, China has little need to import coking coal, despite a gradual growth in future demand. China would continue to export coal, mainly to Chinese

Taipei, Korea and Japan, which are closer to the big Chinese export terminals than some of the coastal provinces in the south. Under this scenario, steam coal exports expand from 66 Mt in 2005 to 178 Mt in 2030 and coking coal exports increase from 5 Mt to 16 Mt during the projection period. Overall, net imports in 2030 would reach 129 Mt, mainly meeting demand in coastal provinces and should further strengthen the price relationship between domestic and internationally traded coal. Key suppliers in 2030 are expected to be Indonesia, Australia, South Africa, Mongolia, Vietnam and Russia.

Coal demand projection from the US EIA *International Energy Outlook 2007*

The US Energy Information Administration (EIA), in its *International Energy Outlook 2007* projects that global coal use will grow in the next few years, with most growth coming from developing Asia, particularly China and India, where coal resources are abundant and economic growth is expected to be strong (averaging 6.5 percent per year in China from 2004 to 2030 in the reference case). It estimates that China and India together will account for 72% of the projected increase in world coal consumption from 2004 to 2030, meeting much of the increase in their demand for energy, particularly in the industrial and electricity sectors (EIA, 2007).

Coal supply is expected to come mostly from China's own coal mines to meet the demand projected in Figure 4.1 which closely matches the IEA Reference Scenario in 2030. The figure also shows the high- and low-macroeconomic growth cases considered by the EIA (±0.5 percentage points from the 6.5% growth rate assumed in the reference case). At the end of 2004, China had an estimated 271 GW of coal-fired capacity in operation. To meet the demand for electricity that is expected to accompany rapid economic growth, this capacity is forecast to reach 768 GW to generate 5 317 TWh (at a higher load factor than assumed by the IEA) by 2030, requiring large financial investments in new coal-fired power plants and associated transmission and distribution systems. The EIA projects that coal use for power generation will increase from 778 Mtce in 2004 to 1 917 Mtce in 2030, at an average annual rate of 3.5%, slightly below the 3.7% projected in the IEA Reference Scenario.[3]

Nearly one-half (45%) of China's coal use in 2004 was in sectors outside of power generation, primarily in the industrial sector. China was the world's leading producer of both steel and pig iron in 2004. Over the projection period, coal demand in China's non-electricity sectors is expected to more than double. Despite such substantial growth, however, the non-electricity share of total coal demand is projected to decline slightly, to 41% of total coal demand in 2030. Coal remains the primary source of energy in China's industrial sector, mostly because the country has only limited reserves of oil and natural gas.

---

3. EIA reports energy production and consumption in quadrillion British thermal units (Btu) on a higher heating value (HHV) basis. EIA data is converted here to million tonnes of coal equivalent (Mtce), lower heating value (LHV) basis, assuming 1 055.056 J/Btu, 29.307 GJ/tce (7 000 kcal/kg) and a 5% difference between the lower and higher heating values of coal.

China's coal imports are projected to total about 121 Mt in 2030, while its exports are projected to total 46 Mt. Thus, the EIA expects China to become a net coal importer of some 75 Mt by 2030, somewhat less than the 129 Mt under the IEA Reference Scenario.

ALTERNATIVES TO COAL

China's rapid economic growth and opening up was initially accompanied by greater diversification in its energy structure, moving away from coal (Figure 3.2). Since 2003, this trend has reversed with the growing use of coal for power generation to meet soaring electricity demand. In 2006, national electricity generation reached 2 866 TWh, having grown at an average annual rate of 13% since 2000. Hydropower output was 436 TWh or 15% of total output, a 12% annual increase since 2000; thermal power output was 2 370 TWh, accounting for 83% of total generation, up 13% per year since 2000 and while dominated by coal, the 11 GW of natural gas-fired power plants had the capacity to produce over 80 TWh; nuclear power output was 54.8 TWh or 1.9% of total output, up 22% per year since 2000; and wind power provided 3.9 TWh (NBS, 2008; Ni, 2007; Li and Gao, 2007).[4] Elsewhere in the energy sector, demand for transport fuels has seen total oil consumption grow by 7.7% per annum over the period 2000-06, compared with 11.3% per annum for coal. This section examines alternative energy sources to reach conclusions on how far these might displace coal use in the future.

Oil

China's total crude oil resources are estimated by the Energy Research Institute (ERI) to be 13-15 Gt (ERI, 2004a). At the end of 2007, the country's proven crude oil reserves were 2.1 Gt, ranking 14th in the world (BP, 2008). Xinjiang and Tibet are key regions for future oil development in China. Between 2000 and 2006, China's oil consumption grew by an annual average of 7.7% to reach 352 Mtoe (Figure 4.2); China overtook Japan to became the world's second largest oil consumer in 2003. In 2006, the country's total production of crude oil was 185 Mt, having grown by just 2.1% per year since 2000 (NBS, 2008). This growing gap between consumption and production means that China's oil imports have risen quickly – China is now the world's third largest oil importer.

China became a net importer of oil in 1993. Since 2000, its net oil imports have grown rapidly from 87 Mtoe to 184 Mtoe in 2006, an average annual growth rate of 13%. In 2006, China's reliance on oil imports was 52% and could reach more than 60% in 2020 (Liu, 2006). Other forecasts confirm this growing import dependence

---

4. China reports "thermal power" output, including coal-fired, natural gas-fired and oil-fired output. There is no reliable data for the share that each of these fossil fuels makes, although coal accounts for more than 95%. Based on the fuel input in 2006, the IEA estimates that 26 TWh were generated from gas-fired plants.

Figure 4.2 .... Oil production, consumption and trade in China since 1971

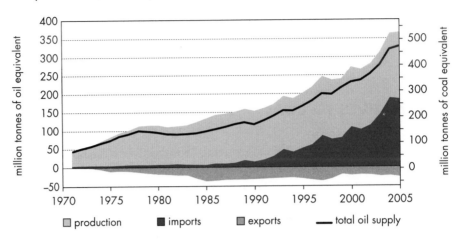

Source: IEA database (wed_bal.ivt available at http://data.iea.org).

(Figure 4.3), with international agencies projecting that China could be around 70-80% dependent on imports by 2030. China's oil imports come mainly from the Middle East and Russia. In the case of the former, they are dependent on safe passage through the Strait of Hormuz and the Strait of Malacca, creating potential risks for China's oil supply security (ERI, 2004a).

Figure 4.3 .... China's growing oil import dependence, 1990 to 2030

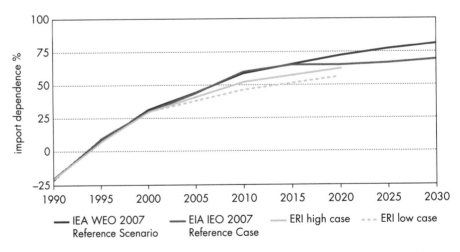

Notes: EIA: US Energy Information Administration; ERI: Energy Research Institute, China; IEA: International Energy Agency; IEO: EIA *International Energy Outlook*; WEO: IEA *World Energy Outlook*.
Sources: EIA (2007); ERI (2004b); and IEA (2007).

In recent years, China has reformed its oil pricing mechanism such that domestic prices are now better linked to international prices. Thus, China's economy is becoming ever more influenced by the higher price of internationally traded oil. The Chinese Academy of Social Sciences' Report on Energy Development 2007, the so-called

Energy Blue Paper, noted that Chinese energy market reforms should gradually accelerate, with price reform once again being the central issue (CASS, 2007). Domestic refined oil product prices will become fully linked to international market prices, a linkage which began in March 2006. To counter the resulting potential for excessive, unearned profits, national oil companies have had to pay a special tax of 20-40% on sales since 1 April 2006 and must also subsidise vulnerable groups affected by the rising prices of oil products.

Despite the potential to develop oil resources in western China, a study on optimising China's energy structure (ERI, 2004a) estimates that China's crude oil production would peak around 2015 at about 200 Mt, falling to 180-200 Mt in 2020. With China's oil demand set to continue growing and with the eventual decline of domestic supply, oil import dependency will grow, leading to increased oil supply risks. Some Chinese experts suggest that China should establish laws, monitoring systems and incentive policies that would encourage or oblige its oil companies to enter the international exploration and production sector (ERI, 2004a).

Natural gas

China's total natural gas resources are estimated to be between 10 and 13 trillion cubic metres (ERI, 2004a). At the end of 2007, China had proven natural gas reserves of just 1.88 trillion cubic metres, ranking 18th in the world (BP, 2008), although other estimates are substantially higher, 3.72 trillion cubic metres in the case of one international association (Cedigaz, 2007), as shown in Figure 4.4. In July 2006, the discovery of new gas resources in the Pearl River Delta was announced, with preliminary reserve estimates of more than 100 bcm, making it China's largest offshore natural gas discovery (Xinhua, 2006). Other discoveries continue to add to reserves.

Figure 4.4 .... Natural gas reserve discoveries in China since 1950 and their current distribution

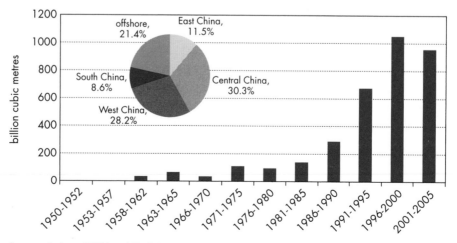

Sources: Cedigaz (2007) and ERI (2004a).

Currently, natural gas represents about 3% of total primary energy consumption (Figure 4.5). China's installed capacity of gas-fired power generation is over 14 GW, including plants under construction. At current domestic coal and gas prices, gas-fired power plants, despite their low-investment cost and high efficiency, cannot offset the price advantage of coal – combined cycle gas turbine power generation is not as competitive as coal-fired generation in China. Nevertheless, in the next decade, with the development of China's natural gas infrastructure and energy price reforms, demand for natural gas is expected to grow faster than that for coal and oil. By 2020, China's total natural gas-fired power generation capacity might reach 60 GW, with strong growth also in the non-power sectors (ERI and SGCC, 2007). This suggests a very slightly faster growth than assumed by the IEA, who forecast gas-fired capacity reaching 31 GW by 2015 and almost 100 GW by 2030 (IEA, 2007a). It is unlikely that China's indigenous gas production will keep pace with rising demand and natural gas imports will become increasingly important.

Figure 4.5 .... Natural gas production, consumption and trade in China

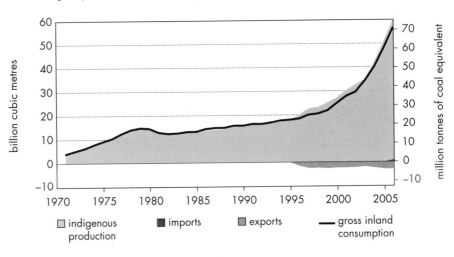

Source: IEA database (ngwbal.ivt available at http://data.iea.org).

To meet Guangdong's and Fujian's rapidly rising demand for natural gas, China began to import liquefied natural gas (LNG) from Australia in 2006 via LNG terminals and pipeline networks within the two provinces. China's natural gas pipeline network lays the foundation for the rapid development of the gas industry – at the end of 2006, the total length of natural gas pipelines reached over 24 090 km (NBS, 2007), mainly in Sichuan Province, the northeast and northern China. The USD 20 billion West-East gas pipeline was completed in 2005 and others are planned. Over the next ten years, new gas pipelines will be laid mainly in and to the north of the Yangtze River Delta region, and in the Sichuan Basin and surrounding areas. The introduction of natural gas from Russia and Kazakhstan will promote gas use in the northeast and parts of North China, while imports of LNG and offshore gas exploitation will provide important supplies to southern regions.

In April 2007, the 11th Five-Year Plan for Energy Development proposed policies on natural gas utilisation and demand-side management (DSM) in order to improve natural gas supply (NDRC, 2007b). DSM can help to optimise energy resource allocation and save energy, leading to lower-cost energy services and environmental benefits. The National Development and Reform Commission issued its gas use policy in August 2007 whereby, for the first time, the government will make use of DSM based on four categories of industrial gas user: preferential, permitted, restricted and prohibited. This should help to control natural gas consumption and prevent unnecessary wastage (NDRC, 2007c). The Chinese Academy of Social Sciences predicts that, in the next 15 years, China's demand for natural gas will grow dramatically, at an average annual growth rate of 11-13%, reaching 100 bcm in 2010, of which 20 bcm will be imported, and 200 bcm in 2020, with imports meeting half this demand (CASS, 2007). This is a faster growth than forecast by the IEA; under the IEA Alternative Policy Scenario and driven by strong environmental policies, gas demand reaches 200 bcm by around 2025 (IEA, 2007).

Because oil and gas supplies from Russia, the Middle East, Central Asia and Latin America are all subject to international market price fluctuations and transport constraints, China's imports of oil and natural gas face risks. Some analysts believe that there is limited growth potential for oil and gas production in the Southeast Asian region adjacent to China. For this reason, some argue that China should not rely on oil and natural gas as alternatives to coal in the future (Liu, 2006).

Nuclear power

China's *in situ* uranium resources totalled 100 000 tonnes (tU) in 2007, with up to 48 800 tU being "reasonably assured" and recoverable[5] (NEA, 2008). Production in 2007 is estimated to have been 750 tU and consumption for power generation was about 1 500 tU (*ibid.*), but will increase significantly with the commissioning of new reactors (Figure 4.6). China imports raw uranium from Kazakhstan and Canada. In January 2005, China and Canada signed a statement on co-operation in the fields of oil, natural gas, nuclear, energy efficiency, clean energy and renewables (NDRC, 2005). Both governments will encourage domestic enterprises to jointly study and develop nuclear energy to reduce costs and improve the security of nuclear power systems. Australia has the world's most abundant uranium resources, with 23% of the world's identified resources. In April 2006, China and Australia signed two agreements providing a legal framework for commercial uranium trade between the two countries (DFAT, 2006). Subsequently, in July 2006, China signed an agreement with the Government of Niger to jointly develop uranium resources. The China Nuclear International Uranium Corporation (SinoU) will lead this development. The Chinese government is developing other policies on uranium exploitation and use. For example, the China National Nuclear Corporation (CNNC) lost its monopoly in uranium development in 2006. In response, CNNC signed a co-operation agreement

---

5. Uranium recoverable using current mining and processing technologies at a cost below USD 130/kgU, based on samples taken from know mineral deposits.

in February 2007 with the Sinosteel Corporation to oversee uranium resource development (Sinosteel, 2007).

Since 2000, China's nuclear power generation output has grown at an average rate of 22% per year to reach 54.8 TWh in 2006 (NBS, 2008). During the 9th Five-Year Plan period (1996-2000), China developed four nuclear power projects with a total of eight units having a combined capacity of 6.7 GW. As a result, China has now mastered PWR nuclear power generation technology at unit sizes of 300 MW and 600 MW, with 60% of components manufactured locally. In August 2007, China began building its largest nuclear power project, the 4 320 MW Red River nuclear power plant in Liaoning, the first in the 11th Five-Year Plan period. China's own CPR1000 nuclear technology will be used for all four units.

Figure 4.6 .... Nuclear power generation capacity growth in China showing reactors under construction and those about to start construction

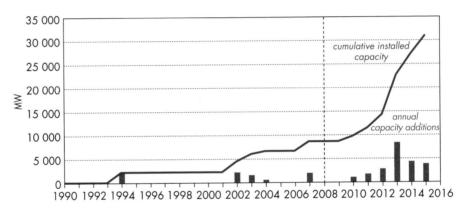

Note: Figure shows the capacity of the 21 reactors currently under construction or about to start construction. A further 20 nuclear power units are planned or proposed with a combined capacity of 20.7 GW. Source: WNA (2008).

China's total nuclear power installed capacity was 8.6 GW in 2008 (Figure 4.6). The Chinese government aims to have 70 GW of nuclear capacity operation by 2020 supplying 5% of the country's electricity needs. To strengthen policy support for nuclear power development, the Energy Research Institute suggests that China needs to establish medium- and long-term nuclear power development plans and financial incentives (ERI, 2004a). Breaking the monopoly on nuclear power plant construction and reducing the high cost of nuclear power are also seen as important, especially the high cost of imported equipment. Large-scale development of nuclear power will also require parallel progress in processing and storing nuclear wastes.

Large-scale hydropower

Water power is the most important renewable energy resource in China. It is broadly distributed, although the majority of resources, about 70%, are located in southwest China. Total installed capacity in southwest China is about 60% of the economically feasible resources (NDRC and ERI, 2006), with the middle and lower reaches of the Yangtze River being most important, followed by the Yellow River, Hongshui River and Min Zhe Gan River (ERI, 2004b). According to the results of the 2003 Nationwide Hydropower Resource Assessment, China's technically exploitable hydropower totals 542 GW, with an annual power generation potential of 2 470 TWh (NDRC and ERI, 2006). The Jinsha River has four dams under construction, site preparation works at three other locations and plans exist for a further five dams. The Lancang (Mekong) River has two completed dams, three under construction and three more proposed. Flowing from the Tibetan Plateau into Myanmar (Burma), the Nu (Salween) River has significant potential and there are plans to build 13 dams. China's total economically feasible capacity is 400 GW, with an annual output potential of 1 750 TWh. By the end of 2006, the national hydropower capacity was 125 GW, accounting for 19% of total generation capacity. Output was 436 TWh, accounting for 15% of the nation's power generation (NBS, 2008).

The Energy Research Institute suggests that China needs to make greater use of its water resources and strengthen hydropower planning (ERI, 2004a). ERI recommends that the electricity market should be gradually opened up to achieve optimal development of the country's largest regional hydropower resources, paying attention to environmental protection. In recent decades, hydropower construction has had to pay up to thirty different taxes and fees, including land occupation tax, mineral resource tax and compensation fees; the hydropower value-added tax has been as high as 17%. These add to the cost of hydropower projects and affect their development (*ibid.*). According to China's long-term renewable energy plan, by 2020, China's hydropower capacity will reach 300 GW, accounting for 75% of economically feasible hydropower resources (NDRC, 2007d). By 2030, China expects to have completed its hydropower development with a total installed capacity of about 350 GW and annual electricity generation of 1 500 TWh. Even with such a large contribution, coal-fired plants will still dominate the generation mix.

Renewable energies

China has abundant and widely distributed renewable energy resources that have the potential to gradually displace coal, oil and natural gas in the nation's energy mix. In recent years, China has strengthened its energy legislation to promote renewable energy. The Electric Power Law 1995, the Energy Conservation Law 2007, the Construction Law 1997, the Atmospheric Pollution Prevention and Control Law 2000 and other laws and regulations have all dealt with renewable energy development issues (NDRC and ERI, 2006). In January 2006, the Renewable Energy Law 2005 was put into effect.[6] The law establishes feed-in tariffs for renewable power and requires utilities to purchase all renewable power generated at attractive, fixed rates,

---

6. According to the Renewable Energy Law 2005, grid companies and petroleum product wholesalers must purchase renewable power and liquid bio-fuels respectively; the energy administrative authorities under the State Council are made responsible for formulating regulations for grid connection, operation and management of renewable power generation; grid companies have responsibility for the construction of transmission lines for renewable power stations; and the organisations responsible for power dispatch must endeavour to dispatch renewable energy sources ahead of others.

thus spreading the additional costs across all consumers. In addition, the law led to the development of legally binding national renewable energy targets, which are among the most aggressive in the world. The national renewable energy plan calls for 15% of all primary energy supply to come from renewables by 2020 (NDRC, 2007d). The plan requires the installation of 120 GW of renewable power generation capacity, including 30 GW each of wind and biomass.

Wind energy

According to estimates by the China Meteorology Research Institute, China's exploitable onshore wind resources are 253 GW, ranking first in the world, with a further offshore potential of 750 GW. Wind power is the most cost-effective renewable energy today. Construction of wind power projects in China began in the 1980s. During the 10th Five-Year Plan period, China's wind power projects have developed rapidly, with total installed capacity increasing from 0.35 GW in 2000 to 2.6 GW in 2006, an average annual growth rate of 40% (Figure 4.7). The installed capacity in 2004 ranked tenth largest in the world, but by the end of 2006, it had risen to sixth (NDRC and ERI, 2006). Despite this rapid growth, China's installed wind power capacity today is just 0.1% of the country's wind resource potential.

Figure 4.7 .... Wind energy installed capacity in China

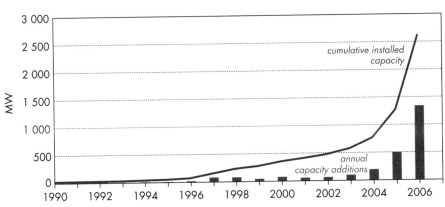

Source: NDRC and ERI (2006).

By the end of 2006, NDRC had awarded 1.85 GW of wind project concessions, leveraging about USD 1.8 billion of private investment. The central government has stated its intention to dramatically expand this concession model in wind-rich areas to attract investment of about USD 3.5 billion and bring the total capacity of wind projects to more than 3.5 GW by 2010. According to the National Medium- and Long-Term Energy Development Plan, by 2010, the total installed wind power capacity will reach 5 GW and 30 GW by 2020 (NDRC and ERI, 2006). An internationally competitive wind power industry is expected to become established by 2010-15 to achieve this longer-term goal.

Biomass energy

China's biomass energy resources include straw and other agricultural wastes such as rice husks, waste from forestry and forest product processing, animal manure, energy crops and plantations, organic effluents from industry, municipal wastewater and municipal solid waste (MSW). Of about 600 Mt of crop straw produced every year, nearly 300 Mt (around 150 Mtce) can be used as fuel. Around 900 Mt of waste from forestry and forest product processing is available each year, and nearly 300 Mt of this (about 200 Mtce) can be used for energy production (NDRC and ERI, 2006). In addition, there are large areas of marginal lands in China that can be used to cultivate energy crops. Presently, the nation's biomass resource that can potentially be converted into energy is about 500 Mtce per year (*ibid.*), less than 20% of current total primary energy consumption. A small proportion of this is used for power generation. By the end of 2005, the installed capacity of biomass power in China reached 2 GW. Bagasse (sugar cane residue) plants totalled 1.7 GW, while MSW incineration and land-fill gas power plants accounted for a further 200 MW; the remainder was agricultural or forestry waste gasification (NDRC, 2007d).

At the end of 2005, the total number of household biogas digesters reached 18 million, with an estimated total annual production of 7 bcm. About 1 500 large-scale biogas plants for livestock waste and organic industrial effluent produced a further 1 bcm (NDRC, 2007d). Biogas is now widely integrated with animal husbandry and has become an important means of waste treatment in the agricultural sector.

China has already begun to produce bio-ethanol for use as a transport fuel. In 2005, the production capacity for bio-ethanol using food grains as a feedstock was just over 1 Mtpa (NDRC, 2007d). The production capacity of bio-diesel made from oil-seed crops, residues from edible oil pressing and waste oil from restaurants reached 50 000 tonnes in 2005 (*ibid.*). The technology for producing bio-ethanol from non-food-grain feedstock has reached the pre-commercial stage in China.

Solar energy

Two-thirds of China's land area has abundant solar energy, particularly in the northwest, Tibet and Yunnan, with average annual radiation levels of over 6 GJ/m^2, an annual surface absorption of solar energy equivalent to approximately 1 300 Gtce (NDRC and ERI, 2006). Solar technologies are used for heating, power generation, lighting and cooling. In China today, solar water heaters are widely used, having been developed commercially. By the end of 2005, the annual production of solar heaters was 15 million square metres (m^2) (NDRC, 2007d). The total heat-collecting area of installed solar water heaters reached 80 million m^2 in 2006, about half the world total, and the national targets are to reach 150 million m^2 by 2010 and 300 million m^3 by 2020 (*ibid.*).

By the end of 2006, the total installed capacity of solar photovoltaic (PV) power in China was about 80 MW (Figure 4.8), about half of which was used for supplying power to residents in remote rural areas and for special applications, such as communications and navigation (NDRC, 2007d). The Township Electrification Program of 2002-03 resulted in the installation of 19 MW of solar PV panels, providing a relatively strong stimulus to solar cell manufacturing in China (NDRC and ERI, 2006). China has now begun to implement building-integrated, grid-connected PV demonstration projects.

The Energy Research Institute regards photovoltaic power generation as a "mature technology" and, by 2020, PV capacity could reach 1 600 MW (NDRC and ERI, 2006; NDRC, 2007d). Some estimates of China's PV manufacturing capacity in 2010 exceed

1 000 MW and would make China the world's largest producer (Li and Shi, 2006). China's long-term development plan shows the solar power industry growing at an annual rate of over 20% between 2006 and 2020, rather slower than ERI's optimistic assessment. Currently, the problem is the higher generation cost of mainstream solar PV power generation technology compared to traditional technologies. In the short term, there may be insufficient supply of polysilicon, and thin-film PV solar cells need further development.

Figure 4.8 .... Photovoltaic power generation capacity in China

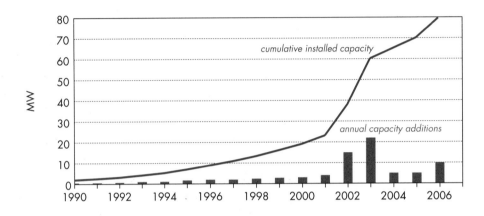

Source: NDRC and ERI (2006).

Small-scale hydropower

Small hydropower generally refers to plants below 50 MW. According to the latest review of small hydropower, China's economically feasible water power resource is estimated to be 400 GW, with an annual generation potential of 1 750 TWh, of which small-scale hydropower accounts for 125 GW, widely distributed throughout the provinces, especially in the southwest (NDRC and ERI, 2006). With supportive policies and incentives, China had more than 40 GW of small-scale hydropower projects by the end of 2006, with a generating output of about 140 TWh. According to the renewable energy development plan, China's hydropower capacity will reach 300 GW by 2020, of which 125 GW will be small-scale hydropower (NDRC, 2007d).

Renewable energy costs and pricing policy

The current high cost of renewable energy restricts its commercialisation. If the cost of coal-fired power generation is 1.0, then the estimated cost of small-scale hydropower is about 1.2, biomass power generation (using methane generated in gasifiers) is about 1.5, wind power is 1.7 and photovoltaic power lies between 11 and 18, thus greatly weakening the economic competitiveness of renewable sources in the electricity market (Oriental Xinbang, 2007).

At the beginning of 2006, the Renewable Energy Law 2005 was officially implemented in China. The State Council and relevant ministries have been supporting its implementation with measures including the *Management Regulations on Renewable Energy Power Generation, Renewable Energy Prices and the Cost-Sharing Pilot Scheme, Renewable Energy*

Tariff Revenue Allocation Pilot Scheme, Guidance on Development of Renewable Energy Industry and *Interim Management Measures for Renewable Energy Funds*. These regulations aim to support the development of renewable energy by providing industrial development guidance, progressing electricity price reform, setting volume goals, implementing favourable taxation policies and subsidies and establishing special funds.

In September 2007, the National Development and Reform Commission issued its Medium- and Long-Term Development Plan for Renewable Energy in China in order to speed up the development of renewable energy, promote energy conservation, reduce pollution, mitigate climate change and better achieve sustainable social and economic development (NDRC, 2007d). According to the plan, China should have raised the share of renewable energy in total primary energy consumption to 10% by 2010 and 15% by 2020, with individual technology capacity targets shown in Table 4.1.

Table 4.1 ....Renewable energy targets in China's Medium- and Long-Term Development Plan

	Total installed capacity/annual production	
	By 2010	**By 2020**
Wind power	5 GW	30 GW
Solar power	0.3 GW	1.8 GW
Hydropower	190 GW	300 GW
Biomass	5.5 GW	30 GW
Biomass use for fuel	1 million tonnes	50 million tonnes
Biogas use	19 billion cubic metres	44 billion cubic metres
Bio-ethanol use	2 million tonnes	10 million tonnes
Bio-diesel use	0.2 million tonnes	2 million tonnes

Source: NDRC (2007).

As elsewhere, pricing policies in China will determine how renewable energy develops. Policy issues include: a lack of internalisation of the environmental and social costs of conventional power generation into power tariffs; an inability to spread the additional costs across all consumers over a wide region; a reluctance to raise wholesale electricity rates, which would allow utilities to pass on the additional costs; a lack of investment support to improve the economics of high capital cost, long-term renewable projects; and a lack of policy support, such as production tax credits and low-interest loans for renewable project developers. Some provinces, such as Fujian and Sichuan, are creating mandatory market share mechanisms to compel utilities to either produce or procure through "green credits" enough renewable energy to meet their 15% obligation. Shanghai is developing a Green Pricing Program, allowing consumers to voluntarily pay more for green electricity, with the utility using the proceeds to build more renewable facilities. By the end of 2006, Shanghai had built 20 MW of wind turbines and has plans to build a further 20 MW.

Hydrogen as an energy carrier

China is pursuing hydrogen fuel cell technology development, with some Chinese experts suggesting that PEM fuel cells should be the main focus. By 2015, hydrogen fuel cell power systems should be competitive with traditional technologies in the main application areas (ERI, 2004c). By 2020, a new generation of hydrogen storage technology might be available and a number of key cities can then contemplate large-scale commercial applications of hydrogen as an energy carrier (*ibid.*). The main strategic tasks are to study hydrogen storage technology, fuel cell technology, hydrogen fuel cell power systems and technical standards for a national hydrogen energy system.

Energy efficiency

Although examples of very efficient appliances, equipment and factories can be found, China's energy efficiency is generally low compared with the advanced levels found in some parts of the world. China's staggering pace of economic growth (averaging 10% per year over the last 25 years) is overwhelming China's ability to invest in energy efficient solutions. For every dollar of economic output, China uses 3 times more energy than the global average, 4.3 more than the United States and 8.3 times more than the super-efficient Japan (IEA, 2007b).[7] China's national development goal for 2020 aims to quadruple the economy from 2000 levels while only doubling energy consumption. Energy efficiency will play a significant role in meeting this challenge. ERI's 2020 energy demand scenario analysis shows that enhanced energy conservation will help China reduce its energy consumption by 15-27% below a baseline forecast (ERI, 2004b).

In the past several years, China has made efforts to develop and implement appliance standards and labelling, and energy efficiency standards for buildings. National appliance standards, covering more efficient refrigerators, air conditioners, lighting, televisions and other consumer appliances, are on track to save 10% of residential electricity consumption by 2010 (ERI, 2004d). In March 2005, China began requiring manufacturers to affix mandatory energy efficiency labels to refrigerators and air conditioners. Then, in September 2006, NDRC approved mandatory energy information labels for central air conditioners and washing machines. China is also improving electric motor efficiency by an average of 4.5% with new motor standards. The Ministry of Construction has completed the groundwork for regulating both residential and commercial building energy consumption. The residential building energy code for China's northern heating zone was updated in 1996, and codes have recently been adopted for the "hot-summer, cold-winter" zone of central China and the "hot-summer, warm-winter" zone of southern China. A national commercial building code with energy efficiency requirements was adopted in 2005 (MOHURD, 2005), although there is insufficient practical guidance to allow contractors to implement the code.

---

7.   Expressed on an exchange rate basis. If expressed on a purchasing power parity basis, then China's energy per unit of GDP is similar to the US and global averages, and around 50% greater than Japan's.

Conclusions

This section has examined the alternatives to coal to determine if these have the potential to meet the Chinese government's goal of long-term, sustainable economic development. China has limited reserves of oil and natural gas, so will need to import more from the international market in the future. This raises energy supply security questions and, with rising oil and natural gas prices, China's economy may suffer. Nuclear and hydropower are important indigenous energy resources that can reduce China's dependence on fossil fuels. In the case of hydropower, its share of primary energy supply is expected to decline, even if China succeeds in harnessing most of the country's technical hydropower potential. China is currently pursuing the world's most ambitious nuclear expansion programme. If by 2020, the country has built and commissioned some of the 40 or so reactors planned and reached the government's 70 GW target, nuclear will still only account for 5% of national power generation. The use of renewable energy sources is growing very quickly in China and current plans will see greater exploitation of wind, biomass and solar energy in the future. The high cost of renewable energy means that its full potential cannot be reached in the short to medium term. Nevertheless, the Chinese government appears willing to channel investment into renewable sources to deliver its target of 15% of primary energy supply by 2020.

Taken together, the promotion of alternatives to coal, coupled with improved energy efficiency, could see coal demand rise less steeply in the future than otherwise forecast. Nothing suggests that the trend of rising coal consumption will flatten or reverse. Indeed, consumption could easily double over the next two decades. Hence, energy conservation should continue to be a top priority, in an energy system dominated by clean and efficient coal technologies, complemented by a strategy to develop alternative energy supplies.

REFERENCES

Berrah, N., F. Feng, R. Priddle and L. Wang (2007), *Sustainable Energy in China: The Closing Window of Opportunity* (中国能源可持续发展报告), The World Bank (世界银行), Washington, DC.

BP (2008), *BP Statistical Review of World Energy,* BP plc, London, June, www.bp.com.

CASS (中国社会科学院 – Chinese Academy of Social Sciences) (2007), 2007 中国能源发展报告 (*China Energy Development Report 2007*), CASS, Beijing.

CCIDRC (China Coal Industry Development Research Centre) (2007), personal communication, CCIDRC, Beijing, October.

CCTMA (China Coal Transport and Marketing Association) (2007), *Forecast of Coal Demand and Growth Rates by Sector During China's 11th Five-Year Plan Period (2006-10)*, CCTMA, Beijing.

Cedigaz (2007), *2006 Natural Gas Year in Review,* Cedigaz, Rueil-Malmaison, France.

DFAT (Department of Foreign Affairs and Trade) (2006), *Agreement Between the Government of Australia and the Government of the People's Republic of China for Cooperation in the Peaceful Uses of Nuclear Energy* and *Agreement Between the Government of Australia and the Government of the People's Republic of China on the Transfer of Nuclear Material*, DFAT, Barton, ACT, Australia, 3 April, www.dfat.gov.au/geo/china/index.html.

EIA (Energy Information Administration) (2007), *International Energy Outlook 2007*, DOE-EIA-0484(2007), EIA, US Department of Energy, Washington, DC.

ERI (能源研究所 – Energy Research Institute) (2004a), 中国国家综合能源战略和政策项目:能源结构调整和优化 (*China's Energy Development Strategy and Policies: Energy Structure Adjustment and Optimisation*), ERI, Beijing.

ERI (2004b), 中国国家综合能源战略和政策项目: 能源需求情景分析, (*China's Energy Development Strategy and Policies: Energy Demand Scenario Analysis to 2020*), ERI, Beijing.

ERI (2004c), 中国能源科技发展战略研究报告, (*Report on China's Energy Technology Development Strategy*), ERI, Beijing.

ERI (2004d), 中国国家综合能源战略和政策项目: 能源效率和节能, (*China's Energy Development Strategy and Policies: Energy Efficiency and Energy Saving*), ERI, Beijing.

ERI and SGCC (Energy Research Institute and State Grid Corporation of China) (2006), *China's Policy on Natural Gas in Power Generation*, 1st draft report of joint research group, ERI and SGCC (as reported by Guangdong Chamber of Commerce, www.oilgas.cc/blog/bo-blog/read.php?174, accessed 14 August 2008).

IEA (International Energy Agency) (2007a), *World Energy Outlook 2007 – China and India Insights*, OECD/IEA, Paris.

IEA (2007b) *Key World Energy Statistics 2007*, OECD/IEA, Paris.

IEA (2008), *Coal Information 2008*, OECD/IEA, Paris.

Li Junfeng and Gao Hu (2007), *China Wind Power Report 2007* (中国风电发展报告 2007), China Environmental Science Press, Beijing, www.gwec.net/uploads/media/wind-power-report.pdf (English) and www.cresp.org.cn/uploadfiles/89/971/wind-power-report.pdf (Chinese).

Li Junfeng and Shi Jingli (李俊峰, 时璟丽) (2006), 我国可再生能源技术的现状与发展 (*Development of Renewable Energy Technologies in China*), 能源研究所 (Energy Research Institute), Beijing.

Liu Ronghua (刘荣华) (2006), 高油价下的中国能源战略选择 ("China's Energy Strategy under a High Oil Price Scenario"), 中国经济周刊，2006年第35期 (*China's Economy Weekly*, Issue 35, 2006).

MOHURD (中华人民共和国住房和城乡建设部 – Ministry of Housing and Urban-Rural Development) (2005), 公共建筑节能设计标准 (*Energy Efficiency Design Standard for Public Buildings*), GB50189-2005, MOHURD, Beijing, 4 April, www.cin.gov.cn/fdcwwj/gg/200610/t20061031_2716.htm.

NBS (2007), "China's Growth in Oil and Gas Network Length", National Bureau of Statistics of China press release (reported at www.china5e.com/news/oil/200710/200710150297. html, accessed 14 August 2008).

NBS (National Bureau of Statistics of China) (2008), *China Statistical Yearbook 2007*, China Statistics Press, Beijing.

NDRC (National Development and Reform Commission) (2005), "Statement on Energy Cooperation in the 21st Century", press release, NDRC, Beijing, 24 January, http://en.ndrc.gov.cn/newsrelease/t20050620_8220.htm.

NDRC (国家发展改革委 – National Development and Reform Commission) (2007a), 煤炭工业发展"十一五"规划 (*11th Five-Year Plan for Coal Industry Development*), press release, NDRC, Beijing, 2007年1月22日 (22 January 2007), www.ndrc.gov.cn/nyjt/zhdt/t20070122_112661.htm (full text at www.ndrc.gov.cn/nyjt/zhdt/W020070306520943990664.doc).

NDRC (2007b), 能源发展"十一五"规划 (*11th Five-Year Plan for Energy Development*), NDRC, Beijing, 四月 (April), www.ccchina.gov.cn/WebSite/CCChina/UpFile/File186.pdf.

NDRC (2007c), 天然气利用政策 (*Natural Gas Utilisation Policy*), NDRC, Beijing.

NDRC (2007d), 可再生能源中长期发展规划 (*Medium- and Long-Term Development Plan for Renewable Energy in China*), NDRC, Beijing.

NDRC and ERI (国家发改委和能源研究所 – National Development and Reform Commission and Energy Research Institute) (2006), 我国可再生能源产业发展报告 (*Report on China's Renewable Energy Industry Development*), NDRC and ERI. Beijing.

NEA (Nuclear Energy Agency) (2008), *Uranium 2007: Resources, Production and Demand*, Report No. 6345, a joint report by the OECD Nuclear Energy Agency and the International Atomic Energy Agency, OECD/NEA, Paris.

Ni, Chun Chun (2007), *China's Natural Gas Industry and Gas to Power Generation*, Institute of Energy Economics, Tokyo, Japan, July.

Oriental Xinbang (东方信邦) (2007), 可再生能源发电仍有制约 ("Constraints Still Exist for Renewable Energy Power Generation"), 东方信邦 (*Oriental Xinbang*, Issue 229).

Pan Qinglin (2005), "China's Coal Demand: Prediction and Prospects", *China Clean Coal Technology*, November, China National Coal Association, Beijing.

Sinosteel (2007), "Sinosteel Signing the Overseas Uranium Resource Development Strategic Cooperation Agreement with CNNC", press release issued 16 February, Sinosteel Corporation, Beijing, http://en.sinosteel.com/xwzx/zgdt/2007-02-16/1518.shtml.

WNA (World Nuclear Association) (2008), *Nuclear Power in China*, country briefing, WNA, London, August, www.world-nuclear.org/info/inf63.html, accessed 31 August 2008.

Xinhua (2006), 南海珠江口盆地发现上千亿方天然气资源, ("Discovery of 100 bcm Natural Gas Resource in Pearl River Mouth Basin of South China Sea"), 张晓松，杜文景 (Zhang Xiaosong and Du Wenjing), 新华网 (Xinhua Net), Beijing, 13 July.

V. A CLEANER FUTURE FOR COAL IN CHINA: THE ROLES OF TECHNOLOGIES AND POLICY

In this chapter, the technologies and policies relating to cleaner coal technologies in China are reviewed. The aim is to present an overview of the technologies in current use, notably for coal mining and coal-fired power generation, and how these might evolve over the coming decades. Government policy will influence this evolution, so extant coal-related policies are described with a critique of their effectiveness. To achieve the goals set out in China's national plans will require the deployment of new technologies, so the final sections of the chapter examine R&D and technology transfer requirements. This introduction continues with a summary of the key issues and how these might influence future international co-operation with China.

From a technology perspective, the dichotomy in China between the advanced and the primitive is stark. There are examples of coal mining techniques and equipment achieving the highest productivities found anywhere in the world. At the same time, there are tens of thousands of small coal mines that perform poorly, by any measure. Appropriate policies are in place to improve the coal mining sector, but they lack traction because other imperatives, such as meeting coal demand and maintaining local economic activity, take precedence. The story is similar in the power generation sector – a clear national policy to improve efficiency and reduce pollution is unevenly implemented at the local level. A large number of FGD plants have been built but could be better used, and low-NOx burners have not been adopted. Chapter 7 reviews how some of these issues have been tackled in IEA member countries, the most challenging one being coal industry restructuring.

This chapter concludes with a review of technology transfer requirements and the role of international co-operation in ensuring that China has access to the technologies it needs for a cleaner future. The IEA believes that the most productive area for international co-operation is in developing policies that create attractive markets around the world for clean technologies. This policy review describes a range of policies relevant to clean coal in China, some mandatory, some not, that, in many cases, are falling short of the well-considered, higher objectives set by State Council. For example, multiple support mechanisms for coalbed and coal mine methane projects are described (in addition to the substantial support under the Kyoto Protocol's Clean Development Mechanism for coal mine methane projects), but with no clarity as to whether the benefits justify this level of support for two techniques that serve very different purposes. In IEA member countries, such targeted support is often the result of industry lobbying, rather than sound policy making. International co-operation on policy development, environmental impact assessment and cost-benefit analysis is perhaps the most rewarding route to pursue since it can result in a framework for commercial activities that have a far greater impact than piecemeal co-operation on individual projects. This theme is returned to in Chapters 9 and 10.

TECHNOLOGY REVIEW

Coal mining

During the 10th Five-Year Plan period (2001-05), China made great progress with coal mining techniques and technologies, improving output and efficiency at mines that encounter a variety of geological conditions, including thin and steeply sloping seams. In thick coal seams, fully mechanised, top-coal caving mining technology, developed in China, allows a greater proportion of the seam to be recovered. A fully mechanised, longwall coal face can now produce three to six million tonnes per annum (Mtpa) and achieve productivities of over 30 000 tonnes per man-year using machinery with rated capacities of 1 500-2 500 tonnes per hour, matching or exceeding the best performance achieved elsewhere in the world.

To implement the National Plan for Medium- to Long-Term Scientific and Technological Development (2006-2020) (MOST, 2006), the 11th Five-Year Plan for Coal Mining Technology Development (NDRC, 2007a) instigates several major projects to develop technologies, including:

■ **Geological surveying technologies.** To guarantee high-efficiency coal mining through the precise and detailed exploration of coal resources, especially in western China where the evaluation of new technologies and information systems to handle geological data will be important.

■ **High-efficiency coal mining technologies.** Mainly equipment for large, deep mines in alluviums with thick coal seams, that are capable of producing 6-10 Mtpa using mechanised and automated mining equipment. Other technologies include those for thin seams, fully mechanised shortwall mining equipment, automatic control systems for coal face equipment, fast drivage techniques for cutting underground roadways, new coal conveying, lifting, blending and loading technologies, and mine information systems. The basic theory and research into technologies for mining deep seams in eastern China and under the sea are also considered important.

■ **Sustainable coal mining technologies.** Including those to safely mine under surface water features and populated areas. Achieving this demands research on: the environmental impacts of coal exploitation and possible mitigation measures; subsidence management; the use of mine water as a resource; the treatment and reduction of coal gangue pollution; mining waste utilisation; and mine site restoration. The industrial-scale demonstration of underground coal gasification is also seen as a priority.

After three distinct periods of development in the People's Republic of China – the transition to socialism following the nation's founding (1949-57), the adjustment period (1958-77) and the period of economic reform and opening up (1978-93) – the coal industry has been gradually restructured as China moved from a planned economy to a socialist market economy. Earlier, cost-based management accounting methods could not be adapted to the demands of a market economy. Therefore, all levels of government, together with coal enterprises, had to explore ways to improve accounting methods. At present, the shadow of China's former planned economy remains and financial management methods still do not meet the requirements of the socialist market economy. The main problems are that reported coal production costs do not reflect true costs, that some charges are unjustifiably high, and that employees' incomes and social security are generally too low. Costs do not reflect the value of coal resources mined or sterilised,[1] or the full cost of pollution, and they do not include the cost of further coal industry rationalisation.

The Chinese government has attached great importance to raising coal resource and environmental compensation fees to fund more sustainable development practices, such as the coal resource tax implemented by the State Administration of Taxation in 2006. Anhui province has set up a new mine construction fund; Shandong province has set up a reform and development fund; Shanxi province has established a fund for coal production reform; and some provinces, including Henan and Anhui, have raised the rate of coal production levies. These reforms have had positive effects, but financial management still lacks an overall, systematic objective. Moreover, coal enterprises are concerned about their ability to develop sustainably in the face of rising wage bills and increased taxes. Further progress should aim to encourage coal enterprises to carry out resource exploration, to protect the environment and treat pollution, and so move the coal industry towards a more sustainable future.

Coalbed methane and coal mine methane

A diverse and practical range of coalbed methane (CBM) and coal mine methane (CMM) drainage methods and technologies are used today in China. CMM involves capturing methane released from the coal seam and surrounding strata before or during mining. In underground mines, ventilation must be sufficient to keep methane levels low enough to avoid the risk of explosion. Since the build up of methane can slow mining, it is common practice to pre-drill coal seams and draw off methane before mining. When this is done to exploit the methane as an energy source, without any intention of mining the coal, it is called CBM. Since the 1950s, China has carried out underground gas drainage trials in coal mines, such as Fushun, Yangquan, Tianfu and Beipiao, when annual drainage was about 60 million cubic metres (mcm). By 2006, the total amount of gas drained annually from the 286 "highly gassy" key state-owned coal mines was 2.6 billion cubic metres (bcm). According to the requirements of the 11th Five-Year Plan, mine gas drainage in China should exceed 5 bcm in 2010, with a drainage rate of over 40%. In 2005, the total quantity of CMM consumed was about 1 bcm. According to the 11th Five-Year Plan, gas utilisation in China is to reach over 3 bcm by 2010, a utilisation rate of over 60%. Today, CMM is used for several purposes: residential use (accounting for over 70%), power generation, industrial fuel, chemical feedstock and fuel for transport. Over 2 000 CBM wells have been drilled in China, 1 700 in the period 2005-07. CBM production in 2006 was less than 0.1 bcm.

---

1. Here, "sterilised" refers to coal resources left underground after mining that will never be recovered because it would be too difficult, too dangerous or too uneconomic to attempt their recovery in a subsequent mining operation. Ideally, all coal resources would be recovered during mining, but this ideal is impossible to achieve (see also footnote 1 in Chapter 2).

CMM technologies

The selection of a suitable CMM gas drainage method is mainly determined by factors such as the source of methane, the type of coal, the coal extraction method and mining procedures employed, and the geological conditions. Currently, the CMM drainage methods used in China can be divided into gas drainage from seams to be mined, gas drainage from adjacent seams, gas drainage from goaf or gob areas, and gas drainage from surrounding rock. They could be also divided into in-seam borehole drainage, ventilation drainage, mixed drainage and drainage from the surface using vertical boreholes. Depending on when the gas is drained, they could be further divided into pre-mining drainage prior to coal extraction, drainage during coal face development, drainage while mining and post-mining drainage. Typical drainage arrangements at underground coal mines include: long-hole drilling parallel to the bed plane to drain gas from the coal seam being worked; cross-seam gas drainage from underground roadways; directional drilling of a long borehole in the roof for gas drainage from an adjacent worked coal seam; goaf drainage from worked areas behind the retreating coal face; and gas drainage from the goaf area via surface boreholes. CMM surface drainage technologies mainly include surface goaf or gob wells in mined areas, and multiple lateral or horizontal wells for gas production from unmined coal seams. All these drainage technologies have been applied as dictated by the characteristics of Chinese coal seams, such as low permeability, high gas content, high risk of gas outburst, multiple seam mining and complex geology.

Alongside CMM drainage technologies, China has developed many CMM utilisation technologies since the 1950s when a black carbon plant using CMM as raw material was established in the Fushun mining area. Since 1982, China formally included CMM utilisation projects in the national energy-saving investment plan (National Planning Committee of China, 1982) and CMM use has grown, mainly in the residential and power sectors. CMM drainage pump stations, secondary booster stations, and municipal gas distribution and storage systems are built to the *Design Code for Town Gas Engineering* (Ministry of Construction, Notice No. 451, 2006). Although many technologies for residential use are mature, some problems remain to be solved. For example, fluctuations in the methane concentration of CMM are relative large and affect the stability of gas supply and combustion equipment.

In recent years, CMM-based power generation in the coal mining areas of China has developed quickly, mainly using reciprocating gas engines and gas turbines. Gas engines are small, light, portable and highly efficient (up to 40%). They can adapt to variations in methane concentration and are often suitable for small-scale power plants. Gas turbines are larger, typically 50-100 MWe from a single unit, with an efficiency of about 30%. Since gas turbines require fuel gas at a high pressure, the safety of compressing CMM is an important consideration to avoid creating a risk of explosion.

CMM costs

The costs for the development and utilisation of CMM in China mainly comprise drainage costs and investment costs for utilisation projects. CMM drainage is a compulsory requirement of state work-safety rules to ensure coal mine safety, so investment in drainage equipment and facilities is part of mining costs. Therefore, the viability of CMM utilisation projects depends mostly on capital investment costs, since the CMM is essentially a zero-cost input.

Since most coal mines in China are located in remote mountainous areas, large CMM utilisation projects are difficult to establish. It would be very challenging to construct long CMM pipelines to cities, regardless of the economics. The additional capital investments needed for residential use are substantial and include secondary booster stations, gas storage and distribution systems in cities, pressure reducing stations and user equipment. CMM projects are affected by factors such as gas price, pipeline construction costs and seasonal gas consumption, such that mining companies have had little enthusiasm to carry out projects for residential use.

For power generation projects, capital and operating costs depend on the technology selected. A gas turbine, has a high capital cost, but low operating costs. Gas engines are more flexible, but have higher operating costs. These power generation technologies are mature in China for use with CMM, having solved problems such as instability of the CMM supply and low methane concentration. A coal mine with an annual CMM drainage of around 1 mcm per year – typical of many coal mines in China – could drive a 500 kW, locally manufactured power generation unit. Enhanced electricity tariffs and priority grid access have greatly encouraged the implementation of power generation projects using CMM (Section 5.2).

CBM and CMM markets

The potential market demand in China for CBM and CMM is huge, since it could be transported and used with natural gas. According to estimates by CCII, natural gas demand in China will exceed 100 bcm by 2010, but there is a serious shortage of supply. The annual market demand in the key natural gas consuming areas around the Bohai Sea, coastal regions of the southeast, the Yangtze River Delta region, central China, and the north-eastern region already exceeds 12 bcm. During the 11th Five-Year Plan period, the preferred markets for CBM/CMM are mainly Shanxi, Beijing, Tianjin, Henan and Hebei. Once this demand is met, the remaining CBM/CMM could supply other provinces in central China or be fed into the West-East gas pipeline and delivered to the Yangtze River Delta region.

The state has issued several related policies on the prices of CBM/CMM for industrial and residential use, stipulating that these should not be lower than the price of natural gas with the same heating value (Section 5.2). This pricing policy stifles demand for CBM/CMM. In mining areas, drainage gas is often discharged to atmosphere because coal mines have no incentive to establish the facilities for reliable pipeline supply, including gas storage tanks and control systems. Local residents are therefore missing the opportunity to be supplied with potentially cheaper CMM.

Future CMM developments

The future development and utilisation of CMM in China depends on the development of drainage technologies and gas utilisation technologies, particularly for mine gas with a low methane concentration. Through the application and improvement of these technologies, CMM which is currently vented to atmosphere could be fully utilised from mines where gas collection and drainage is difficult.

With stricter environmental protection policies for GHG emissions, the reduction and utilisation of low-concentration CMM emissions are imperatives. At present, the concentration of methane in the drainage gas from many coal mines in China lies between 10% and 20%, below the 30% requirement stipulated for safety reasons in the *Coal Mine Safety Regulations* (SACMS, 2005). Although some domestic manufacturers have developed and field tested new products that could use low-concentration gas, they lack regulatory approval; safety issues need to be further addressed. Technologies with potential include gas purification equipment and power generation equipment that can directly use gas with low methane content.

Purifying mine gas could expand its utilisation possibilities and dramatically enhance CMM utilisation rates. At present, CMM can be purified by either absorption separation or permeation separation. Absorption separation depends on the adsorption properties of different substances. When gas flows through these substances, some molecules are preferentially adsorbed; when this reaches saturation, the molecules can be released in a regeneration step to complete the separation. This is the principle behind molecular sieves and pressure swing adsorption. Pressure swing adsorption is currently a mature CMM purification technology where pressure differences lead to differences in adsorption property, allowing different gases to be absorbed selectively and hence separated. Permeation separation purifies gas through a device with a membrane. Gases with smaller molecules penetrate the membrane, while gases with larger molecules do not pass through, allowing the separated gases to be collected from either side of the membrane.

The economic utilisation of low-concentration CMM, with a methane content in the range of 6-25%, depends upon solving two problems. Firstly, a safe gas transport system must be in place; secondly, power generation units should be able to operate with the low-concentration gas. China's domestic enterprises have developed power generation technologies for low-pressure, low-concentration CMM, and these have been put into commercial operation. Operating costs are low with efficiencies of 32-40% (higher heating value). Technologies such as automatic air-fuel ratio control, lean combustion, sump temperature control, digital ignition and cooled fuel pressure boosting are all employed.

In order to satisfy the requirements of work safety during mining activities, a ventilation system is usually employed. Research by CCII indicates that most of China's coal mines have not adopted CMM drainage systems. Instead, mine gas is discharged to atmosphere via the ventilation system. According to conservative estimates, the annual emission of ventilation air methane (VAM) is over 12 bcm. The volume of VAM will increase with growing coal output, attracting ever more attention. Currently, VAM utilisation technologies include: the thermal flow-reversal reactor (TFRR) from MEGTEC System Technology Company of Sweden and the catalytic flow-reversal reactor (CFRR) from the Canadian Energy Technology Research Institute. These are both at the demonstration stage in China.

The TFRR can be used to oxidise low-concentration gas and recover the heat energy released during oxidation. The heat may be used to produce hot water or hot oil, or to raise steam to drive a turbine for power generation. The system has no open flame, so no emissions of nitrogen oxides. It has high energy conversion efficiency and the methane oxidation efficiency can reach 98%, turning the methane almost entirely into carbon dioxide and water.

The CFRR has been developed in the laboratory and a field test has been carried out in Canada with a methane processing capacity of 600 m^3/min. This reactor has a recirculation time longer than that of the TFRR, so is a larger device. The temperature at its centre can reach 350-800°C and it does not produce nitrogen oxides or carbon monoxide. The heat retention is good and dust in the ventilation air is unlikely to damage the system.

The application of VAM oxidation power generation technology has been successful at West Cliff mine in Australia, and it has great significance for China's coal mines. It could more fully utilise the CMM resource, help relieve energy shortages and effectively reduce emissions of a greenhouse gas to protect the environment. A feasibility study for China's first power generation project using VAM has been completed in the Huainan mining area, providing a model for other mining areas in China.

Coal preparation

Current status and future development of coal preparation technologies

Raising the proportion of coal that is screened and washed has long been promoted as an important step to improve combustion efficiency, minimise ash production and reduce pollution, but much work remains. In 2006, of the 961 coal preparation plants with washing capacities of over 150 kt raw coal per year (total 837 Mtpa), 318 were at key state-owned mines, with a total capacity of 658 Mtpa, 164 at local state-owned mines (88 Mtpa), and 479 at TVE and private mines (91 Mtpa). Out of the total, 504 were coking coal preparation plants (316 Mtpa) and 457 were steam coal washing plants (521 Mtpa). In 2005 (the latest year for which figures are available), 703 Mt of raw coal were washed, 33% of raw coal production, including 405 Mt of raw steam coal and the rest coking coal. Key state-owned mines washed 590 Mt or 58% of their raw output, local state-owned mines 56 Mt or 19%, and TVE and private mines washed 58 Mt or 7%.

Figure 5.1 shows that jig, heavy-medium and flotation are the main coal washing methods in China. In 2005, coal washed by jigs was 316 Mt, accounting for 45% of total coal cleaned; 278 Mt of coal were cleaned in heavy medium (40%); 67 Mt by flotation (10%); and 42 Mt or 6% by other methods, such as dry cleaning, moving-screen jigs, shaking tables and spirals.

Figure 5.1 .... Percentage shares of different coal cleaning methods (left scale) and total washed tonnage (right scale) in China, 1975-2005 and projections to 2020

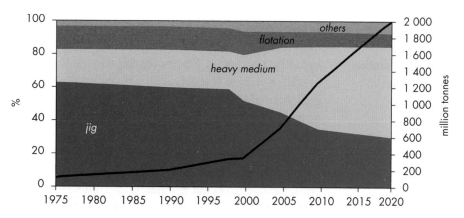

Sources: NDRC (2006a) and Yu (2006).

With reductions in capital and operating costs, the overall cost of heavy-medium washing has fallen close to that of jigs. In the near future, the heavy-medium method will dominate coal cleaning, given its higher separation efficiency. Jigs are still the preferred methods for easy-to-wash coal because of lower operating costs, especially at steam coal preparation plants. In the coming years, conventional flotation methods will remain the main method for cleaning fine coal at coking coal preparation plants, although dry cleaning and moving-screen jigs will have a role in arid areas.

Costs of coal preparation technologies

As coal preparation plants have become larger, investment and operating costs have fallen. The capital investment is now between RMB 30-80 per tonne of raw coal for coking coal preparation plants and RMB 20-50 for steam coal plants (Table 5.1). The investment cost for new steam coal preparation plants with capacities above 10 Mtpa is at the lower end of the range.

Operating costs are about RMB 10-25 per tonne of raw coal at coking coal preparation plants and about RMB 5-15 at steam coal preparation plants. Consumables and power costs account for 25-30% of the operating costs for steam coal preparation, with labour accounting for 40-50% and depreciation the remainder. In the future, lower capital costs should see the depreciation cost fall.

Table 5.1 ....Capital and operating costs of steam and coking coal preparation plants (RMB per tonne of raw coal)

	Steam coal preparation		Coking coal preparation	
	Current	**Future (2030)**	**Current**	**Future (2030)**
Capital cost	20 – 50	15 – 30	30 – 80	20 – 50
Operating cost	5 – 15	3 – 10	10 – 25	8 – 15

Source: Estimates by CCRI based on trends in coal preparation technology development and costs.

Coal-fired power generation

Current trends in coal-fired power generation technologies

In recent years, power generation and electricity consumption in China have increased more than in any other country – the average annual growth was 13% during the 10th Five-Year Plan period. Power consumption was 2 866 TWh in 2006 and 3 256 TWh in 2007, a year-on-year growth of 14%, while the installed capacity of power plants grew from 624 GW to 713 GW, a year-on-year growth of 14%. Electricity is mainly generated from coal (Figure 5.2). In 2007, the installed capacity of coal-fired power generation accounted for about 75% of total capacity and the actual power generated from coal accounts for more than 80% of total production. Coal use for power and heat generation accounted for about 58% of total coal consumption. In the long term, renewable, nuclear and new energy sources will gain in importance, but coal-fired generation will still account for more than 70% of total power generation over the next 20 to 30 years.

In China, approximately 740 units have a capacity of 300 MW or more among a total of approximately 1 550 units with a capacity above 50 MW (there are about six thousand units with a capacity above 6 MW). Among new power generation projects, 600 MW units now dominate. About 95 supercritical or ultra-supercritical (USC) units with a capacity of 600 MW or more had been put into operation by mid-2007, with another 70 units under construction, due for commissioning before 2010. The share in total coal-fired generation capacity of all types of supercritical plant is about 12% (Table 5.2). This is growing gradually as about 60% of the new-builds in China are large, supercritical units. In comparison, the share of supercritical units in coal-fired generation capacity is about 70% in Japan and 30% in the US.

Figure 5.2 .... Power generation in China – output and capacity by energy source, 2005

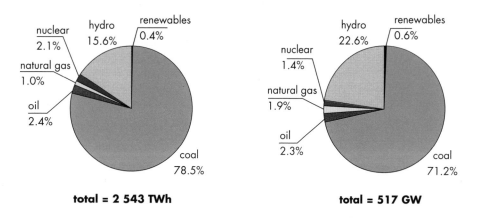

total = 2 543 TWh   total = 517 GW

Source: IEA (2007).

Table 5.2 .... Coal-fired power generation technologies used in China, 2005-30 (gigawatts)

	2005	2006	2007	2020	2030
Total generation capacity	**517**	**624**	**713**	**~1 500**	**2 000 – 2 300**
Coal-fired	368	454	524	1 040	1 200
subcritical	355	419	464	700	440
supercritical	13	32	~50	200 – 220	300 – 330
ultra-supercritical	0	3	10	80 – 90	270 – 280
IGCC	0	0	0	44	170
Gas-fired and oil-fired	22	30	40	60	200
Total thermal capacity	**390**	**484**	**564**	**1 100**	**1 400**
Installed FGD capacity	53	162	270	700 – 800	1 000 – 1 100

Sources: NBS (2008); SERC (2008); and unpublished projections by MOST expert group and CCRI.

By the end of 2006, coal-fired power generation units with FGD accounted for over 30% of the total installed thermal capacity (State Council, 2007). About 40% of new units are fitted with FGD, while the retrofitting of FGD to existing power plants is progressing slowly (NDRC and SEPA, 2007a). Overall, there is an upward trend in SO_2 emissions as coal consumption increases, adding urgency to the need to improve the management and enforcement of existing pollution control legislation (Section 5.2). For example, units put into operation recently are mostly equipped with FGD, but the rate of FGD use is low, resulting in total SO_2 emissions of 25.9 Mt in 2006 (Figure 6.2).[2] Management of NOx emissions also needs strengthening – on-line monitoring and control should be improved.

During 2006, the first of four 1 000 MW ultra-supercritical units at China Huaneng Group's Yuhuan power plant was put into operation. With a generation efficiency of up to 43%, coal consumption is around 290 gce/kWh. By the end of 2007, seven 1 000 MW USC units were in operation, including those at China Huadian Corporation's Zhouxian power plant and China Guodian Corporation's Taizhou power plant. In addition, there are a number of 600 MW USC units in operation, including at China Huaneng Group's Yingkou power plant and China Power Investment Corporation's Kanshan power plant. Other USC units are under construction and will be put into operation in 2009. At present, 8.2 GW of USC units have been put into operation and 100 GW are under construction.

Also in 2006, China's first 210 MW circulating fluidised bed combustor (CFBC) boiler was successfully put into commercial operation in Jiangxi and construction of 300 MW CFBC boilers have just begun in Yunnan and Sichuan. Huaneng Group has started to build a 250 MW IGCC demonstration power station in Tianjin, being the first-stage of the GreenGen project,[3] while Huadian Group and Dongguan City both have IGCC projects under construction.

Costs of power generation technologies

Capital costs of power generation units are shown in Table 5.3. The capital costs of recently constructed supercritical units have been higher, at up to RMB 1.65 billion for a 300 MW unit (RMB 5 500/kW or USD 775/kW). This is around 30% more than the cost of sub-critical units which remain the favoured choice among many project developers. Production costs are shown in Table 5.4.

2. NBS reported an annual increase in SO_2 emissions of 1.6% to 25.9 Mt in 2006. A revised figure of 24.6 Mt was published by the new Ministry of Environmental Protection on 4 June 2008. This means that SO_2 emissions actually fell in 2006.
3. GreenGen Ltd Co is a consortium of China Huaneng Group (majority shareholder), China Datang Group, China Huadian Corporation, China Guodian Corporation, China Power Investment Corporation, Shenhua Group, State Development and Investment Corporation and China National Coal Group established in December 2005 and supported by NDRC, MOST and other ministries to develop and commercially demonstrate a 250 MW IGCC with hydrogen production and CO_2 capture and storage by about 2015 (www.greengen.com.cn). See also Chapter 9.

Table 5.3 ....Capital costs of power generation technologies in China
(RMB per kilowatt)

	Current capital cost	Capital cost in 2030
FGD	200 – 300	200
SCR (fitted to new units)	200 – 300	200
Subcritical + FGD	3 300 – 4 200	3 200
Subcritical + FGD + SCR	3 600 – 4 400	3 450
Supercritical + FGD	3 500 – 4 200	3 500
Supercritical + FGD + SCR	3 700 – 4 400	3 600
USC + FGD	3 800 – 5 000	3 650
USC + FGD + SCR	4 000 – 5 200	3 800
IGCC	6 500 (900 USD/kW)	5 800 (800 USD/kW)

Sources: Henderson (2005); Li (2006); MOST (2004); MOST (2005); Nalbandian (2006); and Xu (2006).

Table 5.4 ....Production costs of 300 MW subcritical and 600 MW supercritical
units with FGD in China (million RMB per year)

	300 MW subcritical		600 MW supercritical	
	Current	**2030**	**Current**	**2030**
Fuel and materials	170 – 180	220 – 230[1]	290 – 310	350 – 380[1]
Depreciation	70 – 80	60 – 70[2]	140 – 150	120 – 130[2]
Labour and others	30 – 40	35 – 44[3]	60 – 70	65 – 77[3]
Total	**270 – 300**	**315 – 344**	**490 – 530**	**535 – 587**
Unit cost (RMB/MWh)	173 – 192	200 – 220	155 – 168	170 – 190

Notes: Costs include FGD. It is projected that:
1. coal prices will rise by 30%;
2. capital costs will decrease by 20%; and
3. labour costs will rise by 8-10%.
Source: CCTM model developed by CCRI.

Coal-fired power plant operators are liable to pay pollutant emission fees
(*e.g.* RMB 0.63 / kgSO$_2$ [NDRC, 2003]), but these are only partially collected and not
effectively recycled, so the aim of encouraging technical progress and strengthening
pollution control is not being achieved. For example, the top-five power generation
groups received RMB 300 million from special environmental protection funds
in 2004, less than 10% of the emission fees they paid (Wang, 2005). In addition,
electricity prices still do not reflect the scarcity of coal resources and the importance
of environmental protection. According to new electricity price regulations, FGD
costs for new power plants can be incorporated into electricity prices at the rate of
RMB 15/MWh (NDRC and SEPA, 2007b). However, it is difficult to do so except at
a few units in some provinces. It is estimated that FGD costs account for about 10%
of the power generation cost. Not allowing these FGD costs to be fully recovered in
electricity prices greatly influences the operation of power plants, to the extent that
installed FGD is not operated (Peng, 2005).

Future power generation technologies

With the national strategy of "replacing small scale by large scale", it is expected that units below 100 MW will be eliminated and, by 2010, units above 300 MW will account for more than 80% of the total number of units. By 2020, units below 200 MW will be eliminated and those above 300 MW will account for more than 90% of the total. Supercritical and USC technologies will be developed and deployed first, followed by IGCC and coal-based polygeneration technologies, as shown in Figure 5.3.

The state has signed agreements for the reduction of SO_2 emissions with provincial governments and five large utilities, so FGD installation and utilisation rates are expected to increase. Taking into account the dramatic growth of the FGD market, as well as the limited specialist manufacturing capacity, shortage of skilled engineers and technical personnel, and insufficient engineering management experience, the FGD installation rate is forecast to reach 80% in 2010, 90% in 2020 and 95% in 2030.

At present, the capital and operating costs of high-efficiency deNOx technologies (selective catalytic reduction or SCR) are relatively high. The government does not have a policy comparable to that for FGD, so SCR will not be installed in large numbers in the foreseeable future. The government will probably impose stringent emission standards and insist on control measures for NOx emissions once the SO_2 problem has been largely solved. At that time, high-efficiency deNOx technologies are expected to develop rapidly. It is projected that the installation of SCR units for power generation will reach 10% in 2010, 30% in 2020 and 50% in 2030.

Figure 5.3 .... Installed capacity forecast for coal-fired units and pollution control technologies in China (gigawatts)

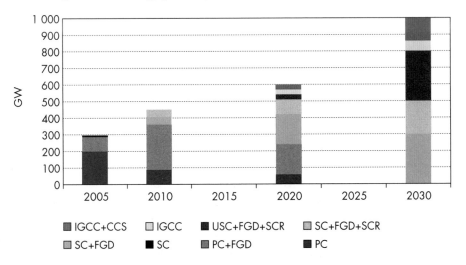

Source: Estimates by CCRI based on development trends.

Coal-to-liquids (CTL)

Current status of CTL

Over 80 potential CTL projects have been announced in the last few years, and at one time the sector was at risk of overheating. The Chinese government announced in 2006 that it would not approve the construction of any further CTL projects before the technology had been successfully demonstrated.[4] Moreover, coal chemical projects without transport arrangements for products will not be authorised. At the same time, the government is encouraging the demonstration of advanced CTL technologies, providing their wider deployment in China is not constrained by ownership of intellectual property. It is expected that a direct coal liquefaction plant, with an annual capacity of one million tonnes of oil products, will complete construction and commissioning in 2009. An indirect coal liquefaction plant, also with a capacity of one million tonnes per year, will commence construction shortly. The plan is to establish a significant coal-based liquid fuels industry by 2020, as projected in Figure 5.4.

Figure 5.4 .... Forecast of oil supply (left scale) and coal liquefaction output (right scale) in China

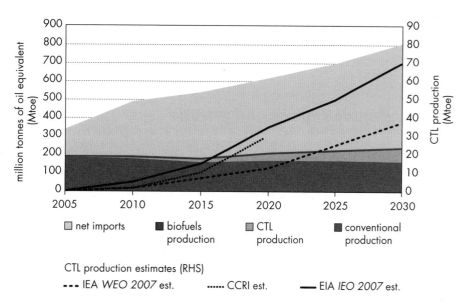

CTL production estimates (RHS)
--- IEA *WEO 2007* est.   ...... CCRI est.   —— EIA *IEO 2007* est.

Notes: CCRI: China Coal Research Institute; EIA: US Energy Information Administration; IEA: International Energy Agency; IEO: EIA *International Energy Outlook*; WEO: IEA *World Energy Outlook*.
Sources: CCRI; EIA (2007); and IEA (2007).

Cost of coal-to-liquids

Capital and production costs for CTL projects in China are shown in Tables 5.5 and 5.6. These do not include the additional costs of CO_2 capture and storage, discussed below.

Table 5.5 .... Capital costs of coal-to-liquids technologies in China (RMB per tonne)

	Current capital costs	Capital costs in 2030*
Direct liquefaction	8 000 – 10 000	6 000 – 8 000
Indirect liquefaction	9 000 – 11 000	7 000 – 9 000

* The cost should decrease by 20% as shown, based on technical maturity and scale.
Source: CCTM model developed by CCRI.

---

4.   In August 2008, the government put a halt on all new CTL project approvals.

Table 5.6 ....Production costs for direct and indirect coal liquefaction in China (million RMB per year, assuming plant capacities of one million tonnes per year)

	Direct liquefaction		Indirect liquefaction	
	Current	**Future (2030)**	**Current**	**Future (2030)**
Fuel and materials	1 046 – 1 620	1 220 – 1 890[1]	1 146 – 1 780	1 340 – 2 080[1]
Depreciation	507 – 633	380 – 475[2]	570 – 633	440 – 500[2]
Labour and others	140 – 200	150 – 220[3]	160 – 220	170 – 240[3]
Total	**1 693 – 2 453**	**1 750 – 2 585[4]**	**1 870 – 2 633**	**1 950 – 2 820[4]**
Unit cost (RMB/tonne)	1 700 – 2 400	1 800 – 2 600[5]	1 900 – 2 600	2 000 – 2 800[5]

1. Total increase will be around 17%, since an energy efficiency improvement of 10% offsets a cost increase of 30%.
2. Depreciation will decrease due to a decline in capital costs.
3. Demand for staff will decrease by 20% due to technical progress and automation, while salaries will increase by 50%. Overall, labour costs will increase by 8-10%.
4. Calculation results.
5. Essentially unchanged, on the basis of the analysis shown.
Source: CCTM model developed by CCRI.

Carbon dioxide capture and storage

In China and elsewhere, the application of CO_2 capture and storage (CCS) is constrained by many factors, including a lack of proven technologies, high costs and limited knowledge of storage potential. Current studies are only at a preliminary stage in China, and some distance from practical application. Based on current trends in technology and policy, it is unlikely that CCS will find large-scale application in China before 2030, as shown by the roadmap in Figure 5.5 developed by CCRI.

Figure 5.5 ....Roadmap of CCS R&D and commercialisation in China

Task	2010 ▶	2020 ▶	2030 ▶	2040 ▶	2050
CO₂ capture	Dissemination of capture technologies for low-concentration CO₂ and cost reduction				
	Demonstration and dissemination of oxygen-rich combustion technologies and cost reduction				
Decarburisation to produce hydrogen	Demonstration of coal-based hydrogen production		Commercialisation of coal-based hydrogen production		Provision of hydrogen energy, including pipelines and hydrogen stations
CO₂ transport	Technical and economic feasibility	Application of CO₂ storage and transport			
CO₂ storage	Research and geological investigation of storage potential	Demonstration and verification	CO₂ capture-transport-storage monitoring plan		

Source: CCRI.

CCS has not been commercialised anywhere in the world, so it is difficult to assess capital and operating costs. Recent estimates made by the IEA suggest that total power generation costs are about 75-100% higher with CCS than for conventional steam cycles (IEA, 2008). This may drop to 30-50% in the long term when capture and storage from coal-fired power plants will approach USD 50 per tonne of CO_2 mitigated. In calculating this figure, pipeline transport of CO_2 over a distance of 100 km is assumed to add USD 1-3 per tonne of CO_2 shipped[5] while storage adds around USD 15 per tonne of CO_2.

Figure 5.6 ....China Huaneng Group's Gaobeidian power plant in Beijing with 3 000 tpa post-combustion CO_2 capture pilot plant under construction

In partnership with CSIRO of Australia and the Thermal Power Research Institute in Xian, China, Huaneng Group commissioned this pilot plant in July 2008. The project is part of the Australia-China clean coal technology initiative (Chapter 9).

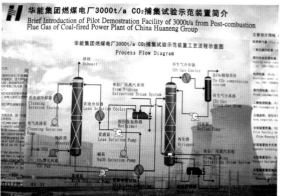

POLICY REVIEW

Policies and regulations for coal mining

In recent years, different Chinese government departments have proposed various policies to promote the sound and sustainable development of the coal industry. The orderly exploitation of coal resources and their rational utilisation are both emphasised. At the same time, coal enterprises are being urged to raise recovery rates, to ensure safety in coal mines and to pay more attention to environmental protection. The main policies concerning coal mining are described in this section.

Opinions of the State Council on coal industry development

The *Opinions of the State Council on Promoting the Sound Development of the Coal Industry* was issued on 7 June 2005 (State Council, 2005a). The State Council proposes intensifying coal resource exploration, strengthening the management of coal resources, perfecting the planning system for coal production and mine development, exploiting coal

---

5. Assuming 10 million tonnes per year is shipped.

resources in a rational and orderly way, and using coal resources efficiently. In terms of exploiting coal resources in a rational and orderly way, it suggests: accelerating revisions to coal mine design criteria; strictly controlling the schedule, method and intensity of coal mining; forbidding mining of non-permitted seams and beyond permit boundaries, as well as other illicit mining; encouraging the adoption of advanced technologies; and mining difficult coal seams, including very thin coal seams. To preserve coal resources, it suggests strictly controlling resource recovery rates, calling for high-yield, high-efficiency mines to be built more quickly. Moreover, it suggests further modernisation of coal production and mining equipment, with greater mechanisation and automation, and use of management information systems, while eliminating the obsolete technologies and techniques found at small- and medium-scale mines (Figure 5.7). The State Council recommends that an all-embracing service organisation for small-scale coal mines should be established quickly, offering relevant technology services. Finally, it suggests enhancing the ability to carry out R&D into important mining technologies, as well as to manufacture mining equipment, by introducing key technologies, integrating technology acquisition with trade, co-operating in manufacturing joint ventures (JVs) and exchanging equipment market share with technology suppliers who promote local manufacture of equipment. Strengthened co-operation between enterprises, scientific research institutes and academia is seen as crucial to advance such technological innovation.

Figure 5.7 .... Datong Group Xiaoyu state-owned coal mine and community, Shanxi

Opened in 1960 with a design capacity of 600 ktpa, Xiaoyu mine was incorporated into Datong Coal Mine Group in 2003 in response to government policy on coal industry consolidation. Production capacity is to be expanded to 2.8 Mtpa using fully mechanised longwall mining with top-coal caving.

11th Five-Year Plan   Many of the State Council's opinions are repeated in the 11th Five-Year Plan, issued in March 2006 and that are intended to guide national socio-economic development from 2006 to 2010. In Chapter 11, Developing the Equipment Manufacturing Industry, equipment for comprehensive coal extraction is regarded as, "a focus for a prosperous equipment manufacturing industry". Specifically, it mentions equipment for large-scale underground mining, transport and hoisting, as well as for coal preparation and large-scale opencast mining. In Chapter 12, Optimising Development of the Energy Industry, it proposes that coal resource exploration should be strengthened, recovery rates should be increased and the impact of coal mining on the environment should be reduced.

11th Five-Year Plan for Coal Industry Development

Under the overarching Five-Year Plan, detailed sector plans were prepared by the relevant government departments. The plan for the coal industry proposes that modernisation, scale and mechanisation should be advanced, together with local manufacture of mining equipment, and that awareness about preserving resources and protecting the environment should be promoted. During the 11th Five-Year Plan period, coal industry development is to be dominated by consolidation, supplemented by construction of new mines. The plan calls for the creation of more coal bases over the coming years, with large-scale, modern opencast coal mines and safe, highly efficient underground mines, with an annual output of over 10 Mt, built in preference to optimise the structure of the coal-mining industry. The aim for coal production is to achieve a mechanisation rate of over 95% at large coal mines, 80% at medium-sized mines, and a mechanisation or semi-mechanisation rate of 40% at small coal mines. Overall, there should be 380 safe and highly efficient coal mines whose output should account for 45% of China's total production; among these, 25 should each produce over 10 Mtpa. The "circular economy" concept should be developed vigorously with the widespread utilisation of coal mine methane, mine water, coal gangue, coal sludge and associated minerals (Figure 5.8).[6]

Figure 5.8 .... Tongmei Datang Tashan coal mine, power plants and industrial complex, Shanxi

Tongmei Datang Tashan industrial park opened in 2003. It comprises a 15 Mtpa coal mine and coal preparation plant, a 50 ktpa kaolin plant, a 200 MW (4 × 50 MW) power plant fired with mine discards, a conventional 1 200 MW (2 × 600 MW) coal-fired power plant, a cement plant, a building materials plant and a dedicated 21 km rail link. Datong Coal Industry Company Ltd, Datang International Power Generation Company Ltd and Datong Coal Mine Group are the industrial park's three shareholders.

_____

6. "Circular economy" (循环经济) is a key concept in current economic policy in China, similar to the ideas of industrial ecology, in which the waste streams from one process are used as material inputs to other processes.

Scientific and technological development

The development of highly efficient coal mining technology and equipment is ranked as a key objective in the National Plan for Medium- to Long-Term Scientific and Technological Development (2006-2020) (MOST, 2006). The Energy Bureau of the National Development and Reform Commission (NDRC),[7] in its preliminary opinions on the plan outline, has proposed that in order to develop highly efficient coal mining, the production capability, technical performance and reliability of domestic equipment for fully mechanised coal mining should be improved. In addition, the application of fully mechanised, longwall equipment sets for thick coal seams, with an output of over 10 000 tonnes per day (tpd), should become widespread, and larger sets, capable of up to 15 000 tpd or 3-5 Mtpa, employed at some mines. Innovative techniques, such as shaft construction in deep and thick alluvium, fast drivage of roadways, modern coal conveying, hoisting, blending and loading systems, and mine automation and information systems should all be employed using domestically manufactured equipment which must be developed and produced quickly. Finally, the Energy Bureau believes that the percentage of coal resources recovered during mining should be increased.

Policies on coal exploration and mining issued by the Ministry of Land and Resources

The *Notice on Further Strengthening the Management of Coal Resource Prospecting and Mining*, issued on 24 January 2006, prescribes that effective measures should be taken to improve the percentage of coal resources recovered. Departments of land and resources at all levels are asked to summarise carefully the results of specialist inspection work on recovery percentages at coal mines, account for different mining conditions, and take expert advice on appropriate measures to address any low recovery rates. These are to include demonstrating the techniques, methods and processes for efficient coal mining in areas with different geologies, and eliminating the catalogue of obsolete coal mining techniques, methods and processes found in China.

The *Notice on Strengthening the Management of Mineral Resources and the Collection of Compensation Fees to Improve Coal Recovery Rates*, issued on 28 April 2006, prescribes that the management and collection of mineral resource compensation fees should be strengthened and that other economic mechanisms for coal resource preservation and rational use by mining enterprises should be established. To this end, steps have been taken to carry out a full review of the approved recovery rates for individual coal mining enterprises, assess their actual recovery rates and ensure that the assessed compensation fees are actually collected.

The *Notice on Strengthening the Management of Comprehensive Exploitation of Coal and CBM Resources*, issued on 17 April 2007, proposes "supporting and encouraging holders of coal mining rights to comprehensively prospect and explore for CBM resources, further strengthening the management of CBM rights and properly resolving the problem of overlap between coal mining and CBM rights". The purpose of the notice is to boost the exploitation and utilisation of both coal and CBM resources.

---

7. The National Energy Administration (NEA) replaced the Energy Bureau in July 2008. It has nine departments, with 112 staff, and its responsibilities include: drafting energy development strategies; offering reform advice; implementing energy sector management; proposing policies on new energy; international co-operation; and energy price management.

Regulations on coal mine safety

Amended by the State Administration of Work Safety in 2005, the *Coal Mine Safety Regulations* provide comprehensive requirements on coal mining technology. The second edition specifies technical requirements for coal mining; the third edition specifies technical requirements for open-pit coal mining (SACMS, 2005).

State Council plans for coal industry restructuring

On 28 September 2006, the State Council published its *Notice on Further Improving the Progress of Coal Mine Restructuring and Closure* prepared by twelve departments, including the State Administration of Work Safety, the State Administration of Coal Mine Safety and NDRC. It makes specific plans concerning the objectives and steps for the closure and restructuring of coal mines. The major reasons for restructuring are: to deal with illegal mining; to close unsafe coal mines and those mining beyond the boundaries of their mining rights; to eliminate coal mines that do not conform with industrial policy; to improve mines that have unacceptable layouts, destroy resources or pollute the environment unnecessarily; and to address those newly built, but illegal coal mines and extension projects that are not authorised or fail to meet safety standards. The objective is to significantly improve coal mining order by 2008. Hence, mining without permission and cross-boundary mining will be effectively stopped, the number of accidents at small coal mines will fall substantially, serious accidents will be curbed, the annual fatality rate at small coal mines will be reduced to below four fatalities per million tonnes, mining practices at small coal mines will be standardised and the number of small coal mines will be dramatically decreased to below 10 000 by 2010.

Policies for the sustainable development of the coal industry

The *Opinions on Issues Concerning the Deployment of Coal Industry Sustainable Development Policies in Shanxi Province* were first put forward jointly by several bodies, including the Shanxi provincial government, NDRC, the Ministry of Finance, the Ministry of Labor and Social Security, the Ministry of Land and Resources, the State-Owned Assets Supervision and Administration Commission, the State Administration of Taxation, the State Environmental Protection Administration, and the State Administration of Coal Mine Safety. On 15 June 2006, the State Council issued its response (Letter 2006, No. 52) in which four economic policies were put forward: a levy to fund the sustainable development of the coal industry, compensation fees for the right to mine coal resources, a levy for environmental restoration of mining areas and a development fund for coal industry restructuring (Section 3.6).

On 30 September 2006, the State Council approved a *Scheme for Deepening the Reform of the System for Equitable Use of Coal Resources*, submitted jointly by the Ministry of Finance, the Ministry of Land and Resources and NDRC. It aims to improve the efficiency of coal resource allocation. The reform was approved for implementation in eight major coal-producing provinces: Shanxi, Inner Mongolia, Heilongjiang, Anhui, Shandong, Henan, Guizhou and Shaanxi. The scheme focuses on extending the use of fees for prospecting and mining rights to reflect the value of coal resources and external costs of their exploitation. In summary, the scheme will increase support for coal prospecting, perfect a policy on coal resource taxation, enhance the management and macro-control of coal resource exploitation, facilitate the rational and orderly development of coal resources, and further increase the percentage of coal resources recovered.

The *General Executive Scheme to Demonstrate Coal Industry Sustainable Development in Shanxi*, approved by the Shanxi provincial government on 31 March 2007, raises the requirements for coal industry development: mine capacities should not be below 300 000 tpa after consolidation; new mines should not be under 600 000 tpa; and coal recovery rates should not be lower than national standards. Work in progress to develop new coal seams and extend boundaries will be terminated for those mines with a capacity under 300 000 tpa and licences for mines with unacceptable layouts will not be renewed.

Revised Coal Law

China's existing Coal Law, enacted on 29 August 1996, aims at the rational exploitation of coal resources and promotion of coal industry development. However, according to key leaders in the Energy Bureau of NDRC, the Coal Law is outdated and this has caused a series of problems in coal resource administration, management of coal operations and coal mine safety, which have resulted in serious waste and loss of resources. At the end of 2003, the China National Coal Association suggested amending the Coal Law and, more recently, representatives at the annual NPC and CPPCC have put forward proposals and bills to revise the Coal Law. Thus, on 14 March 2005, NDRC announced the formal start of a process to amend the Coal Law. From that time until 15 October 2005, comments were collected from interested parties and the public. One of the more controversial topics in the amendment was entry requirements for coal industry participants, including tightening the issue, management and review of prospecting licences, production licences, business licences and other licences related to the development of coal resources. An umbrella Energy Law is similarly to be drafted, with the aim of setting out broad principles for each energy sector, including the coal sector.

Emissions reduction and energy saving in the coal industry

Efforts to build a resource-saving and environment-friendly coal industry, and to promote its sustainable development, were put forward in the *Opinions on Emission Reductions and Energy Saving in the Coal Industry* (NDRC, 2007b). It requests that, when designing coal mines, priority should be given to clean production techniques, processes with low emissions, and equipment with high resource-recovery rates. In accordance with the requirements on energy saving and emission reductions, coal mines must adopt state-of-the-art equipment and up-to-date production processes.

Policies on coal preparation

The Coal Law 1996, the Atmospheric Pollution Prevention and Control Law 2000 (revision), the *Opinions on Promoting the Sound Development of the Coal Industry* (State Council, 2005a), the *Technical Policies for SO_2 Emission Prevention and Control* (SEPA, 2002) and some local government regulations and policies all encourage the development and application of coal preparation technologies and the use of washed coal. However, current policies for coal preparation lack coherence. For example, governments at different levels encourage the use of washed coal for boilers and kilns, but some designers and manufacturers of boilers and kilns still design and supply equipment for low-quality coal. In addition, retrofits are not available to convert existing boilers from low-quality coal to prepared coal. Thus, users of washed coal face increased fuel costs, but might see no obvious gain in efficiency – lessening their enthusiasm for prepared coal.

There are no national coal standards for different industries and no mandatory regulations. Existing standards do not properly reflect the need for environmental protection and do not reflect the varied requirements of users. For example, the type and quality of coal available for industrial boilers does not meet the quality requirement of stoker-fired boilers, which results in many combustion problems. Weak implementation of environmental policies have also hampered the deployment of coal preparation technologies and the use of washed products. Low emission fees for smaller, decentralised users and the difficulty of enforcing current laws mean that low-quality, cheaper coal finds a ready market.

Policies and regulations for CBM/CMM development and utilisation

Since the 1990s, the Chinese government has offered various forms of assistance to the coalbed methane and coal mine methane (CBM/CMM) sectors through its development plans and industrial policies, and the government continues to strengthen its support of the sector. Together, these policies reduce CBM/CMM project development costs and reduce investment risks for CBM/CMM utilisation project developers.

11th Five-Year Plan for CBM/CMM Development and Utilisation

The 11th Five-Year Plan for CBM/CMM Development and Utilisation (NDRC, 2006b) proposes that by 2010, four objectives should be achieved. Firstly, the national CBM/CMM output should reach 10 bcm (5 bcm from surface borehole extraction and 5 bcm from underground extraction). Secondly, the volume of CBM/CMM utilised should be 8 bcm (5 bcm from surface extraction and 3 bcm from underground extraction). Thirdly, proven reserves of CBM should be expanded to 300 bcm. Fourthly, a CBM/CMM industry should be established to develop and utilise this resource.

Specific CBM/CMM development and utilisation policies

In June 2006, the State Council's *Opinions on Accelerating CBM/CMM Extraction and Utilisation* (State Council, 2006a) were made public. In addition to clarifying the guiding principle of gas extraction prior to coal mining, integrated with gas control and utilisation, sixteen suggestions were made, as summarised here. Restrictive measures include the following: local administrative departments of land and resources should supervise and control CBM/CMM exploration and exploitation activities according to the law; coal mining can only commence when the gas content of the coal to be mined has been reduced below a specified standard (measured in m^3/t); gas extraction must be integrated with coal mining; and specific standards for emissions of CMM to atmosphere must be applied. Supportive polices include the following: land needed to establish CBM/CMM extraction and utilisation projects should enjoy priority according to state regulations; fees otherwise payable by enterprises for work safety can be used for constructing surface and underground CBM/CMM extraction systems; suppliers of in-specification CBM/CMM can enjoy priority connection to natural gas pipelines and public town gas pipelines; coal enterprises can generate electricity with

CBM/CMM for self use and any surplus electricity is given priority access to the grid; the price of gas can be fixed through negotiation between the seller and buyer; and enterprises directly engaged in CBM exploration and development from surface facilities can apply for a reduction of, or exemption from charges for exploration and mining rights up to 2020. CBM/CMM extraction and utilisation equipment can benefit from accelerated depreciation – the Ministry of Finance is currently formulating other preferential tax policies together with NDRC. Finally, all levels of government can offer subsidies and discounted loans to developers of CBM/CMM extraction and utilisation projects.

In November 2006, the State Administration of Coal Mine Safety issued specific minimum standards for CMM extraction (AQ1026-2006). Before tunnelling and mining in coal seams subject to gas outburst, the gas content or pressure within the control region must be reduced to safe values or to below 8 m^3/t or 0.74 MPa if safe values cannot be assessed.

In April 2007, NDRC issued a *Notice on CBM/CMM Price Management*. It specifies that the price of gas not distributed via city pipeline networks can be determined freely through negotiations between suppliers and buyers, while the price of gas distributed via city pipeline networks and those categories under government control should be determined in a timely manner according to its heating value compared to substitute fuels such as natural gas, coal gas and liquefied gas.

Also in April 2007, NDRC issued a *Notice on Executing Opinions on Generating Electricity with CBM/CMM*. This encourages the deployment of power generation sets (gensets) with an installed capacity of 500 kW and above to generate electricity from CBM/CMM, and the development of reciprocating gas engine gensets of 1 000 kW and above, as well as high-efficiency gas turbine gensets of larger power ratings. The notice requires that electricity generated by CBM/CMM power plants should be given priority by grid operators who should purchase electricity surplus to on-site needs and make timely payments equal to the subsidised price of electricity generated from biomass, as specified by NDRC in the *Trial Management Method for Electricity Prices and Sharing Expenses for Electricity Generated with Renewable Energy* (the price had previously been the same as that paid to local coal-fired power plants fitted with FGD). Lastly, CBM/CMM power plant owners do not face market price competition and do not undertake any responsibilities for grid stability.

Tax policies and subsidies

With the approval of the State Council, the Ministry of Finance and the State Administration of Taxation jointly issued a *Notice on Issues of Tax Policies to Speed up CBM/CMM Extraction* on 7 February 2007. It introduces a VAT-refund policy for CBM/CMM extraction enterprises who develop technologies and expand production. It specifies in the notice that special equipment purchases for drilling and logging, CBM/CMM extraction pumps, CBM/CMM monitoring systems and gensets can benefit from accelerated depreciation. For example, if enterprises engaged in CBM/CMM extraction separately account for projects using domestic equipment, 40% of their investment is deductible from taxable income in the first year. Development expenses for new technologies are 100% deductible, with 50% allowable in the first year. A temporary suspension of resource taxes for enterprises engaged in surface CBM extraction is also specified in the notice. Finally, there is the five-year income tax exemption for CMM drainage operations and to enterprises who use CMM as a raw material, starting from their first profitable year.

In April 2007, the Ministry of Finance issued *Executing Opinions on Subsidising CBM/ CMM Development and Utilisation Enterprises* whereby any enterprise engaged in CBM/ CMM extraction within China is entitled to financial subsidies. Specific conditions include that CBM extracted by enterprises should be used on site or marketed for residential use or as a chemical feedstock. The volume of extracted CBM used to generate electricity by enterprises does not enjoy financial subsidies (but see above). It is specified in the opinions that the central financial authority would subsidise properly metered CBM/CMM extraction at a standard 0.2 RMB/m^3 of pure methane, although local financial departments can determine the subsidy level and calculation methods independently.

Other policies relevant to CBM/ CMM

CMM, as a source of clean energy and a means to reduce a key greenhouse gas, has been granted much policy support from central government. For example, enterprises involved in CMM exploitation enjoy: a reduction of or an exemption from mine prospecting fees and mining fees; lower customs duties on imported equipment, parts and special tools used for CMM technology improvement projects; accelerated depreciated allowances on drainage utilisation equipment; and VAT allowances for drained CMM when used as a raw material before 2020.

CMM power generation projects listed in the *Catalogue on the Comprehensive Utilisation of Resources* (2003 revision) can enjoy certain preferences (NDRC, 2004a). Power authorities will grant enterprises generating electricity and heat from CBM/CMM, *i.e.* cogeneration or combined heat and power plants, a grid connection as long as the individual units have an installed capacity of above 500 kW and meet required standards. They are exempted from paying the connection fee normally charged to small-scale, coal-fired power plants, are excluded from quotas under the national distribution plan, and benefit from priority electricity sales to the grid at wholesale prices, or even at higher prices if approved by the provincial authorities. Those with an installed capacity of 1.2 MW or less are not required to support the grid by load following; plants above this capacity can deliver their full output during periods of peak demand, but will never be required to drop below 85% output.

The former State Development Planning Commission and the former State Economic and Trade Commission jointly issued the *Catalogue of Key Industries, Products and Technologies whose Development is Encouraged by the State* (2000 revision) which includes equipment for mine gas control, for CBM exploration and development, and for low heating value fuels. Domestic investors are exempted from paying tariffs and VAT on such imported equipment for CMM projects.

NDRC and the Ministry of Commerce jointly issued the *Guidance Catalogue for Foreign Industrial Investment* in which equipment for the exploration and development of coal and associated resources is listed as "encouraged content for foreign investors"

(NDRC and MOFCOM, 2007). CBM/CMM development projects are classified into two types. Type 1 relates to projects exploiting CBM as the major resource with large-scale exploration and development conducted in unmined regions of coalfields, outside of coal production areas. Such projects are managed according to the rules governing oil and natural gas exploration and development. Type 2 relates to projects extracting CMM within a licensed mining area. Such projects are managed according to policies stipulated by the state. Although the state encourages foreign investment in projects of both types, there are differences in terms of management and policies.

Policies on power generation

Recent Chinese policies have focussed on accelerating the closure of smaller power generation units and encouraging the construction of large, high-efficiency, clean units. Current laws concerning energy conservation, price formulation, emission fees and emission controls are mostly of a guidance nature; further detail is found in regulations and implementation rules. For example, the Electric Power Law 1995 forbids the use of certain power equipment and technologies, while the Energy Conservation Law 2007 (revision) encourages the development and use of clean coal technologies, such as CFBC, suitable for application in China. The revised Atmospheric Pollution Prevention and Control Law 2000 requires the adoption of technical and economic policies and measures that promote the clean use of coal, enable the collection of emission fees and ensure all emissions are controlled. Thus, FGD and de-dusting equipment must be installed where emissions exceed emission concentration standards or absolute emission standards. In general, SO_2, NOx and dust concentration standards for new plants are around double those applying in the EU.

Detailed regulations to encourage clean, high-efficiency coal-fired power generation and environmental protection are contained in documents issued by government departments. These are mostly concerned with environmental protection and efficient utilisation, such as the *Management Methods to Encourage Comprehensive Resource Utilisation* (NDRC and Ministry of Finance, 2006), the *Notice on SO$_2$ Emission Control at Coal-Fired Power Plants* (SEPA, 2003), the *Notice on Planning and Construction of Coal-Fired Power Generation Plants* (NDRC Energy 2004, No. 864), and the *Preliminary Administrative Measures for the Assessment of FGD Projects* (NDRC, 2006c). Other mandatory laws and standards also promote the application of clean, high-efficiency coal-fired power generation technologies, such as the *Emission Standard for Air Pollutants from Thermal Power Plants* (GB13223-2003), the *Technical Code for Designing Fossil Fuel-Fired Power Plants* (DL5000-2000), the *Technical Code for the Design of Environmental Protection Systems at Fossil Fuel-Fired Power Plants* (DLGJ102-91), the *Technical Code for the Design of Waste-Water Treatment Systems at Fossil Fuel-Fired Power Plants* (DL/T 5046-2006), the *Technical Code for the Design of De-dusting Equipment at Fossil Fuel-Fired Power Plants* (DL/T 5142-2002) and others. Government departments have also issued relevant industrial policies such as the *National Industrial Technology Policy* (SETC, 2002), the *Industrial Restructuring*

Guidance List (2005 revision) published by NDRC, and the *Notice on a Catalogue of Outdated Production Capacities, Processes and Products to be Eliminated* (SEPA, 2000) and its subsequent updates.

These policies, laws and regulations encourage the development of high-specification power generation technologies with large-capacities, high-efficiency, low water usage and effective environmental controls. These technologies include supercritical and USC units above 600 MW, and CFBC and IGCC units above 300 MW. Approvals of units with a coal consumption of more than 300 gce/kWh (305 gce/kWh when air cooled) and conventional coal-fired units with a capacity of less than 300 MW are restricted. Derogations can be made for smaller units burning waste coal (Figure 5.9) and for CHP plants (Figure 5.10).

Figure 5.9 .... Xishan Coal and Electricity Group Gujiao power plant, Shanxi

The diversified Shanxi Coking Coal Group is engaged in coal mining, mining equipment manufacture, coal chemical production and power generation and distribution. Its subsidiary, Xishan Coal and Electricity Group, owns the 6 Mtpa Tunlan coking coal mine and adjacent 600 MW Gujiao power plant fuelled with blended rejects from the mine's 4 Mtpa wash plant. Construction is under way to increase the generation capacity at Gujiao to 3 000 MW.

In order to promote the robust development of China's power industry, NDRC has required the closure of small-scale thermal power units with high energy consumption and poor pollution control, delegating this task to provincial governments and to the power and grid companies. During the 11th Five-Year Plan period, coal-fired power generation units facing closure in areas with large power grids include: those below 50 MW; those below 100 MW and having operating for over 20 years; those below 200 MW and having reached their design lives; those with a coal consumption 10% higher than the provincial average or 15% higher than the national average; and those that fail to meet environmental standards. The total capacity that needs to be shutdown is 50 GW (Xinhua, 2007) and the State Council has issued a *Notice on Some Joint Opinions of NDRC and the Office of the National Energy Leading Group on Accelerating the Closure of Small Thermal Power Generation Units* (NDRC, 2007c) requiring local governments and enterprises to make specific implementation plans. Significant progress has reportedly been made towards this goal.

Figure 5.10 .... Datong Tong Mei power plant, Shanxi

This 4 × 50 MW combined heat and power (CHP) plant near Datong is Datong Coal Group's first CHP plant. It was jointly developed with Datang Power, a shareholder in the project. The plant consumes 1.2 Mtpa of high-ash (typically 28%) coal that would otherwise have a very low saleable value or be discarded.

For power generation units with FGD, higher electricity prices are allowed. The increased investment cost for FGD is calculated and reflected in regulated prices according to policies such as the *Notice on Standardising Electricity Tariff Management* (SDPC, 2001a) and the *Notice on Pilot Electricity Price Reform in the Northeast Regional Power Market* (NDRC, 2004b). Some policies provide preferential funding for clean coal power generation projects, such as the *Preferential Policy on Power Generation Demonstration Projects with Clean Coal Technologies* (SDPC, 2001b). Some local policies demand priority dispatching of power generated by high-efficiency and environment-friendly technologies, such as in Shandong and Shanxi provinces. However, despite this financial and policy support, realising China's desulphurisation target remains challenging.

The above policies and regulations are mostly directive rather than mandatory and are subject to change. Where policies are mandatory, they are sometimes difficult to implement because policies from different government departments are not always consistent. Even when policies are clear, inadequate enforcement can mean that they are not effective. As a result, many new power plants have been built since 2001 without approval, often in response to power shortages. A considerable number of these are smaller coal-fired units under 135 MW with outdated technologies, high energy consumption and serious pollution. The reliance on coal-fired power generation units below 300 MW has increased, even with the construction of larger, more advanced units, resulting in an increase in specific coal consumption to an average of 366 gce/kWh of electricity generated in 2006. The policy of closing smaller and older plants will help, but with growing electricity demand this policy will be difficult to enforce if power shortages are to be avoided. More attention needs to be given to existing emission standards; enforcing, extending and tightening these remains a viable route to achieving China's overall pollution control objectives.

Coal chemicals and CTL policies

This subsection examines the formulation of standards to raise the permitted capacity thresholds, energy efficiency standards and emission standards for the coal chemical industry, including coal-to-liquids (CTL). Many national programmes and plans, such as the 11th Five-Year Plan for National Economic and Social Development, the National 863 Program, the Technical Development Program, the 11th Five-Year Plan for Coal Industry Development, and the Medium- and Long-Term Energy Conservation Plan aim to develop the coal chemical industry.

In order to scientifically develop the coal chemical industry, the state issued a *Notice on Strengthening the Management of Coal Chemical Projects and Promoting Sound Development of the Industry* (NDRC Industry 2006, No. 1350). The notice regulates the minimum capacity of coal chemical projects: 3 million tonnes per year for CTL projects; 1 million tonnes per year for coal-based methanol and DME projects; and 600 000 tonnes per year for coal-based olefins projects. In addition, since coal liquefaction is still at the demonstration stage, the technology should be proven before wider application. To ensure the rational use of resources, lignite and sub-bituminous coal with low calorific value should be used for coal liquefaction in preference to higher quality bituminous coal. The planning of the coal chemical industry should be strengthened across all regions. Finally, the notice suggests that requirements for the rational use of resources and environmental protection should be proposed as part of an alternative oil strategy that avoids disorderly project construction.

However, current laws and regulations lag behind recent developments in the coal chemical industry. Development objectives appear to be concerned more with project scale, technology choice and subsequent product processing, and less with existing policies, laws and regulations. There are many plans at enterprise level but little policy guidance at regional level, and no national policy framework that could bring about efficient allocation and utilisation of resources, and influence the orderly development of this emerging energy industry. Without an industrial policy to establish an effective coal chemical industry structure and market access, the development of a potentially valuable industry is hampered. In addition, there is a lack of co-ordination between the chemical, oil and equipment supply industries, which may lead to problems in key technical support areas and poor market development. Finally, no specific regulations exist for the construction of CTL plants by foreign companies. In conclusion, further study is needed to understand exactly how oil-substitution technologies will influence China's national energy strategy.

Some policies and regulations encourage national investment grants, subsidies and discounted loans alongside preferential policies for foreign investment. These policies and regulations include the *Industrial Restructuring Guidance List* (2005 revision) published by NDRC, the *Decision on Reform of the Investment System* (State Council, 2004), the *Provisional Measures for the Management of Subsidies and Discounted Loans for Project Investment within the Central Government Budget* (NDRC, 2005), the *Provisional Measures for Authorisation of Foreign Investment Projects* (NDRC, 2004c), the *Guidance Catalogue for Foreign Industrial Investment* (NDRC and MOFCOM, 2007), and the *List of High Technologies and Products to be Encouraged by Foreign Investment* (MOST and MOFCOM, 2006).

However, for coal chemicals and CTL, there have been no special economic incentive policies. Therefore, project developers seek preferential treatment under the above regulations. On 28 August 2008, in response to techno-economic and environmental concerns, NDRC announced a moratorium on CTL project approvals.

Other relevant policies

Emission trading

Government environment departments are responsible for overseeing the construction of FGD and the management of SO_2 emissions. The *Reply to the Control Plan for the Total Emissions of Major Pollutants during the 11th Five-Year Plan Period* (State Council, 2006b) required that total annual SO_2 emissions be controlled below 9.5 million tonnes by 2010. SEPA signed letters of agreement for SO_2 emission reduction objectives with provincial governments and the five largest power companies, and in 2006 issued *Guidelines for Total SO_2 Emission Allocation*. The *Decisions on Implementing the Scientific Development Concept to Strengthen Environmental Protection* (State Council, 2005b) proposed establishing SO_2 emission trading in some regions with certain conditions, although no schemes have been established to date.

CO_2 capture and storage

In December 2005 and February 2006, the Ministry of Science and Technology (MOST) signed memoranda on carbon dioxide capture and storage, marking the formal start of innovative research led and organised by the Chinese government.

China's National Climate Change Programme (NDRC, 2007d) proposes that China should vigorously develop CTL, coal gasification, coal-to-chemicals conversion, coal-based polygeneration and CO_2 capture, utilisation and storage to increase the capacity of its response to climate change.

Figure 5.11 .... Henan Coal Gas (Group) Co. Ltd. Yima coal gasification plant, Henan

Five Lurgi gasifiers produce 2.6 mcm/day of town gas, supplying mainly households in Zhengzhou, Yima and other cities along the route of the company's 200-km pipeline. Coal is trucked from three local mines. Construction of the plant was a key state project under the 9th Five-Year Plan (1996-2000) and attracted investment support from the Australian government. Phase I of the project was completed in 2001 and Phase II in 2006. Although town gas sales are not profitable, because prices are set by the state at c.RMB 1.1/m³ (c.USD 9/mmBtu), the company produces a growing range of profitable by-products. Further expansion is planned, including a 500 ktpa methanol plant.

Water issues   At power plants, the *Technical Code for the Design of Waste-Water Treatment at Fossil Fuel-Fired Power Plants* (DL/T 5046-2006) defines water treatment requirements. The *Notice on Planning and Construction of Coal-Fired Power Generation Plants* (NDRC Energy 2004, No. 864) indicates that only environmentally friendly power plants with low water consumption are to be permitted. For coal chemical projects, the *Notice on Strengthening the Management of Coal Chemical Projects and Promoting Sound Development of the Industry* (NDRC Industry 2006, No. 1350) imposes strict controls on their construction in arid regions. However, insufficient water supply and inadequate ecological and environmental impact assessments remain issues to be resolved in many regions.

R&D REQUIREMENTS

China has long recognised the need to establish an effective R&D platform through the high-level co-ordination of activities by government, enterprises and research institutes. Government- and industry-supported R&D organisations, some with significant budgets, have evolved during the course of China's economic restructuring. Nevertheless, domestic RD&D should be strengthened, building on China's strengths in basic scientific study, applied research, component technology development, systems development, industrial demonstration and technical support. As in other countries, the engineering for commercial applications and widespread deployment should refine and perfect technologies already demonstrated.

Coal preparation

Restricted by the general development level of this sector of Chinese industry, the manufacturing ability, reliability of equipment and degree of automatic control all fall below current ideals. In recent years, many large-scale vibrating screens, centrifuges and cyclones have been imported; domestic equipment suppliers have suffered as a result of this competition.

Chinese experts have recommended that coal processing should develop towards greater automation, larger scales and improved throughput. Effective waste utilisation and pollution control measures should be employed to avoid the secondary pollution associated with spoil disposal and water discharges. Finally, they recommend that R&D into equipment reliability should be strengthened.

Power generation

In the case of conventional power generation technologies, China's own high-efficiency, clean coal combustion technologies lag behind the best available technologies from elsewhere. The principal reasons for this are inadequate R&D funding and a shortage of suitably qualified personnel. A lack of suitable, high-strength materials is also an important factor that restricts the development of supercritical and ultra-supercritical units at large scale. Insufficient R&D of key components and technologies is the main obstacle to domestic manufacture of USC units. For example, the core technologies of system design and component manufacture have not been attained for boilers, steam turbines or electrical generators.

Regarding FGD for units above 200 MW, technologies with Chinese-owned intellectual property fall behind the most advanced technologies found elsewhere, in terms of maturity and technical performance, but are nevertheless very cost competitive.

For circulating fluidised bed combustion, R&D of domestically produced, large-scale installations is not adequate. Thus, key equipment depends on production through joint ventures or under licence.

For IGCC and coal-based polygeneration, there is an even bigger gap between domestic and foreign technologies in coal gasification, gas turbines and large-scale Fischer-Tropsch synthesis.

Chinese experts have recommended that the key technologies for ultra-supercritical units should be studied to grasp the core technologies, with research on hydro-dynamics, heat transfer, steam flow distribution, air flows and combustion dynamics, coupled with system simulation and combustion tests. They recommend that several key technologies for IGCC and polygeneration be developed, such as advanced large-scale gasification, high-temperature gas clean-up, large-scale air separation and system integration and optimisation.

Coal gasification and CTL

For coal gasification, Chinese experts suggest focussed R&D should be carried out on the gasification process itself, mass and heat transfer, gas clean-up processes, automatic control and the reaction processes in coal gasification systems operating under high pressure and high temperature. For indirect coal liquefaction, work should focus on technical integration, engineering design, selection of key equipment and safe operation at full load and during start up and shut down. Large-scale, indirect liquefaction requires the successful scale up of Fischer-Tropsch synthesis technologies.

Direct liquefaction depends on technologies for coal gasification, hydrogen production, catalyst preparation, oil-coal slurry heating and transport, solvent recycling and hydrogenation, liquefaction and product separation, residue treatment and refining of the liquefied oil product. Many of these are new and face the test of commercial operation at scale – an engineering risk that is being taken today in China (Figure 5.12).

Figure 5.12.......... Shenhua CTL plant under construction in May 2008 near Erdos, Inner Mongolia

Scheduled for commissioning in September 2008, this USD 1.5 billion direct coal liquefaction (DCL) demonstration train will produce 25 000 b/d (1.1 Mtoe per year) of mainly diesel product, but also naphtha, LPG and phenol. Its annual coal requirement of 3.4 Mt will be brought from the nearby Shangwan mine where there is considerable scope to expand production. The train forms part of a planned 3.2 Mtpa plant which will be expanded to 5.0 Mtpa in a second phase of construction. Shenhua uses imported equipment alongside its own proprietary liquefaction technology in what is the world's first commercial demonstration of DCL.

CO_2 capture and storage

China already participates in the global R&D effort to develop CCS technologies. New technologies for CO_2 capture and purification should be developed actively as well as completing a study of China's geological sequestration potential for large-scale CO_2 storage. Overall, there is a need to explore the technical feasibility and economic rationality of CCS. The ultimate aim is to realise near-zero emissions of CO_2 during the whole process of coal utilisation.

TECHNOLOGY TRANSFER AND INTERNATIONAL CO-OPERATION

There is no clear definition of what "technology transfer" means, although it commonly takes the form of imported goods, technology licensing and foreign direct investment. When China joined the World Trade Organisation in 2001, it became part of a global agreement that encourages technology transfer. More specifically, parties to the UNFCCC Kyoto Protocol have agreed to co-operate in the transfer of technologies pertinent to climate change.[8] The IEA supports this principle, but is concerned that some believe it requires privately owned technology to be transferred freely, in other words, gifted to developing countries. Privately funded R&D depends on investors being able to earn a return from successful results, at least greater overall than the cost of R&D failures. Government-funded R&D rarely creates technology with immediate commercial application, since this would not be permitted under various state-aid rules. Commercial companies do offer technologies that could assist China to improve its coal supply and use; the route to transferring these on a material scale is through trade and commerce. The immediate incentive for this is the sheer size of the market for clean coal technologies in China. An equally attractive incentive, in the medium to longer term, is the potential to develop and manufacture products in China that meet the global demand for cleaner coal technologies. Certainly, the capital costs quoted above are half those for similar equipment manufactured in OECD countries. The opportunities for international partnerships, joint ventures, production-sharing agreements and other commercial relationships are enormous and provide a sustainable way forward that benefits China and the world at large. The state-managed system of technology transfer described in this chapter, with approvals based on guidance lists and barriers in the shape of import tariffs, inhibits progress and restricts the power of commercial enterprises to deliver clean coal technology solutions.

Experience with technology transfer to date

Chinese companies would like to import technologies and equipment that have been successfully demonstrated, not those that need time to develop and improve before productivity benefits are achieved. These concerns influence the decisions of companies wishing to introduce new energy technologies.

---

8. Under Article 10 of the 1998 Kyoto Protocol to the United Nations Framework Convention on Climate Change, "All parties, ... shall: ... (c) Cooperate in ... the transfer of, or access to, environmentally sound technologies, know-how practices and processes pertinent to climate change, in particular to developing countries, including the formulation of policies and programmes for the effective transfer of environmentally sound technologies that are publicly owned or in the public domain and the creation of an enabling environment for the private sector, to promote and enhance the transfer of, and access to, environmentally sound technologies."

Some foreign technologies are not suitable for use with Chinese coal, so design parameters must be adjusted to suit particular coals. For example, special, long-residence time burners are needed for the low-volatile coals of southern China that do not burn readily, and coal mills must be designed based on a coal's Hardgrove grindability index which varies widely between coals, even coals within the same country. Other foreign technologies may not be appropriate for use in regions where water is scarce, so designs must be adapted. For example, air cooling would never be considered for the temperate climate of northern Europe, but is a major environmental benefit in the arid regions of South Africa and Shanxi. These design issues are well understood by power equipment suppliers around the world, who adapt their designs on a project-by-project basis. This means that no two coal-fired power stations are alike and, more importantly, no single design solution exists for China or any other region or country. Coal-fired power stations are bespoke products from an industry where technical experience is as valuable as manufacturing capability. However, the number and ability of Chinese technical personnel needed for effective co-operation with foreign technical personnel is insufficient (and *vice versa*), resulting in difficulties when attempting to transfer technology to local regions. This is a barrier to the adoption of conventional clean coal technologies, before even considering how China might "leapfrog" to state-of-the-art technologies.

Furthermore, Chinese companies have imported multiple examples of similar foreign technologies, often without careful consideration of their suitability for use in China. This has impaired China's ability to assimilate any particular technology and see it through to widespread deployment. For example, China has imported more than twenty variants of coal gasification technology from abroad.

Today, Chinese government policies encourage imported technologies, but require a level of foreign investment in their demonstration to share risk. Incentives for the demonstration of clean coal technologies include preferential import tariffs and lower value added taxes.

A commercial framework for future international co-operation and technology transfer

An alternative route to technology transfer is the introduction of advanced and established technologies in commercial projects. Key equipment and materials can be imported with the bulk manufactured locally, where possible. When importing foreign advanced technologies, factors considered during the authorisation process include the maturity, performance specifications and materials of key components. At the same time, partnerships with domestic original equipment manufacturers are encouraged by the Chinese government. At present, some key clean coal technologies are imported. For example, FGD technologies for units above 200 MW are imported from Germany, Japan, the US and Austria. Some other technologies have been developed in co-operation with foreign companies, such as once-through boilers.

In the coal-mining sector, greater international participation in coal prospecting, mine planning, equipment selection, coal-mining operations and supervision, and environmental management would speed the application of new knowledge and skills in the industry. While current Chinese legislation provides for private and foreign ownership of coal resources, the government does not encourage majority foreign investment in coal production, thus denying itself lasting benefits.

Foreign technologies and experience provide a reference point to promote the adoption of advanced technologies in China. Co-operative development, joint investment and joint operations can all strengthen international exchange and co-operation. With economic globalisation, the development of new technologies often depends on the establishment of such international partnerships, joint ventures and production-sharing agreements. For example, limited by domestic materials and processing technologies, China looks to international assistance for gas turbine manufacturing. In response to the growing market for gas turbines in China, GE Corporation of the US has joint venture agreements with Harbin Power Equipment Company and Nanjing Turbine and Electric Machinery (Group) Company Ltd. for the assembly of large-frame gas turbines using a combination of locally produced and imported components. At the same time as seeking technology input through such joint ventures, China should actively participate in the international R&D effort to establish new clean coal technology options for the future.

REFERENCES

EIA (Energy Information Administration) (2007), *International Energy Outlook 2007*, DOE-EIA-0484(2007), EIA, US Department of Energy, Washington DC.

Henderson, C. (2005), *Towards Zero Emission Coal-Fired Power Plant*, CCC/101, IEA Clean Coal Centre, London.

IEA (International Energy Agency) (2007), *World Energy Outlook 2007 – China and India Insights*, OECD/IEA, Paris.

IEA (2008), *CO_2 Capture and Storage: A Key Carbon Abatement Option*, OECD/IEA, Paris.

Li Zhenzhong (2006), *R&D and Commercialisation of Denitrification for Coal-Fired Power Plants*, China Science and Technology Industry.

MOST (中华人民共和国科学技术部 – Ministry of Science and Technology) (2004), *Study on New Clean Coal Power Generation Technology*, The National High Technology Research and Development Program of China (863 Program) project, MOST, Beijing.

MOST (2005), *Evaluation Method and Technique of Clean Coal Technology*, The National High Technology Research and Development Program of China (863 Program) project, MOST, Beijing.

MOST (2006), 国家中长期科学和技术发展规划纲要 (2006-2020年) (*National Plan for Medium and Long-Term Scientific and Technological Development [2006-2020]*), MOST, Beijing, 9 February, www.gov.cn/jrzg/2006-02/09/content_183787.htmand www.most.gov.cn/yw/t20060209_28601_0.doc.

MOST and MOFCOM (Ministry of Science and Technology and Ministry of Commerce) (2006), 鼓励外商投资高新技术产品目录 (*List of High Technologies and Products to be Encouraged by Foreign Investment*), 商资发[2006]652号 (Notice No. 652), MOST and MOFCOM, Beijing, 31 December.

Nalbandian, H. (2006), *Economics of Retrofit Air Pollution Control Technologies*, CCC/111, IEA Clean Coal Centre, London.

NBS (National Bureau of Statistics of China) (2008), *China Statistical Yearbook 2008*, China Statistics Press, Beijing.

National Planning Committee of China (1982), *National Energy-Saving Investment Plan*.

NDRC (国家发改委 – National Development and Reform Commission) (2003), *Management Measures and Collection Standards for Emission Fees*, NDRC, MOF, SEPA and former SETC.

NDRC (2004a), (国家发展改革委 – National Development and Reform Commission) (2004a), 《资源综合利用目录（2003年修订）》 ("Catalogue on the Comprehensive Utilisation of Resources (2003 revision)"), 发改环资 [2004]73号 (Fa Gai Huan Zi [2004] No. 73), NDRC / Ministry of Finance / State Administration of Taxation, Beijing, 12 January.

NDRC (2004b), 《关于东北区域电力市场上网电价改革试点有关问题的通知》 ("Notice on Pilot Electricity Price Reform in Northeast Regional Power Market"), 发改价格[2004]709号 (Fa Gai Jiage [2004] No. 709), NDRC, Beijing, 27 April.

NDRC (2004c), 《外商投资项目核准暂行管理办法》 ("Provisional Measures for Authorisation of Foreign Investment Projects"), 令第22号 (Order No. 22), NDRC, Beijing, 2004年10月9日 (9 October 2004), www.ndrc.gov.cn/zcfb/zcfbl/zcfbl2004/t20050601_5142.htm.

NDRC (2005), 《中央预算内投资补助和贴息项目管理暂行办法》 ("Provisional Measures for the Management of Subsidies and Discounted Loans for Project Investment within the Central Government Budget"), 令第31号 (Order No. 31), NDRC, Beijing, 2005年6月8日 (8 June 2005), www.ndrc.gov.cn/zcfb/zcfbl/zcfbl2005/t20050630_26952.htm.

NDRC (2006a), *National 11th Five-Year Plan for Coal Washing and Processing*, NDRC Energy Bureau and China Coal Industry Development and Research Center, Beijing.

NDRC (2006b), 煤层气(煤矿瓦斯)开发利用"十一五"规划 (*National 11th Five-Year Plan for CBM and CMM Development and Utilisation*), NDRC, Beijing, 2006年6 月26日 (26 June 2006), www.ndrc.gov.cn/nyjt/nyzywx/t20060626_74591.htm.

NDRC (2006c), 《火电厂烟气脱硫工程后评估管理暂行办法》 ("Preliminary Administrative Measures for the Assessment of FGD Projects"), NDRC, Beijing, 21 March, www.cec.org.cn/news/showc.asp?id=28784.

NDRC (2007a), 煤炭产业政策 (Coal Industry Policy), NDRC, Beijing, 23 November, www.ndrc.gov.cn/zcfb/zcfbgg/2007gonggao/t20071128_175124.htm and www.ndrc.gov.cn/zcfb/zcfbgg/2007gonggao W020071128508929704690.pdf.

NDRC (2007b), 《关于印发煤炭工业节能减排工作意见的通知》 ("Opinions on Emission Reductions and Energy Saving in the Coal Industry"), 发改能源[2007]1456号 (NDRC Energy [2007] No. 1456), NDRC and State Environmental Protection Administration, Beijing, 2007年7月3日 (3 July 2007), www.ndrc.gov.cn/zcfb/zcfbtz/2007tongzhi/t20070705_146315.htm and www.gov.cn/ztzl/jnjp/content_683310.htm.

NDRC (2007c), 《关于加快关停小火电机组的若干意见》 ("Opinions on Accelerating the Closure of Small Thermal Power Generation Units"), 国发[2007]2号 (Guo Fa [2007] No. 2), NDRC and Office of the National Energy Leading Group, Beijing, 2007年1月20日, (20 January 2007), www.ndrc.gov.cn/zcfb/zcfbqt/2007qita/t20070131_115037.htm.

NDRC (2007d), *China's National Climate Change Programme*, NDRC, Beijing, June, http://en.ndrc.gov.cn/newsrelease/P020070604561191006823.pdf.

NDRC and Ministry of Finance (2006), 《国家鼓励的资源综合利用认定管理办法》 ("Management Methods to Encourage Comprehensive Resource Utilisation"), 发改环资[2006]1864号 (Notice [2006] No. 1864), NDRC and Ministry of Finance, Beijing, 2006年9月7日 (7 September 2006), www.gov.cn/zwgk/2006-09/13/content_387619.htm.

NDRC and MOFCOM (Ministry of Commerce) (2007), 《外商投资产业指导目录 (2007年修订)》 ("Guidance Catalogue for Foreign Industrial Investment – 2007 revision"), 令第57号 (Order No. 57), NDRC and MOFCOM, Beijing, 2007年10月31日 (31 October 2007), replacing 2004 revision, www.ndrc.gov.cn/zcfb/zcfbl/2007ling/W020071107537750156652.pdf.

NDRC and SEPA (State Environmental Protection Administration) (2007a), 《现有燃煤电厂二氧化硫治理"十一五"规划》 ("SO_2 Control at Existing Coal-Fired Power Plants under the 11th Five-Year Plan"), 发改环资[2007]592号 (Notice [2007] No. 592), NDRC and SEPA, Beijing, www.mep.gov.cn/law/gz/bmhb/gwygf/200804/t20080410_120980.htm.

NDRC and SEPA (2007b), 《燃煤发电机组脱硫电价及脱硫设施运行管理办法》（试行） ("Operation and Management of Flue Gas Desulphurisation Equipment at Coal-Fired Electricity Generating Plants" [trial implementation]), 发改价格[2007]1176号 (Fa Gai Jiage [2007] No. 1176), NDRC and SEPA, Beijing, 29 May, www.ndrc.gov.cn/zcfb/zcfbtz/2007tongzhi/W020070612625366316898.pdf.

Peng Jingang (彭全刚) (2005), 火力发电企业面临五大风险 ("Five Top Risks Facing Thermal Power Companies"), 中国电力企业管理，2005年10月 (*China Power Enterprise Management*, October 2005).

SACMS (国家煤矿安全监察局 – State Administration of Coal Mine Safety) (2005), 《煤矿安全规程》 ("Coal Mine Safety Regulations"), Decree No. 16, SACMS, Beijing, promulgated 30 November 2004, effective 1 January 2005, replaces 2001 edition, www.chinasafety.gov.cn/files/2004-12/09/F_42cd456f6a924f7f8d36815edaa3e531.pdf and www.chinasafety.gov.cn/files/2004-11/30/F_b123bd5548c341e29fa45c9351cea44e_gc.doc (table of amendments).

SEPA (State Environmental Protection Administration) (2000), 《淘汰落後生產能力、工藝和產品的目錄》 ("Catalogue of Outdated Production Capacities, Processes and Products to be Eliminated"), 第二批 (2nd Batch), 环发[2000]35号 (Huan Fa [2000] No. 35), SEPA, Beijing, 2000年2月21日 (21 February 2000), www.mep.gov.cn/info/gw/huanfa/200701/t20070105_99299.htm.

SEPA (2002), 《燃煤二氧化硫排放污染防治技术政策》 ("Technical Policies for SO_2 Emission Prevention and Control"), SEPA, State Economic and Trade Commission and Ministry of Science and Technology, Beijing, 30 January, www.mep.gov.cn/xcjy/zwhb/200203/t20020319_78974.htm.

SEPA (2003), 《关于加强燃煤电厂二氧化硫污染防治工作的通知》 ("Notice on SO_2 Emission Control at Coal-Fired Power Plants"), 环发[2003]159号 (Huan Fa [2003] No. 159), SEPA, 15 September, www.mep.gov.cn/info/gw/huangfa/200309/t20030915_86600.htm.

SERC (State Electricity Regulatory Commission) (2008), *China Electric Power Statistical Yearbook 2007*, SERC and the China Enterprise Confederation, China Statistical Publishing House, Beijing.

SETC (former State Economic and Trade Commission) (2002), 国家产业技术政策 *(National Industrial Technology Policy)*, SETC / Ministry of Finance / Ministry of Science and Technology / State Administration of Taxation, Beijing, 21 June.

SDPC (国家计委 – State Development Planning Commission) (2001a), 《关于规范电价管理有关问题的通知》 ("Notice on Standardising Electricity Tariff Management"), 计价格[2001]701号 (Ji Jiage [2001] No. 701), SDPC, Beijing, 23 April.

SDPC (2001b), 《对洁净煤技术发电示范工程项目实行优惠政策》 ("Preferential Policy on Power Generation Demonstration Projects with Clean Coal Technologies"), 计价格[2001]1370号 (Ji Jiage [2001] No. 1370), SDPC, Beijing, 25 July.

State Council (国务院) (2004), 《国务院关于投资体制改革的决定》 ("Decision on Reform of the Investment System"), 国发[2004]20号 (Guo Fa [2004] No. 20), Office of the State Council, Beijing, 2004年11月4日1 (16 July 2004).

State Council (国务院) (2005a), 《国务院关于促进煤炭工业健康发展的若干意见》 ("Opinions of the State Council on Promoting the Sound Development of the Coal Industry"), 国发[2005]18号 (Guo Fa [2005] No. 18), Office of the State Council, Beijing, 2005年6月7日 (7 June 2005), www.gov.cn/zwgk/2005-09/08/content_30251.htm.

State Council (国务院) (2005b), 《国务院关于落实科学发展观加强环境保护的决定》 ("Decisions on Implementing the Scientific Development Concept to Strengthen Environmental Protection"), 国发[2005]39号 (Guo Fa [2005] No. 39), Office of the State Council, Beijing, 2005年12月3日 (3 December 2005), www.gov.cn/zwgk/2005-12/13/content_125680.htm.

State Council (国务院) (2006a), 《国务院办公厅关于加快煤层气（煤矿瓦斯）抽采利用的若干意见》 ("Opinions of the State Council on Accelerating CBM/CMM Extraction and Utilisation"), 国办发[2006]47号 (Guo Fa Ban [2006] No. 47), Office of the State Council, Beijing, 2006年6月15日 (15 June 2006), www.gov.cn/zwgk/2006-06/19/content_314623.htm.

State Council (国务院) (2006b), 《国务院关于"十一五"期间全国主要污染物排放总量控制计划的批复》 ("Reply to the Control Plan for the Total Emissions of Major Pollutants during the 11th Five-Year Plan Period"), 国函[2006]70号 (State Letter [2006] No.70), Office of the State Council, Beijing, 2006年8月5日 (5 August 2006), www.ndrc.gov.cn/zcfb/zcfbqt/2007qita/t20070131_115037.htm.

State Council (2007), *China's Energy Conditions and Policies*, White Paper, Office of the State Council, Beijing, 26 December.

Wang Zhixuan (王志轩等) (2005), 现有火电厂二氧化硫治理十一五规划的基本思路 ("Thinking on 11th Five-Year Plan for SO_2 Control at Existing Thermal Power Plants"), 中国电力，2005年11月 (China Power, November 2005).

Xinhua (2007), China to Close Small Units over 50 GW to Ensure Energy Saving and Emission Reduction Aims, Xinhua News Agency, 29 January 2007, http://news3. xinhuanet.com/fortune/2007-01/29/content_5670863.htm.

Xu Shisen (2006), "The Status and Development Trends of IGCC in China", presented at EU China NZEC Workshop on Near-Zero Emissions Coal: Power Generation with Carbon Capture and Storage in China, Beijing, 4-5 July, www.chinaesco.net/PDF_ppt_lt/pdf_dir/xushisen.pdf.

Yu Ertie (2006), "Application and Reorientation of Jigs", *Coal Preparation Technology*, 2006(2).

VI. A CLEANER FUTURE FOR COAL IN CHINA: THE ROLES OF MARKETS AND INSTITUTIONS

The application of clean coal technologies in China cannot be divorced from the dynamics of its domestic coal market and government oversight of the coal-producing and coal-using sectors. Technology choices are largely determined by these two factors, driven by economic and regulatory considerations. This chapter looks ahead to examine how the coal market in China might evolve in the short to medium term. A stable coal supply is fundamental to achieving other goals. Over the last decade, coal shortages, volatile prices, poor product quality, transport bottlenecks, financial losses and other, near-term issues have all, at times, distracted leaders, government officials and enterprise management from giving their full attention to longer-term, sustainability issues. A properly functioning coal market, with effective supply and demand responses, has clear and immediate benefits that give the space and freedom to address the more difficult problems associated with coal use. A legal and regulatory outlook summarises the latest thinking and proposals for new coal-related laws. These will set the institutional framework for coal production and use for some years to come. Such a framework will influence the development of the secondary legislation needed to implement and enforce China's growing volume of broad-based statute law. More than anything else, enforcement of well-crafted legislation will determine how clean coal technology is deployed in China. An effective framework creates a demand for technology solutions, which can often be best delivered in a competitive market place.

MARKET OUTLOOK

The Chinese coal market has been characterised by alternating periods of under and over supply. Until the mid 1990s, coal had been in short supply – a situation that reversed as plentiful supplies became available towards the end of the decade. The rapid demand growth since 2001 has led, once again, to supply-side tightness in a sometimes volatile market. The various coal sector reforms, described in Chapter 3, have contributed significantly to these observed changes. In this section, a short-term (to 2010) and medium-term (to 2020) outlook is presented with an underpinning analysis of the factors that will influence future coal supply and demand patterns, prices and the levels of imports and exports. This complements the long-term (to 2030) outlook in the IEA *World Energy Outlook* (Section 4.1).

Influence of government policy on short- and medium-term outlook

After 1998, investment in key state-owned mines dwindled; new projects were put on hold as the government pursued its policy to close smaller TVE and private mines. Coal production declined, falling to 1.23 Gt in 2000. Since then, Chinese coal output has grown robustly, by an annual average of 11% through to 2006, thus alleviating shortages (Table 6.1). During this period, the Chinese coal industry was transformed from its loss-making position of the past to one of profitability, with particularly strong profits reported since 2004. Fundamental to this improvement was the increase in coal prices – from 2000 to 2006 the price of coal for electricity generation rose 80% while prices rose 116% for other users.[1]

Table 6.1 ....Coal industry production, investments, profits and prices, 1998-2006

	Coal production, million tonnes	Investment, RMB billion	Profit, RMB billion	Gross margin, %	Asset-liability ratio, %	Coal prices, RMB/tonne	
						Non-utility use	Utility use
1998	1 305					160	133
1999	1 238		–1.7			144	121
2000	1 231	18.8	0			140	121
2001	1 268	21.8	4.2	2.9	64.8	151	124
2002	1 398	28.6	8.6	4.7	57.0	168	137
2003	1 670	41.4	13.8	6.0	56.7	174	139
2004	1 956	70.2	30.7	8.9	60.2	206	163
2005	2 159	114.4	54.0	10.9	61.5	270	213
2006	2 320	147.9	67.7	10.6	60.4	302	218

Notes: Investment, profit, gross margin, asset-liability ratio and price (ex-mine) data are for coal mining enterprises large enough to be included in official statistics. Profits include subsidies.
Sources: production data: IEA (2008); investment, profit, gross margin and asset-liability ratio data: CEMAC (2006); coal prices: CCII (2006).

Increased profits and the general ease of raising capital allowed the coal sector to increase its investment massively from just RMB 19 billion in 2000 to RMB 148 billion in 2006. With the industry now enjoying its best period ever, it is expected that, in the short term, production capacity will be further expanded and supply boosted. Total production capacity in 2006 was around 2.4 Gt, of which 1.35 Gt was at medium- to large-scale coal mines and 1.05 Gt at small coal mines. 1 560 new coal mines were under construction in 2006, with a production capacity of 653 Mtpa, some of which will replace mines that close due to depletion or for policy reasons – an estimated capacity of 170 Mtpa. Hence, production capacity might be 2.8 to 2.9 Gtpa by 2010.

Three factors influence the medium-term outlook: reform of state-owned mines; a deterioration of the coal resource base; and constraints on coal distribution. Tackling these in the next 10 to 15 years is crucial if coal supply is to expand. Reforms aimed at improving the efficiency of key state-owned coal mines have not yet been completed. So while gross margins have improved with rising coal prices, production costs have also risen markedly over the past few years, absorbing much of the increase in selling prices (Figure 3.10). There are many factors behind the increase of production costs. Firstly, coal mine workers across the country, particularly in northeast China, enjoyed pay rises and looked to recover unpaid wages totalling RMB 3.2 billion in 2003

---

1. See Section 3.6 for a discussion of why coal prices for electricity generation were held lower.

(Huang *et al.*, 2004). Other input costs also increased. For example, from 2000 to 2006, the prices of fuel and electricity rose by 52%, although steel prices rose by a more modest 36%.[2] More fundamentally, costs are increasing as production moves to deeper seams. Finally, mining enterprises face higher costs for environmental protection and production safety. Looking to the future, the average wage of coal miners lags behind that of workers in the manufacturing sector, and this is likely to put further pressure on wage costs. It remains to be seen whether coal mining companies will be able to manage costs when prices are less favourable. With gross margins still improving, there is little concern about operational efficiency. During the recent investment boom, retained profits allowed the coal sector's asset-liability ratio to jump from 56.7% in 2003 to 60.4% in 2006; and yet, many coal companies are concerned that profits might not be sustainable under a low-price scenario, and this could lead to defaults on the substantial bank loans that have helped fund investment. If such a scenario became reality, investments in the coal industry could shrink again, leading to supply shortages.

Many coal mines in China have been exploited for more than half a century and others have closed due to exhaustion. Even where older mines have not closed, production costs have increased as deeper seams are worked. In addition, following the supply shortages beginning in 2002, many key state-owned coal mines produced beyond their nominal capacity, shortening their remaining life.[3] Furthermore, in regions with excellent resources, such as Shanxi, TVE and private coal mines account for a significant share of output, but only mine the most accessible coal seams, avoiding those susceptible to flooding and gas risks as these would be more costly to exploit. As a result, the resource recovery rate is very low, at around 10-15%, and the remaining coal resource is effectively sterilised. Consequently, there are concerns that the primary coal production areas will face exhaustion in the future (Box 2.1).

Today, coal production remains concentrated in a few provinces where there is still a large potential to expand production, of which the most significant is Shanxi. However, the capacity of China's railways to transport more coal is limited, despite the trend since the 1980s of an increasing share for coal in total rail freight. As noted above, by 2010, coal mines with a combined capacity of 170 Mtpa might close due to resource exhaustion. By 2020, an additional 380 Mtpa of capacity might be lost. To replace this capacity, coal production will have to shift further inland to western regions. Under such a scenario, without massive investment in new railway infrastructure, transport bottlenecks could restrain the expansion of supply.

Effective reform measures could avoid the risk of supply restrictions. A concern is that, despite the supportive policies enacted in 2004 to create large coal production bases and the subsequent investment in key state-owned mines, operational efficiency has not improved universally. In many places, poorly performing mines, previously

---

2. By comparison, coal prices for utility users rose by 80% over the same period and by 116% for non-utility users.

3. With coal in short supply, new coal mines have been commissioned within much shorter timeframes than normal. Usually, it takes 5 to 6 years before a new coal mine starts commercial production; however, there have been opportunistic cases with lead times of less than 3 years. Such compression of development time might result in lower resource recovery rates and affect resource utilisation in the long run.

scheduled for closure, remain in production to meet demand. Market-based reforms, as experimented with since the mid-1990s, should have driven out inefficiency – either through closures or productivity improvements. Instead, when faced with coal supply shortages in 2002, the government responded with what might be viewed as protectionist policies for key state-owned mining companies. These policies have created some very large mining enterprises, both through consolidation and investment in new capacity – nowhere more visible than at the thirteen large coal bases described in Chapter 3. This is certainly the right strategic direction, but has required a level of investment that might have been hard to justify on purely economic grounds, and while consolidation can create larger, more influential coal companies, it does not in itself lead to improved operational efficiency. Worse still, it might even delay reforms.

Coal demand outlook

From 2000 to 2006, coal demand grew by an average of 11.3% each year, restrained at times by a lack of supply following a period of under investment in coal mining. Two major factors that may dampen future coal demand are regulations to cool an overheated economy and the energy conservation and environmental protection measures outlined in the 11th Five-Year Plan.

Power, steel, chemicals and construction materials are the four principal, coal-using sectors. Of these, steel and construction materials have been identified as "overheated sectors" in the last few years, suggesting that the massive growth in demand for coal over recent years is not considered normal. The thermal power sector has also been affected by the overheated economy with annual electricity demand growth averaging 13% from 2000 to 2006. Since 2004, the government has adopted a series of regulatory measures to cool the economy and these began to have an effect from around 2006 such that coal demand, particularly in the steel and construction materials sectors, may moderate. Energy conservation and environmental policies will also exert an impact on future coal demand. According to the 11th Five-Year Plan, energy consumption per unit of GDP will be reduced by 20% between 2005 and 2010. In one initiative designed to deliver this ambitious goal, 998 energy-intensive enterprises have signed letters of commitment with local governments that make senior executives personally responsible for implementing energy conservation programmes to achieve assigned targets.[4] By tackling low energy efficiency in industry, the recent sharp rise in coal demand need not be sustained in the future.

In the coal-fired power sector, where efficiency improvements were already being made, the government has set an overall target to reduce coal consumption per unit of electricity supplied to 355 gce/kWh by 2010 and to 320 gce/kWh by 2020. Over the 10th Five-Year Plan period, specific consumption fell 4.6% from 392 gce/kWh to 374 gce/kWh in 2005, and fell again to 365 gce/kWh in 2006, suggesting that the 2010 target should be easily achieved; many utilities have signed commitment letters. In 2006, the average size of coal-fired unit was 60.9 MWe, giving much further

---

4. Closures and industry consolidation had reduced the number of enterprises in the programme to 953 by mid-2008.

scope for efficiency gains as the policy to close units below 100 MWe takes effect and new projects with supercritical or ultra-supercritical units of 600-1 000 MWe are commissioned. One explanation for the relatively lenient specific coal consumption targets in the power sector is the priority given to air pollution control, such as FGD which reduces power plant efficiency.

Given all these factors influencing demand in the major, coal-consuming sectors, the China Coal Industry Development Research Centre projects that coal demand may reach 2.8 Gt by 2010, an average annual growth of 6.2% from 2005, and 3.2 Gt by 2020 following more moderate growth of just 1.3% per year over the preceding decade (Table 6.2 and Figure 4.1). The projection shows the rising shares taken by power and chemicals as coal demand continues to grow in these sectors, while falling in other sectors sometime after 2010. Overall, the outlook is similar to the IEA *World Energy Outlook* Alternative Policy Scenario (Section 4.1).

Table 6.2............Coal demand outlook (million tonnes) and sector shares (%) in 2010 and 2020

	2005	Share	2010	Share	2020	Share
Power	1 110	53.4%	1 700	60.7%	2 100	66.2%
Steel	255	12.3%	350	12.5%	320	10.1%
Construction materials	285	13.7%	340	12.1%	300	9.5%
Chemicals	96	4.6%	180	6.4%	250	7.9%
Residential and others	332	16.0%	230	8.2%	200	6.3%
Total	**2 078**	**100.0%**	**2 800**	**100.0%**	**3 170**	**100.0%**

Source: CCIDRC (2007).

Coal supply and price outlook

This section examines the impact on coal supply and prices of the above demand projection which can be summarised as soaring demand in the short term followed by more restrained growth in the medium term. The 2.8 Gt demand in 2010 can be met by the production capacity of 2.8-2.9 Gtpa that will then be in operation, as identified above. This suggests a reasonably balanced market – so prices might be little changed from recent levels, at least in the short term.

In the medium term, restrictions to supply growth may be mirrored by slower demand growth leading to a supply-demand balance with continuing price stability. However, there are still uncertainties. In the short term, higher prices encourage over-production that, in the past, has led to gluts. If coal prices remain at their current high level, the investment boom could be prolonged such that production capacity outstrips demand in the medium term. The production overhang of the late 1990s serves as a warning – over supply led to falling prices; many companies saw revenues decline dramatically, so scrambled to sell more coal; there was further over supply and prices dropped even further. Table 6.1 shows that mines are now profitable and the government sees this as helping the reform process. The current goal is to reduce the number of small coal mines to below 10 000 by 2010, from 24 000 in 2005, and to limit total production from small coal mines to 700 Mtpa. If output from these mines can be managed, the government hopes to balance supply and demand without a significant drop in coal prices. Yet controlling output from TVE and private coal mines could have unpredictable outcomes, including price volatility, since these are the more responsive, swing producers. There are other factors that might cause coal prices to increase. Coal prices for power use are held relatively low compared to prices for non-power use and this price gap had grown to a 28% discount in 2006 (Table 3.4). Such price distortions not only affect the economics of coal production, but also reduce the incentive for energy conservation measures in the power industry. This situation will be gradually improved during and after the 11th Five-Year Plan period as coal prices move more freely, and presumably relatively higher for electricity generation.

In Section 3.6, the price reforms that are currently being implemented in Shanxi province are described. Various external costs have been imposed on coal producers, adding pressure to raise prices. Although, at the moment, this pricing reform is limited to Shanxi, the fact that this province accounts for one-quarter of national production means that reform there should have a big influence on coal pricing across China. If reforms are extended to other provinces, coal prices could rise significantly. Based on all the surcharges listed in Table 3.5, covering resource taxes, environmental fees and safety levies, it is estimated that coal production costs in Shanxi will eventually increase by RMB 60-80/t. The reforms were introduced in January 2007 and, in the first quarter of 2007, ex-mine coal prices rose by RMB 40/t, or approximately 25%. End users will experience proportionately smaller price rises since transport and other costs must be added to the ex-mine price; in coastal regions, the ex-mine price might account for less than half of the delivered price. Such pricing reform is imperative if the coal industry in China is to develop sustainably. With external costs increasingly reflected in coal prices, it is expected that market responses will play a greater role in reducing externalities.

Outlook for imports and exports

China's coal exports increased rapidly from 32 Mt in 1998 to 94 Mt in 2003 (Figure 6.1) when it briefly became the world's second largest coal exporter after Australia, meeting strong demand in the East Asia region where imports from China reached 80 Mt compared with 137 Mt from Australia. Behind this rapid growth of exports was oversupply in the domestic market. Key state-owned mines faced sluggish demand in the late 1990s and were concerned about massive stock builds – exports into the international market were seen by many companies as a means to survival and the Chinese government allowed exports to grow. However, the government announced that a quota system for coal export licences would be implemented, with an 80 Mtpa cap in 2004. The outturn for 2004 was 86 Mt as previously issued export licences were carried forward. This sudden cut in exports had a profound impact on many coal importing countries.

Figure 6.1 ....China's coal imports and exports

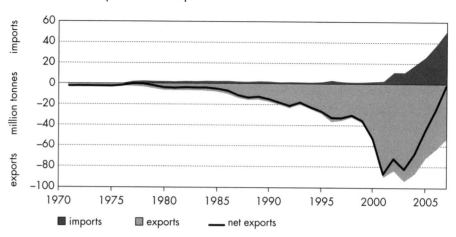

Source: IEA database (World Coal Statistics.ivt available at http://data.iea.org).

China's coal exports are small compared with domestic consumption, peaking at just 6% in 2003, but, being at the margin, they are heavily influenced by the domestic supply-demand balance. By 2006, they had fallen to 63 Mt and, in 2007, fell further to 54 Mt, only 6 Mt greater than imports. China's coal imports have also seen a remarkable change since 2002. Before then, imports were an insignificant 2 Mtpa, but grew abruptly to 11 Mt in 2002, 37 Mt in 2006 and 48 Mt in 2007. In the past, imported coal had played a minor role in the Chinese coal market; only those areas far from production bases relied on imports, for example Guangzhou province. However, due to tight supply in the domestic market, China now imports coal from overseas in order to meet more general demand.

Since 2004, China has exercised export controls on coking coal, resulting in a sharp cut in coking coal exports which fell 56% in 2004 to 5.8 Mt. At the same time, coking coal imports rose 162% to 6.8 Mt, meaning that China became a net importer in 2004, which has had a major impact on the world market. International prices doubled between 2003 and 2004 to USD 110/t (CIF) and have followed a rising trend since then to over USD 300/t in 2008. In 2007, China remained a net importer of around 3.2 Mt of coking coal.

China is the largest exporter of coke in the world. In 1985, China exported less than 370 kt, but, by 2000, coke exports reached a peak of 15.2 Mt. One reason for this growth was the tightened environmental protection standards in the EU and elsewhere which pushed up the costs of coke production. Coke exported from China dominated the global market in 2003, when China exported 14.7 Mt or 56% of global coke trade. In 2006, coke exports stood at 14.5 Mt. Due to the rapid growth of domestic steel production and ever-increasing domestic demand for coke, exports from China might decrease in the future. The Chinese government has outlined in its 11th Five-Year Plan the need to exercise export controls on coking coal and coke; these will likely have a marked impact on global trade in these commodities.

Coal mining enterprises in China are not free to import or export coal, they must go through one of four licensed trading companies: China National Coal Corporation, Shanxi Coal Import and Export Group Corporation, Shenhua Group Corporation and China Minmetals Corporation. Although many other coal mining companies, including Yanzhou Coal Group Corporation (ranked sixth by production in 2006), applied for licences, they did not receive approval. By only allowing four enterprises to import and export coal, the state can exercise its control over coal supply, allowing exports in times of surplus and reining these back in times of shortage.

The following measures have also been used to influence the level of coal exports from China since 2000: value added tax (VAT) levied at 13% was refundable on exports; exports were exempt from the Railway Construction Fund surcharge on primary routes; exports were similarly exempted from the Port Construction Fund surcharge at major ports; and priority was given to export coal when allocating rail transport quotas. To an extent, these preferential policies constituted a subsidy to key state-owned coal mines plagued by poor domestic demand in the late 1990s, since they allowed the mines to earn a higher income by exporting. However, this changed around 2002 when domestic prices became higher than those on the international market. Certain coal mines were still able to make good profits from exports, due to the preferential policies, but at the same time, coal imports were increasing. So coal that was being exported with government subsidy could have been sold on the domestic market without subsidy, and China could have avoided the foreign currency payments being made for coal imports. As coal supply tightened, the VAT refund rate on coal exports was gradually reduced and other preferential policies eliminated. On 1 May 2004, the Railway and Port Construction Fund exemptions ended, despite the earlier announcement on 15 February 2003 that these exemptions would last until the end of 2005. Also in May 2004, the VAT refunds for coke and coking coal ended and, in September 2006, VAT refunds were similarly abolished for steam coal. To further dampen coal exports, a 5% export tax has been imposed since November 2006, while coking coal and coke are now listed as restricted products for export.

In the past, the Chinese government has said that the country would maintain a certain volume of coal exports. However, in the 11th Five-Year Plan, this position has shifted to preserve domestic resources and the government now encourages Chinese coal companies to invest in mining rights overseas. Overall, coal exports became less profitable or even unprofitable and, since 2004, coal exports from China have declined sharply. It is quite likely that Chinese coal exports will continue to decrease, and China might become a major net coal importer in the future, mirroring what has already happened with oil and iron ore. Import duties on 26 commodities, including coal, were reduced from a 3-6% range to between zero and 3% in November 2006, a measure designed to increase domestic supply.

LEGAL AND REGULATORY OUTLOOK

Evolution of government policy

Ever since China's economic reform and opening up, the coal industry has enjoyed rapid growth, meeting the energy demands of a growing economy. Today, China is the world's largest coal producer and home to several coal corporations of international standing. However, in the course of its development, the coal sector has encountered many problems: an imperfect market economy; costly production expansion; a low level of scientific and technological development; too many accidents; severe resource wastage; and lagging environmental protection. With the growing demand for coal, the pressure to resolve these problems will only increase. In response, the Chinese government is constantly refining its coal industry policies to ensure sufficient resources, protect the environment and improve work safety.

In its *Opinions on Promoting the Sound Development of the Coal Industry* (State Council, 2005a), the State Council summarised the Chinese government's coal industry policy with the principal aim of relieving supply tightness. The opinions set out several measures – advanced technology being the central pillar, guided by the overarching "scientific concept of development". By adopting scientific and technological development principles, the State Council expects the coal industry to follow a sustainable development path over the coming years. This path is characterised by high rates of resource utilisation, safe working conditions, wide economic benefits and low environmental pollution. Several goals are identified in the opinions. Firstly, the government should administer coal resource prospecting, mine development and industry management in a way that protects and preserves resources for rational and orderly exploitation. Secondly, the industry should establish a coal-supply system with large-scale coal bases and enterprises as its backbone. Thirdly, a system to guarantee work safety with strong management and sufficient investment is demanded. Finally, the comprehensive utilisation of coal resources is promoted through the "circular economy" concept which applies to all stages from mining, through processing and transformation, to waste management.[5]

Regulations, rules and monitoring under the Coal Law 1996 and the Mineral Resources Law 1996 will be important in achieving these goals. The State Council also lays down a number of principles that the coal industry should adhere to: commissioning advanced production capacity while closing obsolete capacity; dealing with the causes of problems, not simply treating the symptoms; putting safety first by taking full account of risks and adopting precautionary measures; blending state guidance and support with the autonomous development of enterprises; combining reform with innovative regulatory mechanisms; and ensuring coal industry development meets economic and social development goals.

The State Council recommends an integrated approach with clear responsibilities to guarantee worker safety in coal mines. "State inspection, local administration and enterprise action" describes China's coal mine work safety system that the State Council says should be further strengthened. Increased investment in coal mine safety is expected to come jointly from central government, local administrations and enterprises according to the principle of "corporate duty with government support". In order to lead and co-ordinate coal mine gas control activities, the State Council has instructed that a leading group be established so that gas control techniques can be improved. To improve coal mine workers' qualifications and skills, enterprises

---

5. See footnote 6 in Chapter 5.

should establish and perfect employee training schemes, covering safety, technical training and continuing education, paying particular attention to miners who operate tunnelling equipment.

In respect of the environment, the State Council sees protection against pollution and restoration of mining damage as paramount and calls for the following. During the exploitation and utilisation of coal resources, the assessment of environmental impacts must be carried out. For major projects, environmental protection measures must be taken strictly in accordance with three principles: "whoever exploits resources should protect them, whoever pollutes the environment should clean it, and whoever destroys the environment should restore it". In order to enhance investment in environmental protection in mining areas, an ecological recovery and compensation mechanism should be designed and implemented, with the responsibilities of enterprises and government clearly defined. For legacy problems, such as mining subsidence caused by the original state-owned coal mines established by central government, viable plans should be drawn up and implemented. The central government should give the necessary funding and policy support, while local governments, at all levels, and coal enterprises should manage and allocate these funds according to the plans.

With regard to coal industry administration, the State Council says that China should enhance the integrated development and utilisation of coal resources, by building up the "circular economic system" and promoting the industrial development of clean coal technologies, including those that use and treat coal gangue, coal slurry, CMM and coal mine water discharges.

Evolution of coal sector administration

Law enforcement in the Chinese coal industry is gradually being perfected and, in terms of resource administration, safety supervision and inspection, environmental protection and industry administration, government from the central to the county levels has set up a relatively complete organisational structure (Table 6.3).

Table 6.3 .... Responsibilities for coal industry administration at different levels of government

| | Resources administration | Safety supervision and inspection | | Environmental protection | Industry administration |
		Safety supervision	Safety inspection		
Central	Ministry of Land and Resources	State Administration of Work Safety	State Administration of Coal Mine Safety	State Environmental Protection Administration	National Development and Reform Commission
Provincial	Provincial Land and Resources Bureaus	Provincial Bureau of Work Safety	District Coal Mine Safety Inspection Bureaus and Regional Coal Mine Safety Inspection Branch Bureaus overseen by the State Administration of Coal Mine Safety	Provincial Environmental Protection Bureaus	At the provincial level and all levels below are Development and Reform Commissions (or Economic and Trade Commissions) or Coal Industry Bureaus, as well as other relevant agencies
Municipal	Municipal Land and Resources Bureaus	Municipal Bureau of Work Safety		Municipal Environmental Protection Bureaus	
County	County (district) Land and Resources Bureaus	County (district) Bureau of Work Safety		County (district) Environmental Protection Bureaus	

Resource administration

In accordance with the Coal Law 1996 and the Mineral Resources Law 1996, the Ministry of Land and Resources is primarily responsible for coal resource administration. It issues mining licences, following tenders, auctions and applications, to enterprises which meet approval criteria. In the first instance, it checks that mining plans submitted by enterprises will achieve full and proper (*i.e.* "comprehensive") utilisation of resources; specifically, it supervises and checks recovery rates. In recent years, other departments of the state have also administered and licensed mineral resources, including coal, according to the Administrative Licence Law 2003.

During September 2004 and in accordance with the Mineral Resources Law 1996, regulations of other relevant statutes and the National Program on Mineral Resources (SCIO, 2001), the Ministry of Land and Resources and the National Development and Reform Commission (NDRC) published a list identifying the first group of nineteen state-planned coal mining areas in Shanxi and Inner Mongolia.

In 2005, the government initiated a special, nationwide examination of coal resource recovery rates. The Ministry of Land and Resources, NDRC and the State Administration of Work Safety formed a supervision group which visited every region to supervise this examination. Enterprises prepared self evaluations, then these were confirmed by local government and inspected by both the Ministry of Land and Resources and NDRC. Conducting such a special inspection of coal resource exploitation was unprecedented in China and is intended to improve the development and utilisation of coal resources. Such an approach would be relevant also to the exploitation of other mineral resources in China.

In some regions of China, illegal mining activities have been discovered, such as collective prospecting and exploitation of coal without licences, exploitation beyond mining-rights boundaries and unapproved, often disorderly mining. These illegal actions do great harm to China's coal resources. In order to realise rational and sustainable development of mineral resources, and to ensure safety, the Executive Meeting of the State Council on 26 July 2005 considered a report by the Ministry of Land and Resources. The State Council requested that the development of mineral resources be uniformly regulated and standardised, with top priority given to the development of coal resources. As a result, each province and region is now regulating mineral resource development in accordance with the spirit of the *Notice Concerning the Comprehensive Rectification and Standardisation of the Regulation of Mineral Resource Development* (State Council, 2005b).

The other major aspect of resource administration is fiscal and tax policy. In respect of coal resource taxes and their relationship with such factors as regional economic development, the Ministry of Finance and State Administration of Taxation announced in 2004 and 2005 regulations concerning the tax collected on coal resources in Shanxi, Shandong, Henan and Guizhou.

Work safety supervision

In accordance with the Office of Legislative Affairs of the State Council, the regulations and inspection work under the Coal Law 1996, the Work Safety Law 2002, the Mine Safety Law 1992 and the *Coal Mine Safety Inspection Regulations of 2000* are principally the responsibility of the State Administration of Work Safety and the State Administration of Coal Mine Safety. These two departments supervise coal mine safety by issuing work safety licences, competency certificates for mine managers and certificates for authorising individuals to carry out special types of work, such as handling explosives, gas inspections and electrical work. They also enforce administrative penalties for violations of the law.

Coal mine safety law enforcement is central to the Coal Law and is an important element of work safety in China. In 1999, the State Council decided to implement coal mine safety inspections, under a vertical management system, to solve the widespread and persistent pursuit of local self-interest in the supervision and administration of coal mine safety. In January 2000, the State Bureau of Coal Mine Safety was formally established, followed, in February 2001, by the State Bureau of Work Safety. In 2001, these were merged as a consequence of a government re-organisation, further demonstrating the importance of coal mine work safety administration and supervision.

In 2005, coal mine safety law enforcement became even more efficient when, in February, the Executive Meeting of the State Council decided to elevate the State Bureau of Work Safety to become the State Administration of Work Safety, a ministerial agency. Meanwhile, the State Administration of Coal Mine Safety under the jurisdiction of the State Administration of Work Safety, was set up to improve the authority of coal mine safety inspectors and strengthen coal mine safety inspections and law enforcement. This elevation of status has improved the authority of local coal mine safety supervision and inspection organisations under local government, so they can now effectively supervise the large, state-owned coal enterprises. When conducting major safety inspections or accident investigations, the State Administration of Work Safety and the State Administration of Coal Mine Safety are able to better co-ordinate their activities under the new structure.

Over time, as regions faced different mining conditions and issues, the competencies and practices of local coal mine work safety administrations became less uniform and, in some cases, the boundary of responsibility between safety inspection and mining supervision became blurred. Therefore, in November 2004, the *Opinions on Perfecting the Coal Mine Safety Inspection System* (State Council, 2004) summarised the then current practices and experiences, and defined a new model for coal mine work safety, based on "corporate responsibility with state and local inspection". The opinions defined the inspection functions of the coal mine safety inspection bodies at national level, and listed many examples to illustrate the major responsibilities of the provincial and local coal mine safety inspection bodies. In addition, the opinions further clarified the inspection mechanism, by delegating the examination and guidance tasks of the coal mine safety inspection bodies to local coal mine safety inspection agencies. At the same time, it optimised the regional spread of inspection bodies, adding coal mine safety inspection bureaus in Hubei, Guangdong, Guangxi, Qinghai and Fujian, and changed the name of coal mine safety inspection offices to regional inspection branch bureaus, further strengthening coal mine safety inspection work in terms of organisational hierarchy. By the end of 2004, 25 provincial coal mine safety inspection bureaus plus the Beijing Coal Mine Safety Inspection Branch had been established in 31 provinces, autonomous regions, province-level municipalities and the Xinjiang Production and Construction Corps. Most provinces have now established subsidiary agencies to the State Administration of Coal Mine Safety.

In order to supervise coal mine safety more effectively, the supervision and inspection bodies attach great importance to co-operation and communication between organisations to avoid any unnecessary disruption arising from enforcement of contradictory laws and regulations by the authorities. In 2004, the State Administration of Coal Mine Safety issued its *Opinions on Further Enhancing Administrative Law Enforcement in Order to Improve Coal Mine Inspection Mechanisms*, which required "independent and uniform law enforcement" between relevant departments of local government, a regular and effective reporting system, and better information exchange with local governments. Through assistance and encouragement, local governments were expected to strive for excellence in coal mine work safety. In August 2005, the State Administration of Work Safety drew up its *Opinions on Further Enhancing the Structure of Work Safety Supervision and Coal Mine Safety Inspection Teams*, proposing that China properly positions the safety function and perfects supervision and inspection systems, while establishing a uniform law enforcement system with clear responsibilities in order to strengthen the legal powers of inspection teams.

Coal mine safety supervision and inspection, and law enforcement differ from the safety administration specifically carried out by coal mine enterprises. In the case of the latter, the law enforcement authorities can direct coal mine enterprises on work safety and offer information and advice to promote mine safety. For instance, in March 2005, in response to the decisions of the 81st Executive Meeting of the State Council and the Videophone Conference for National Coal Mine Safety Reform and Gas Treatment Work to reduce coal mine gas emissions and the frequency of coal mine gas accidents, NDRC, the State Administration of Work Safety and the State Administration of Coal Mine Safety together prepared Fifty Proposals on Coal Mine Gas Treatment based on the experience of gas treatment at Huainan, Pingdingshan and Songzao coal mines. The three agencies then distributed the fifty proposals to coal mining enterprises to serve as a reference when dealing with similar situations.

Coal mine safety law enforcement continues to improve as corruption and collusion are eliminated, especially since the State Council decreed that local government or state-owned enterprise officials with shares in coal mines had to relinquish these by 22 September 2005 (SAWS, 2005). Some coal mines not meeting work safety requirements and other illegal coal mines have been shut down (State Council, 2005c). However, some mines might appear shut when inspected during the day, only to re-open each night, continuing to produce coal. This difficulty with enforcement illustrates just one of the many grave problems that China faces as it moves to improve coal mine safety.

Environmental protection

Coal production and use leads to atmospheric, water and land pollution. As well as its ambitious target to reduce energy intensity by 20%, the Chinese government has set targets to reduce annual sulphur dioxide and dust emissions by 10% below their 2005 levels by 2010 (Figure 6.2). Achieving these targets will require special attention be paid to pollution from coal; the combustion of coal accounts for most of China's sulphur dioxide (SO_2) emissions. Between 1990 and 2005, China's SO_2 emissions grew from around 19 Mtpa to 26 Mtpa, making it the world's largest emitter of this noxious gas that can cause breathing difficulties and acid rain. Emissions of oxides of nitrogen (NOx) grew from 7 Mtpa to 15 Mtpa and particulate emissions ($PM_{2.5}$) grew from 12 Mtpa to 14 Mtpa (IEA, 2007).

Figure 6.2 ....China's energy intensity, emissions of sulphur dioxide and particulates with targets for 2010

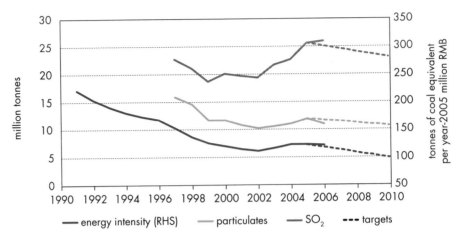

Sources: energy intensity data 1991-2006: NBS (2008); emissions data 1997-2001: SEPA (2001 and 2005); emissions data 2002-06: NBS (2007).

In accordance with the Coal Law 1996, the Environmental Protection Law 1989 and the Water and Soil Conservation Law 1991, environmental inspection, supervision, treatment and protection in the coal sector are mainly the responsibility of the environmental protection agencies, although water and soil conservation schemes are subject to approval by the water resource administrations. In 1998, the State Environmental Protection Bureau was elevated to the State Environmental Protection Administration (SEPA), symbolising the further strengthening of the legal powers held by the environmental protection administrations. In March 2008, SEPA acquired full ministerial status to become the Ministry of Environmental Protection (MEP).

In recent years, environmental protection bodies have enhanced their supervision and treatment of environmental pollution caused by coal development and utilisation. In 2004 and 2005, SEPA successively issued: a *Reply to the Problem of Effluent Discharge from Coal Mine Enterprises,* a *Reply to the Problem of Airborne Coal Dust Pollution,* a *Reply to the Problems of Effluent Discharge and Dust Produced by Opencast Coal Mines and to Proposed Levy,* and a *Reply to the Problem of Effluent Discharge from Coal Gangue on Stripping Soil.* In these replies, SEPA has more specifically defined the levels of effluent discharges from coal production that are subject to levies, thus enabling better legal control over the level of environmental pollution. In addition, in order to implement the Environmental Protection Law, to protect the environment, prevent pollution, assure human health and promote sustainable development of the economy and society, SEPA proposed setting pollutant-emission standards for the coal industry. An opinion-seeking draft was completed and, in April 2005, was distributed to relevant units for study and written comment. In 2006, the standard was publicised by SEPA and its implementation will certainly play an active role in enhancing the enforcement of environmental protection law (SEPA, 2006).

Industry administration

As a result of the changes to the economic planning system and institutional restructuring after China's reform and opening up, the central government's administration of the coal industry has evolved through many stages. In 1988, the Ministry of Coal was dissolved and incorporated into the Energy Ministry. Then, in 1993, the Energy Ministry itself was abolished and the Ministry of Coal reinstated. In 1998, the Ministry of Coal was replaced by the State Coal Industry Bureau under the supervision of the State Economic and Trade Commission. All 94 key state-owned coal mines, established under the direction and jurisdiction of the original Ministry of Coal, were transferred to local governments. In 2001, the State Coal Industry Bureau was incorporated into the State Economic and Trade Commission. Two years later, in 2003, the State Economic and Trade Commission was disbanded and an Energy Bureau was set up under NDRC. According to the Coal Law 1996 and to the State Council's decision on institutional reform, coal industry administration shall be undertaken by a dedicated arm of government. At present, NDRC's National Energy Administration holds the legal powers of the original Ministry of Coal, except for resource administration, safety inspection and environmental protection, which respectively fall under the Ministry of Land and Resources, the State Administration of Work Safety and the Ministry of Environmental Protection.

After the abolition of the Ministry of Coal, NDRC became the central administrative agency for the coal sector. At local government level, coal industry administration is undertaken by a variety of different agencies, with frequent changes and little consistency when naming these agencies. Hence, administration might be the responsibility of local State Economic and Trade Commissions, local Development and Reform Commissions or Coal Industry Bureaus. Since 2004, central government and local governments at all levels have recognised the need to streamline coal industry administration and have embarked on structural changes to this end.

Several provinces have agencies dedicated to coal sector administration, among them Liaoning and Shaanxi. In June 2005, the Liaoning provincial government issued a *Notice on Setting up the Liaoning Coal Industry Administration Bureau* to establish a new provincial coal bureau on the basis of the original provincial Coal Industry Bureau, but with an elevated status. Its staff trebled from 20 to 60, with increases in the number of managers, units and offices – visibly strengthening the management. The functions performed by the new bureau include overseeing the management of provincial state-owned coal enterprises and reviewing management salaries, administering the special coal industry fund, planning coal industry development and carrying out rectification works at small coal mines where these have been compulsorily acquired by the State-Owned Assets Supervision and Administration Commission.

Institutional reforms in Shaanxi province meant that the provincial coal bureau was at one time reduced to just 15 staff, which was grossly inadequate to manage this important industrial sector. In June 2005, Shaanxi provincial government adopted a programme to enhance the coal industry bureau by expanding the bureau and employing more staff. It integrated the provincial coal bureau and coalfield geological bureaus, set up a single office administration system and increased the number of staff to around one hundred while requiring that key coal-producing cities and counties establish independent coal bureaus. New powers related to industry administration were delegated to the provincial bureau, including the management of coalfield prospecting and selection of technologies employed, coal resource management, mine safety inspections and the examination of qualification certificates that certain mine workers are required to hold.

In June 2005, *Opinions on Promoting the Sound Development of the Coal Industry* (State Council, 2005a) included a special section to "strengthen the planning and management, and perfect the supervision system for coal resource development", requiring that Development and Reform Commissions at all levels apply planning, industrial policies, laws and regulations in a comprehensive way to strengthen the supervision and management of coal industry development. At the same time, governments at all levels in the coal-producing regions are instructed by the State Council to strive towards strengthening and perfecting coal industry management and to enhance the administration of coal resources, mine development and coal production.

Future revisions to coal-related legislation and regulations

In its *Opinions on Promoting the Sound Development of the Coal Industry* (State Council, 2005a) the State Council declared that the central government will endeavour to improve the legal and regulatory system for the coal sector. Relevant departments of the Chinese administration have started legislative programmes to draft new laws and regulations, and revise some of the current ones. Major work includes the following.

Drafting of the new Energy Law

Energy-related issues and problems affect the economic development and social stability of China, as well as its national security. Since the reform and opening up of the sector began in the 1980s, the energy industry has developed by leaps and bounds. However, with the rapid growth of energy production and consumption, and with the on-going reform of administrative mechanisms and systems, new problems and contradictions arise in energy supply and demand, safety, efficiency, environmental protection and market structure. China aims to resolve these problems and contradictions to ensure sustainable economic and social development through sustainable energy development.

In developed countries, it is common to solve energy-related problems through the adoption of laws and other legal and regulatory measures. China is the second biggest energy consumer after the US and it aims to establish a legal system for energy matters proportionate to this status. However, the Chinese energy legal system is currently in its infancy; the Electric Power Law 1995, the Coal Law 1996, the Renewable Energy Law 2005 and the Energy Conservation Law 2007 are the main energy-related laws, each adopted to cover legal aspects in particular areas. The country lacks a basic energy law covering the sector as a whole. Hence, the State Council requested that the Office of the National Energy Leading Group establish a group to draft a new Energy Law.[6] This law should aim to solve fundamental, overarching and specific problems in the energy sector as it develops. The legislation should also correct the absence of an

---

6. The Office of the National Energy Leading Group was absorbed into the new National Energy Administration during 2008.

overall structure and add some missing elements to the current legal system. In order to establish administrative mechanisms that meet the new challenges in the energy sector, the drafters of the proposed Energy Law aim to establish a comprehensive legal system that addresses the following:

■ **Energy and economic security.** The Energy Law should give a legislative response to the growing importance to China of the international energy market against a background of economic globalisation, guarantee the stable and reliable supply of energy, and ensure stable national economic and social development.

■ **Energy efficiency and environmental protection.** China is a large producer and consumer of energy, but it should balance development with the need to avoid energy profligacy, giving priority to energy efficiency and environmental protection. Thus, the proposed Energy Law starts from the concept of scientific development to build a harmonious society and proposes a comprehensive legal system to create an environmentally friendly society.

■ **Energy structure and economic growth.** The optimisation of the energy sector and patterns of energy consumption will directly influence the economic structure of China and will require cross-region, cross-industry and cross-department development policies. The government believes that such policies and related systems cannot be implemented effectively until after the new Energy Law is formally adopted.

■ **Energy sector reform and government supervision.** Energy supply and use influence national development, so the government believes it must supervise the behaviour of enterprises and consumers. The scope, extent and means of this government supervision of the energy market should be defined by the new law.

■ **Market access and investment structure.** Proposals demand that the Energy Law should be able to impose requirements on the scope and extent of energy markets, level of energy sector investment, funding for technology development and deployment, environmental protection, work safety and enhancing international energy co-operation.

■ **Technology innovation.** The Energy Law will also boost technical innovation and evolution in the energy sector. Regulations are proposed on energy exploitation and utilisation equipment, on patent protection and for an energy technology development fund that would enhance the ability of the energy sector to innovate independently.

Revision of the Mineral Resources Law

Coal is China's most important mineral resource. The proposed revision of the Mineral Resources Law 1996 will exert a major influence on coal industry development. The law was promulgated in 1986, during the planned economic period, and revised in 1996, strengthening the administration and security of exploration and mining rights. However, more than a decade of intense change has passed since then and the Mineral Resources Law has become inappropriate for the mining sector in China today. The industry faces a series of problems, such as access thresholds that are set too low, thus allowing too many small-scale enterprises with poor mine layouts, antiquated equipment and large overheads that result in severe ecological destruction and environmental pollution. In response, the Ministry of Land and Resources sought opinions on a revision of the Mineral Resources Law in 2003 and is now working to complete the revision and to revise other supporting laws and regulations. These steps are expected to encourage exploration activity and give protection under property rights law to holders of mineral rights.

Revision of the Coal Law

The Coal Law, promulgated in 1996, no longer meets the requirements for sound development of the coal industry. Many prescriptions in the law fall short of current needs, aggravating problems in areas such as coal resource management, mine operation and management, protection of miners' rights and environmental protection. Other problems, such as significant resource wastage and sterilisation, frequent accidents and confused market signals, severely hinder the sound development of the coal industry in China. Therefore, efforts are being made to revise and perfect the Coal Law.

In February 2005, following directions from the State Council and NDRC, revision of the Coal Law began. Those tasked with this work include the Office of Legislative Affairs of the State Council, the Ministry of Land and Resources, the State Administration of Coal Mine Safety, China National Coal Association, China Coal Information Institute and China National Coal Geological Bureau, led by NDRC's Energy Bureau. A drafting group was established under the joint chairmanship of Mr Wu Yin, Vice Director of the Energy Bureau, and Dr Huang Shengchu, President of the China Coal Information Institute – the institute responsible for the drafting process.

The revision of the Coal Law is not only notable for the coal industry, but is also important for national economic and social development as a whole, since it will influence the security of energy supply in China. The revision addresses resource management, work safety, mine operation and management, and environmental protection. After an initial meeting of the drafting group, NDRC issued a bulletin through the internet and held seminars, collecting opinions and suggestions from government, special interest groups, enterprises, non-profit organisations and citizens. The drafting group responded with numerous amendments before completing the draft revision during the first half of 2007. In late 2008, the process leading to a revised Coal Law was still in progress.

Revision of the Mine Safety Law

The Mine Safety Law 1992 is implemented through the *Regulations for the Implementation of the Mine Safety Law* (State Council, 1996). Together, the law and regulations play a significant and positive role in mining safety, accident prevention, personal safety protection and promotion of sound industry development. However, along with economic reforms and the establishment of a socialist market economy, the context of mine safety regulation has changed profoundly. New situations with new problems have arisen. The organisational form of mining enterprises is changing radically, as are the administrative mechanisms for mine safety. Results of investigation and research by the Chinese government reveal that the Mine Safety Law and its implementation rules are dated and no longer satisfy the needs of work safety, so a revision was called for.

In September 2004, leaders of the State Bureau of Work Safety and its State Bureau of Coal Mine Safety held a conference to make initial arrangements for research work and revision of the Mine Safety Law. Before the end of December 2004, the State Bureau of Work Safety issued a *Notice on Carrying out the Research and Revision of the Mine Safety Law* to each province, autonomous region and municipality, and to the Xinjiang Production and Construction Corps' Administration of Work Safety, thus formally starting the revision of the law. From April to August 2005, the Standing Committee of the National People's Congress reviewed enforcement of the Work Safety Law 2002 and, in his review report, the vice-chairman of the Standing Committee suggested reviewing the Mine Safety Law. This assured further attention from relevant departments.

By late 2008, the research work and revision of the Mine Safety Law was progressing, incorporating suggestions and opinions received from across the nation, lessons from international mine safety legislation and practical experience. The work includes a summary of the development of mine work safety legislation in China over more than a decade.

Revision of the *Coal Mine Safety Inspection Regulations*

The *Coal Mine Safety Inspection Regulations of 2000* define the requirements of inspections administered under a vertical management system and play an important role in enhancing coal mine work safety. However, in the process of implementing the regulations, problems arose, mainly in the overlap of authority between central government and local governments, but also because of inconsistency with other laws and regulations. In order to solve these problems, the State Administration of Coal Mine Safety began a revision of the regulations in 2007, with the China Coal Information Institute in charge of drafting work.

An initial survey and research work concluded that the revision should further define the functions of inspection bodies in order to properly establish a coal mine safety management system with a clear division of responsibilities between the state, local inspectors and enterprises, and a separation of inspection and management functions. Meanwhile, China is moving to enhance safety management systems, safety training, detection and elimination of potential dangers, preparation of underground working procedures, adoption of safety technologies and safety inspections – all in accordance with the *Special Regulation on Preventing Coal Mine Work Safety Accidents* (State Council, 2005d). Since there are many conflicting laws, regulations and rules concerning coal mine safety inspection, the revision should consolidate these into a single set of consistent regulations with uniformly applied penalties for non-compliance.

Towards cleaner energy

The energy sector is very much in the spotlight in China, with wide recognition of the need to improve legal texts, regulatory methods and institutional capacity. In addition to the legislative developments in the coal sector and broader energy sector, described above, ongoing reform in the power sector and in environmental protection will determine how rapidly better technologies and practices are adopted by coal-supplying and coal-using enterprises. In the course of designing and carrying out market and regulatory changes, China has sought to absorb the experience of other countries, and is itself becoming a test bed for innovative laws and practices. The continued focus on energy issues by top national and local leaders, along with a growing, grassroots awareness of health, safety and environmental issues seem certain to keep the pressure on for the improvements that new and revised laws promise.

REFERENCES

CCIDRC (中国煤炭工业发展研究中心 – China Coal Industry Development Research Centre) (2007), personal communication, CCIDRC, 国家安全生产监督管理总局研究中心 (State Administration of Work Safety - Supervision and Management Research Center), Beijing, October.

CCII (China Coal Information Institute) (2006), *China Coal Development Report*, CCII, Beijing.

CEMAC (中国经济景气监测中心 – China Economic Monitoring and Analysis Center) (2006), 中国能源产业地图2006-2007 - 发展战略凸现 节能环保备受关注 (*The Industrial Map of China Energy 2006-2007 – Highlighting Development Strategies, Energy Conservation and Environmental Protection*), CEMAC – National Bureau of Statistics, Social Sciences Academic Press, Beijing.

Huang Shengchu, Hu Yuhong, Liu Wenge and Lan Xiaomeri (2004), *China Coal Outlook 2004*, China Coal Information Institute, Beijing.

IEA (International Energy Agency) (2007), *World Energy Outlook 2007 – China and India Insights*, OECD/IEA, Paris.

IEA (2008), *Coal Information 2008*, OECD/IEA, Paris.

NBS (National Bureau of Statistics of China) (2007), *China Statistical Yearbook 2007*, China Statistics Press, Beijing.

NBS (2008), *China Energy Statistical Yearbook 2007*, China Statistics Press, Beijing.

SAWS (State Administration of Work Safety – 国家安全生产监督管理总局) (2005), 《关于清理纠正国家机关工作人员和国有企业负责人投资入股煤矿问题的通知》 ("Circular on the Elimination of Shareholdings in Coal Mines by Officials of State Bodies and State-Owned Enterprises"), 中共中央纪委 (CPC Central Commission for Discipline Inspection) / 监察部 (Ministry of Supervision) / 国务院国有资产监督管理委员会 (State-Owned Assets Supervision and Administration Commission) / SAWS, Beijing, 2005年8月30日 (30 August 2005), www.chinasafety.gov.cn/2005-12/26/content_150918.htm.

SCIO (国务院新闻办公室 – State Council Information Office) (2001), 全国矿产资源规划 2001年4月11日 (*National Program on Mineral Resources*, 11 April 2001), reported in 《中国的矿产资源政策》 ("China's Policy on Mineral Resources"), SCIO, Beijing, www.scio.gov.cn/zfbps/gqbps/2003/200601/t86414.htm.

SEPA (国家环境保护总局局长 – State Environmental Protection Administration) (2001), *Annual Environmental Statistics Report 2000*, China Environmental Science Press, Beijing.

SEPA (2005), *China Environmental Statistics 2004*, China Environmental Science Press, Beijing.

SEPA (2006), 煤炭工业污染物排放标准 (*Emission Standard for Pollutants from the Coal Industry*), GB 20426-2006 部分代替：GB 8978-1996, GB 16297-1996, 2006-10-01

实施 (GB 20426-2006, partly replaces: GB 8978-1996, GB 16297-1996, effective 1 October 2006), 中华人民共和国环境保护部 (Ministry of Environmental Protection - formerly SEPA), Beijing, 1 September, www.sepa.gov.cn/tech/hjbz/bzwb/dqhjbh/dqgdwrywrwpfbz/200609/W020070319523989759335.pdf.

State Council (国务院) (1996), 《中华人民共和国矿山安全法实施条例》 ("Regulations for the Implementation of the Law of the People's Republic of China on Safety in Mines"), 已于1996年10月11日经国务院批准.(approved by the State Council on 11 October 1996), 劳动部令（第4号）, 发布日期：1996年10月30日 (promulgated by Decree No. 4 of the Ministry of Labour on 30 October 1996).

State Council (国务院) (2004), 《国务院办公厅关于完善煤矿安全监察体制的意见》 ("Opinions on Perfecting the Coal Mine Safety Inspection System"), 国办发[2004]79号 (Guo Fa Ban [2004] No. 79), Office of the State Council, Beijing, 2004年11月4日 (4 November 2004), www.gov.cn/gongbao/content/2004/content_63049.htm.

State Council (国务院) (2005a), 《国务院关于促进煤炭工业健康发展的若干意见》 ("Opinions of the State Council on Promoting the Sound Development of the Coal Industry"), 国发[2005]18号 (Guo Fa [2005] No. 18), Office of the State Council, Beijing, 2005年6月7日 (7 June 2005), www.gov.cn/zwgk/2005-09/08/content_30251.htm.

State Council (国务院) (2005b), 《国务院关于全面整顿和规范矿产资源开发秩序的通知》 ("Notice Concerning the Comprehensive Rectification and Standardisation of the Regulation of Mineral Resource Development"), 国发[2005]28号 (Guo Fa [2005] No. 28), Office of the State Council, Beijing, 2005年8月18日 (18 August 2005).

State Council (国务院) (2005c), 《国务院办公厅关于坚决整顿关闭不具备安全生产条件和非法煤矿的紧急通知》 ("Emergency Notice on Effecting the Urgent Closure of Unsafe and Illegal Coal Mines"), 国办发明电[2005]21号 (Ming Guo Banfa [2005] 21), Office of the State Council, Beijing, 2005年8月24日 (24 August 2005), www.chinasafety.gov.cn/2005-08/24/content_140879.htm.

State Council (国务院) (2005d), 《国务院关于预防煤矿生产安全事故的特别规定》, ("Special Regulation on Preventing Coal Mine Work Safety Accidents"), 446号国务院令 (State Council Decree No. 446), Beijing, 于2005年9月3日起施行 (effective 3 September 2005), http://202.123.110.5/zwgk/2005-09/06/content_29440.htm.

VII. COAL INDUSTRY EXPERIENCE IN OTHER COUNTRIES

The adoption of clean coal technologies (CCTs) must encompass the entire coal-supply chain from production to end use, including resource recovery, health and safety, environmental protection and waste disposal. Without progress in all these areas, the continued use of coal as an energy source cannot be considered sustainable. In addition, there are many non-technical issues that must be considered if coal is to retain a stable, long-term role as part of the world's primary energy mix. This chapter examines socio-economic aspects of coal production and coal-sector restructuring in several major coal-producing and -using countries, the legislative frameworks relating to resource acquisition, security of tenure and coal use, the importance of a strong and independent safety-enforcement system, and the need for effective environmental protection measures along the coal production, transport and utilisation chain.

Section 7.1 addresses historical coal production and use in selected countries to demonstrate how coal production and demand grows during industrialisation, then enters a "post-industrialisation" stage in which coal becomes more important for electricity generation than for heat production. Although domestic coal production has been reduced or has even ceased in some of these countries, coal continues to provide a significant proportion of their primary energy needs. The construction of new power plants to meet the electricity demand of China's booming industrial economy and increasingly urban society is comparable to the industrialisation seen in Europe and the United States (US) over the last two centuries, but at an unprecedented speed and scale. With the Chinese government's policy of consolidating coal production in mind, Section 7.2 looks at the experience of coal-sector restructuring in IEA member countries. Since the 1980s, state-owned coal industries that were once deemed critical to ensuring security of energy supplies have been restructured in Western Europe, and those in the transition economies of Eastern Europe followed during the 1990s. In all cases, national government, European or international assistance was needed to stimulate economic regeneration of the coalfields once mines closed. Section 7.2 provides an overview of the measures adopted in Western Europe, in Poland (representing one of the transitional economies in which coal mines have been closed and many jobs lost) and in the eastern US, where the closure of small coal mines during the 1980s and 1990s has great relevance to the situation today in China. Section 7.3 reviews legislative issues relating to coal production and use in IEA member countries, including revenue generation in the form of royalties and resource-acquisition payments, and compliance with environmental standards. Section 7.4 addresses energy policy issues and how these issues can affect the more widespread adoption of cleaner practices. Overall, the chapter aims to show how China can benefit from the wider introduction of clean coal technologies and avoid some of the mistakes that have been made elsewhere in the past.

TRENDS IN LARGE COAL-USING COUNTRIES: PAST AND PRESENT

Coal has provided energy for industrial development worldwide. The replacement of wood by coal provided the foundation for the Industrial Revolution of the 18th century, as a fuel and reductant for iron-making, and for raising steam. This led to major cost savings in the production of metals and allowed the land and sea transport of manufactured goods at much lower cost and with greater reliability than was possible before. Since then, virtually every industrialised country has relied on coal for economic development. During the 20th century, the emphasis swung away from the direct use of coal in industry and transport towards its use as fuel for electricity generation, helped by the growing availability of oil as a transport fuel. The low thermal and mechanical efficiencies of steam engines and small boilers was a major factor in the shift towards the greater use of electricity and alternative fuels in industrial and commercial applications. Since the mid-20th century, coal consumption for residential heating and cooking use has also waned as a result of the greater availability of electricity and natural gas, together with tighter restrictions on atmospheric pollution in urban areas. While household uses were not large in tonnage terms, the change to alternative fuels was significant in terms of the public's perception of coal, as fewer people came into regular contact with it.

This pattern has been common to many industrialised countries, with those industrialising more recently enjoying faster development by adopting technologies already developed elsewhere. A distinction can be made between the older generation of industrialised countries, such as France, Germany, the United Kingdom (UK) and the US, where major industrial development occurred during the 19th century, and those such as Japan and South Korea that achieved spectacular industrial growth in the second half of the 20th century. A third category includes countries where the industrial infrastructure has been shaped by central planning. Examples include the countries of the former Soviet Union and its satellite states in Eastern Europe where, since the early 1990s, the focus has been on modernisation and the replacement of energy-intensive processes with those that are more efficient and environmentally acceptable. The transition from centrally planned to market economies has had major impacts on both industrial output and energy demand. Marked reductions in coal demand were accompanied by the need for coal-mining enterprises to become financially viable. The coal industries in many of these countries today bear little resemblance to the high-cost integrated enterprises of the 1980s. In a fourth group of less-industrialised countries, coal is underpinning strong economic growth. In nations such as South Africa and Australia, established industrial capabilities have been augmented by rapidly rising coal exports. India's burgeoning domestic coal demand reflects the country's growing use of coal in industry and for power generation. Colombia and Indonesia have less well-developed industrial capabilities, but benefit from major coal exports. China itself is following a path comparable to that taken by other nations, in terms of its industrial development and coal demand. What sets China apart is the sheer speed of its recent economic growth and continued heavy reliance on coal.

The remainder of this section reviews coal production and use in selected countries from the groupings described above. The United Kingdom, Germany and the United

States represent long-established industrialised countries, Japan and South Korea the more recent group, South Africa is a developing economy with both domestic and export coal markets, while Poland and eastern Germany are used to illustrate industrial development and coal use in former centrally planned economies. Finally, conclusions are drawn and comparisons made to better understand the situation in China today.

Long-established industrialised countries

United Kingdom

Coal production in the United Kingdom dates from the 16th century or before, when most of the output was used as residential fuel. Wood and charcoal remained the principal sources of energy for industry until coke was first used successfully for iron smelting in the early 18th century. Industrial expansion led to a rapid increase in demand for coal throughout the 19th century; by 1913, output had risen to a peak of 292 million tonnes per annum (Mtpa). Main uses at that time were as a fuel for industry, and for rail and sea transport. There was also a significant export trade, with British coal being shipped worldwide, mainly for use as bunker fuel for ships (Walker, 2000). UK coal production subsequently fell, with an output of around 200 Mtpa by the time the industry was nationalised in the late 1940s (Figure 7.1). Export markets dwindled as coal production began or expanded in other countries, coal was replaced by oil in naval bunkering, and industrial use fell as greater reliance was placed on electricity. While the decline in output accelerated during the second half of the 20th century, demand for coal has remained in the order of 65-70 Mtpa, with imports meeting a growing share of this demand. Changes in the structure of coal use between 1980 and 2005 are shown in Figure 7.1.

Figure 7.1 .... UK coal production and consumption, 1880-2006

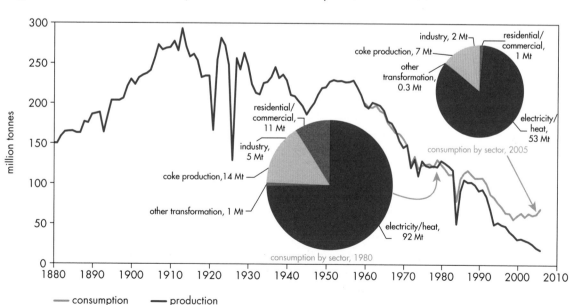

Sources: production data, 1880-1912: Mitchell and Deane (1962); production data, 1913-1959 (DTI, 2001); production and consumption data, 1960-2006: IEA database (World Coal Statistics.ivt available at http://data.iea.org).

According to Vaux (2004), a Hubbert curve can be fitted to historical UK coal production, illustrating the general increasing and decreasing trends in output. Between the production peak in 1913 and the late 1940s, wars and economic depression resulted in lower output than would have been predicted by the curve, which suggests that without these influences production would have reached a maximum in the early 1930s.

Germany

Historically, German hard coal production has been centred on the Ruhr coalfield in the west of the country. Industrial-scale mining began in the mid-18th century, allowing the region to become a major iron-making centre. Small-scale, near-surface operations were gradually replaced by higher-capacity mines at greater depths (Walker, 2000). Production from the Ruhr coalfield stood at around 2 Mt in 1850, 60 Mt in 1900 and 114 Mt in 1913. Production fell during the First World War and during the 1920s, with 73 Mt being produced in 1932, but subsequently resurged to reach 129 Mt in 1940 (Statistik der Kohlenwirtschaft, 2008).[1] Following the Second World War, German hard coal production as a whole (including other coalfields) reached a peak of around 150 Mt in 1956 (125 Mt from the Ruhr), after which there has been a steady reduction as shown in Figure 7.2, which also shows changes in the structure of coal use between 1980 and 2005.

Figure 7.2 .... German hard coal production and consumption, 1900-2006

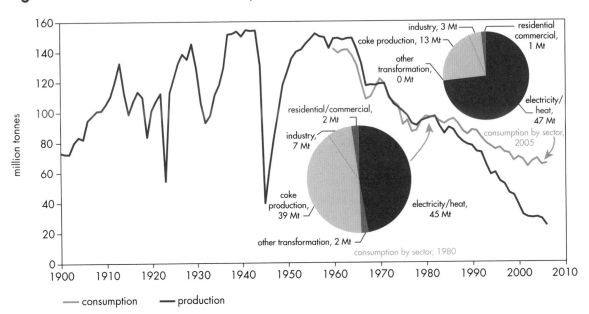

Sources: production data, 1900-1959: Bogalla (2008) and Statistik der Kohlenwirtschaft (2008); production and consumption data, 1960-2006: IEA database (World Coal Statistics.ivt available at http://data.iea.org).

Coal was seen as a strategic energy resource in the Federal Republic of Germany while the country was partitioned, with production being heavily subsidised. Following German reunification in 1990, this need has diminished, with plans now in hand to end subsidies by 2018.

---

1. Data reported in tvF (*Tonne verwertbare Förderung* – tonnes of saleable production).

United States

Industrial coal production began with the exploitation of anthracite resources in northern Appalachia; output exceeded 1 Mtpa by the 1840s, serving local residential and industrial users in the eastern states. Rapid industrialisation and the expansion of the railway network were drivers for substantial growth in coal production during the second half of the 19th century. Westward expansion of the railway system brought with it the need for coal supplies across the nation; production sourced from the Midwest and western coalfields met this need and was also used to fuel local industry (Walker, 2000). US output rose rapidly from around 8 Mt in 1850 to 614 Mt in 1918, after which it fell back as industrial demand slumped during the Great Depression of the 1920s and early 1930s. Recovery from the late 1930s to the 1950s was followed by decline as cheap oil took over a large share of coal's former industrial and transport markets. However, growing electricity demand and the oil price crises of the 1970s led to a resurgence in demand for coal, and although there have been geographical shifts in the centres of production since then, demand has continued to increase. Figure 7.3 shows US output of hard coal and lignite since 1900. Vaux (2004) notes that a Hubbert curve fitted to US coal production would suggest that peak output will be reached in 2030-40, although other factors could alter this significantly. For example, Appalachian coal production is already in decline, while production from the west continues to grow.

Figure 7.3 .... US production and consumption of hard coal and lignite, and coal industry employment, 1900-2006

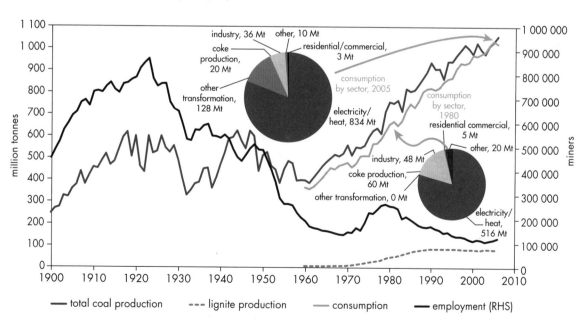

Sources: production data, 1900-1959: EIA (2006a); production and consumption data, 1960-2006: IEA database (World Coal Statistics.ivt available at http://data.iea.org); employment data: MSHA (2007a).

Box 7.1 .... Coalbed methane production in the United States

In 2006, CBM accounted for 50 bcm or 9% of annual natural gas production in the US; proven reserves are estimated to total more than 556 bcm with 4 500 bcm of resources (EIA, 2008). Tax credits received by major energy companies for non-conventional fuel production under the Crude Oil Windfall Profits Tax Act of 1980 were a key factor in the growth of CBM production during the 1990s. Internal Revenue Code Section 29 provided a tax credit of about USD 1/mmBtu for the sale of CBM from wells drilled between 1980 and 1992, and to production from those wells through to 2002. Since then, CBM production has grown on a purely commercial footing and, by the end of 2007, over 60 000 CBM wells had been drilled in the US, of which 19 647 were drilled in the three-year period 2005-07. CBM production is not without technical and environmental difficulties. Dewatering of coal seams, which is necessary to produce the methane gas, means that large volumes of often polluted water must be disposed of safely. To maintain gas flow from the coal seam, it is sometimes necessary to use hydraulic fracturing (hydrofracing) techniques to open up flow paths. Water treatment and hydrofracing add cost, and have a major bearing on the economics of CBM production.

Newer industrialised countries

Japan

Japan industrialised in the second half of the 20th century, but the country's industrial roots stretch back further. By 1900, its coal production was already 10 Mtpa, with a peak production of over 56 Mt being achieved in 1940. Post-Second World War economic recovery was fuelled by coal, although by the 1950s it was being replaced by cheaper imported oil. Japan's coal output in 1960 was 58 Mt, in 1980 18 Mt and in 2000 it had fallen to 3 Mt. However, Japan remains one of the world's major coal consumers, both for industrial uses such as cement and steel production, and for electricity generation, with virtually all of its needs now met by imports. The country has been the world's largest net coal importer since the 1960s, with hard coal imports in 2006 totalling 178 Mt. With such a high reliance on imported coal for electricity generation and coke production, Japan has been at the forefront of developing new technologies to improve efficiency. It has also been a leader in the use of pulverised coal injection (PCI) in iron making, which has helped to limit its import requirement for coking coal. Key features of coal use trends over the 25 years between 1980 and 2005 are the rapid increase in demand for steam coal for power generation (Figure 7.4), and the increase in the use of steam coal for PCI applications and as a blend in coke ovens.

Republic of Korea

Since the mid-20th century, South Korea's growing energy demand had been typical of an industrialising country. However, since it has few indigenous energy resources, its energy needs must be met mainly by imports. Having peaked at 24 Mt in 1988, domestic coal production is now limited to around 3 Mtpa of generally poor-quality anthracite. The country became the world's second largest net importer in the late 1980s, and has retained this position, with coal imports in 2006 totalling 80 Mt. Major

Figure 7.4 ....Japanese hard coal production and consumption, 1960-2006

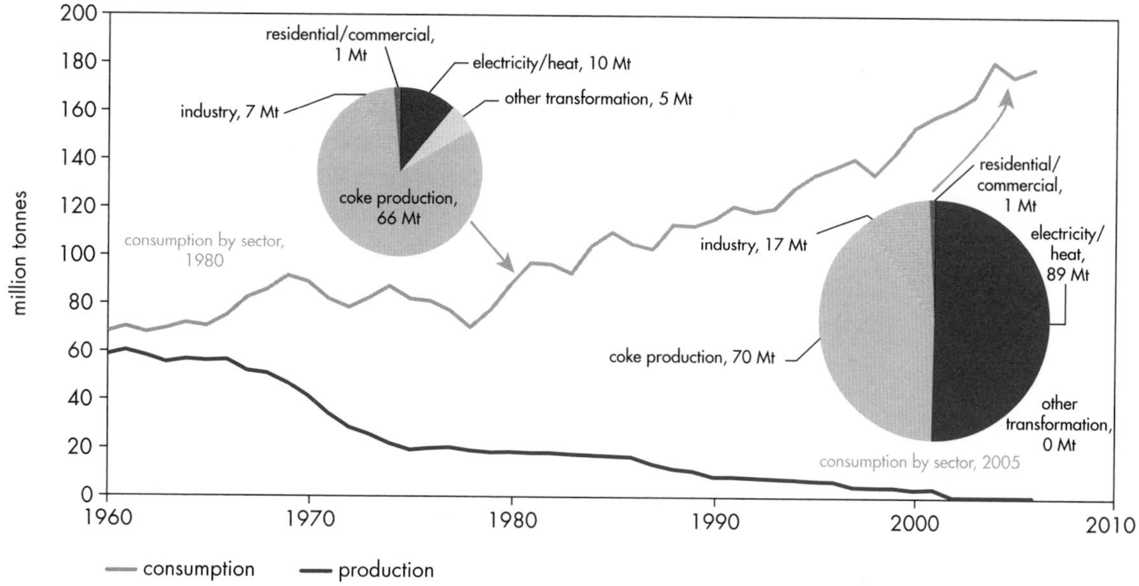

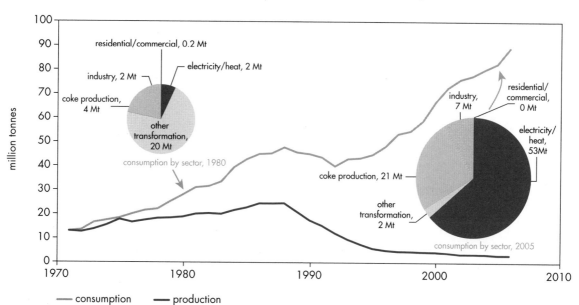

Source: IEA database (World Coal Statistics.ivt available at http://data.iea.org).

trends in coal consumption since 1980 have been the increase in demand for electricity generation and for industrial use, with cement and steel production having grown substantially during this period (Figure 7.5).

Figure 7.5 .... South Korean hard coal production and consumption, 1971-2006

Source: IEA database (World Coal Statistics.ivt available at http://data.iea.org).

A developing industrial economy

South Africa

Unique in Africa in terms of its industrial development, South Africa is both a major coal user and a major steam coal exporter. Development of the country's coal resources followed diamond and gold prospecting; coal production was already over 3 Mtpa at the start of the 20th century. Output rose to around 25 Mtpa by 1950 as demand increased from both the industrial and power generation sectors (Walker, 2000). The country's lack of commercial oil resources coupled with 30 years of political isolation, from the 1960s to the 1990s, provided the stimulus for large-scale coal development, including the use of coal as a feedstock for synthetic oil production. Coal production reached around 50 Mtpa by the mid-1960s, doubled during the 1970s and continued to soar in the 1980s and 1990s as exports increased. Staged development of its port facilities allowed coal exports to expand while domestic demand slowed in response to external political pressures during the 1980s, leaving the country with surplus coal-fired electricity generation capacity. Since the resumption of multi-party politics in the 1990s, coal initially faced increased competition from natural gas and imported hydroelectric power from Mozambique, although general economic growth and the expansion of the power supply network have since led to greater coal demand. Historical production and coal exports since 1900 are shown in Figure 7.6.

Figure 7.6 .... South African coal production and exports, 1900-2006

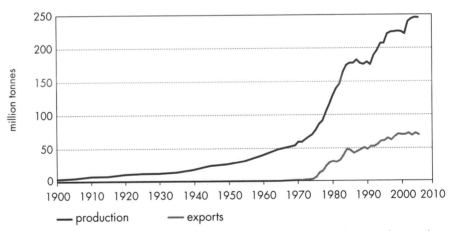

Sources: production and export data, 1900-1970: Minerals Bureau (1988); production and export data, 1971-2006: IEA database (World Coal Statistics.ivt available at http://data.iea.org).

Transition economies

Poland

From the late 1940s until the end of the 1980s, Poland was a Soviet satellite state within the Eastern Bloc. Socialist economic development during this period placed an emphasis on heavy industrial growth fuelled by domestic coal production. Coal resource exploitation centred on the Upper Silesia coalfield, with output of both steam and coking coal. Coal exports to other Eastern Bloc countries were later expanded to include sales to consumers in Western Europe and beyond. In the 1970s, the country was exporting around 40 Mtpa and exports briefly peaked again in 1984 at 43 Mt before falling (Figure 7.7). Poland remains heavily reliant on coal for electricity generation (92% in 2005), with very limited alternative resources and no nuclear generation. Privatisation of its steel industry led to major modernisation programmes, with reduced demand for coking coal. Energy demand fell sharply during the early 1990s as industrial output slumped, but subsequently recovered in response to a general economic upturn. The country's accession to the European Union in 2004 requires stricter controls on emissions from all large coal-fired plants, and thus will speed the introduction of clean coal technologies over an agreed period – full implementation of the EC Large Combustion Plants Directive (2001/80/EC) being required by 2017, with specific dates agreed for particular installations and pollutants.[2]

Figure 7.7 ············ Polish production of hard coal and lignite, hard coal exports, and coal industry employment 1945-2006

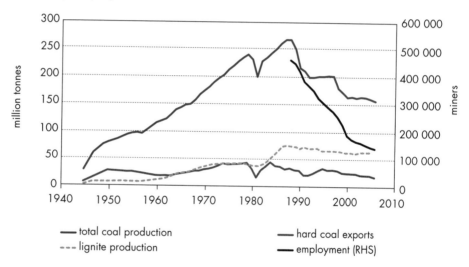

Sources: hard coal production and exports, 1945-1957: NCB (1958); lignite production, 1945-1959: Kasztelewicz (2006); production and export data, 1960-2006: IEA database (World Coal Statistics.ivt available at http://data. iea.org); employment data: Piekorz (2004); Mining Annual Review (2000-2007); and, Kasztelewicz (2006).

Eastern Germany Although eastern Germany produces lignite, not hard coal, it is included here as a further example of the impact of economic restructuring on energy demand. Environmental considerations played a major role in forcing the adoption of new emission standards on mining and power generation operations that had been established under the former centrally planned economic system (Walker, 2001). The former German Democratic Republic (East Germany) was heavily reliant on lignite for power generation. Following a reduction in oil supplies from the Soviet Union in the early 1980s, lignite production was expanded substantially, reaching a peak output of 312 Mt in 1985. Reunification of the two parts of post-Second World War Germany in 1990 resulted in the collapse of eastern Germany's industrial economy. The decline in lignite output in this part of the country did not end until 1999, by which time it

---

2. Official Journal of the European Union, OJ L 236, 23 September 2003, pp. 900-903.

was producing around 65 Mtpa of power station fuel (Figure 7.8). Privatisation of the region's lignite mining and electricity generation industries has led to major investment in new technologies, including the introduction of clean coal power plants to meet European Union emission standards.

Figure 7.8 .... German lignite production, 1950-2006

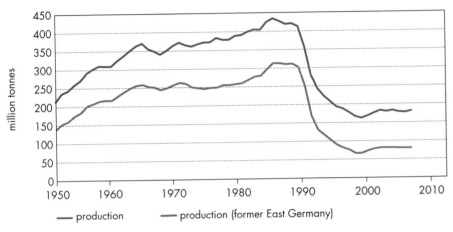

Sources: production data, 1950-1959 and for former East Germany: Statistik der Kohlenwirtschaft (2006); production data, 1960-2006: IEA database (World Coal Statistics.ivt available at http://data.iea.org).

Conclusions and relevance for China

Across the world, economic development and coal use within industrialised countries have followed a largely predictable path: a steep increase in coal demand, followed by a gradual decline in industrial use and growth in coal use for electricity generation. Countries such as Germany and the UK show a clear shift from industrial and residential use of coal to its use for power production, with imported coal meeting an increasing share of demand. In the US, growing electricity consumption creates demand for domestic coal production, although the focus of coal mining has shifted from the mature eastern coalfields to those in the Midwest. The overall change from direct coal use to indirect use via electricity is a common trend. In countries such as the Republic of Korea and Japan, an almost total reliance on imported coal has led to innovations to improve utilisation efficiency and minimise coal demand. Transition economies, such as Poland and the eastern part of Germany, illustrate how stability can return to coal production after the initial shock of economic restructuring. Rationalisation of production capacity into cost-effective units plays a key role in this process, as does the removal of non-core activities from the coal-mining companies. As a developing economy with an established industrial base, South Africa shares several common features with China, such as its reliance on coal for power generation. Its experience also illustrates how coal can be used to produce synthetic oil, a route that China is itself developing, although the process generates greater carbon dioxide emissions than does conventional petroleum refining.

In comparison with the long-established industrialised countries, where coal use has risen, peaked and then often fallen back, China is in the strong growth phase of a recognisable pattern (Figure 7.9). The closest comparison can be drawn with long-term trends in the US, where, since the post-Second World War period, coal demand has risen steadily, especially for power generation, largely met by indigenous production. In China, similar trends can be seen: industrial use is continuing to grow while household use remains important, especially in rural areas, but it is the strong coal demand for power generation and steelmaking that has led to unprecedented growth. From a starting point of 1 563 Mtce consumption in 2005, the IEA *World Energy Outlook 2007* Reference Scenario shows China's coal consumption rising to 3 427 Mtce in 2030 (over 4 500 Mt). Given the country's large coal resources and the doubling of output to 2 482 Mtpa between 2000 and 2006, it is not unrealistic that this level of demand could be met by indigenous production, although the Reference Scenario also shows rising net imports.

Figure 7.9 ....Chinese hard coal production, 1949-2006

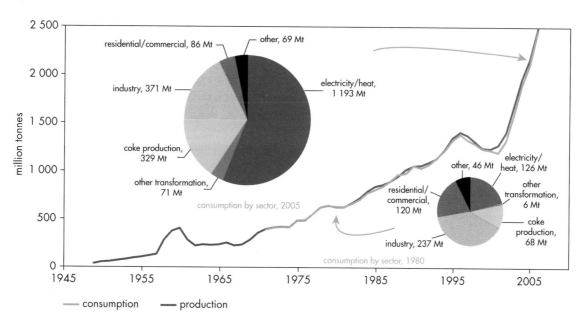

Sources: production data, 1949-1970: CCII; production and consumption data, 1971-2006: IEA database (World Coal Statistics.ivt available at http://data.iea.org).

Post-industrial countries in Western Europe and North America in particular, together with countries with transitional economies, have shown that primary energy intensity can be reduced. With demand for coal continuing to increase, and with coal projected to retain a dominant position within its primary fuel mix, there are huge opportunities for China to bypass many of the developmental stages through which other countries have passed. China's ambitious target of a 20% reduction in energy intensity between 2005 and 2010 is an important response. The transfer of clean coal technologies from elsewhere, and their development within China, offer both the prospect of better resource utilisation and greater conversion efficiencies. Although R&D must continue, there is no reason why China should have to wait for the traditional development cycle to run its course when practical shortcuts exist. The adoption of existing state-of-the-art technologies and their adaption to Chinese conditions will enable China to benefit much sooner than previous experience elsewhere would suggest.

TACKLING TECHNICAL AND SOCIAL ISSUES DURING COAL INDUSTRY RESTRUCTURING

As described above, by the mid-20th century the direct use of coal in industrial processes and for powering transport became less important, and electricity generation became the main use of coal. In addition, coal faced increasing competition from oil, nuclear power generation and natural gas. Where the coal and power generation sectors were in state control, as was often the case in Western Europe during the second half of the 20th century, markets for coal were assured. This was less apparent in the US, where both coal production and power generation have traditionally been in private-sector ownership, although long-term supply contracts helped to maintain market stability for both producers and consumers of coal. In Western Europe, fuel competition and the move to large-scale, coal-fired power generation led to the consolidation of coal production into larger, more efficient units, with fewer employees. Thus, during the second half of the 20th century direct employment in hard coal mining in Western Europe fell from around 1.8 million to fewer than 100 000. In the US, the introduction of new health and safety legislation, together with falling coal prices as large producers became more efficient, forced many smaller operations out of business. The effect was to reduce the number of jobs to around 105 000 in 2003, a fall of around 156 000 over a 25-year period, with few alternative employment opportunities for redundant miners. Another major factor in Western Europe was the change in governments' perceptions over the security of energy supply following the collapse of the Soviet Union. In Germany, heavily subsidised hard coal and inefficient lignite production suffered production cuts once alternative energy supplies could be assured. Imports of both natural gas and coal now account for a greater proportion of the country's primary energy supply than was the case from the 1950s to the 1980s.

Thus, major coal industry restructuring has taken place in a number of IEA member countries over recent decades, notably in Western Europe, in response to a political unwillingness to continue public-sector support of coal production. From the 1980s, another key factor was the gradual liberalisation of the European electricity supply industry, which by then had become the principal consumer of coal and favoured cheaper imported coal over higher-cost domestic coal. Coal sector restructuring has by no means been confined to Western Europe. In Japan, for example, domestic coal production has been phased out, state-supported coal mining in eastern Canada has ended, and the focus in India has been on replacing largely un-mechanised, labour-intensive underground mines with new, more productive surface operations. In addition, since the early 1990s the transition economies of Eastern Europe have reduced state support for their coal industries, resulting in widespread reductions in labour requirements.

In each case, restructuring has had major impacts on the social and economic conditions of the communities that formerly depended on coal mining, both directly and indirectly, for employment. The approaches taken by national governments have varied widely, ranging from extremely structured, strongly paternalistic, long-term programmes of social and economic support, to the use of redundancy payments with little provision for the future well-being of former mining communities. Yet, there are strong economic and social reasons for promoting regeneration in economically depressed regions, such as former coal mining regions.

Governments have to strike a balance between maintaining high-cost coal production and funding long-term economic regeneration programmes. In the case of Western Europe, European Union policy has been to force national governments to reduce or eliminate production subsidies as one element of a broader policy of creating free-market conditions across the EU. Many governments have had no choice but to close uneconomic mining capacity. Where high-cost mining has continued, as in northern Spain, this has been seen as a better long-term option for the communities involved than providing social support in areas with few prospects for alternative employment.

The remainder of this section reviews coal-industry restructuring in IEA member countries, with a focus on the hard coal sector, since this has most relevance to the situation in China today and hence to the conclusions drawn here. The reported experience in Germany, France, Poland and the United Kingdom during the second half of the 20th century highlights differences in the policies that were adopted. A summary of experience in Belgium, the Netherlands and Spain is also included. Experience from the United States offers considerable insight into the social impacts of small mine closures when there is less state involvement than seen in Europe. Programmes that have been implemented in all the countries to support communities affected by mine closures are examined, including their costs.[3] It should be noted that in Western Europe, support was provided by the European Community through its regional development programmes of targeted investment in areas affected by major industrial restructuring. The accession of a group of Eastern European countries to the European Union in 2004 led to a refocusing of this support to other regions within the EU. Aid to the Western European coalfields, such as those in South Wales and northern France, was redirected to poorer, more deserving regions such as Upper Silesia in Poland and Ostrava in the Czech Republic.

Coal industry restructuring in European IEA member countries

Germany

Between the end of the Second World War and 1990, the two halves of Germany followed fundamentally different political and economic paths. In the Federal Republic (West Germany), energy security concerns initially ensured that hard coal production was given high priority, although at a substantial economic cost. In the Democratic Republic (East Germany), early dependence on oil from the Soviet Union was replaced

---

3. The investment required to implement regeneration programmes is shown in US dollars. These amounts have been converted from local currencies at exchange rates of the day (historical dollars).

by increasing reliance on domestic lignite, especially once Soviet oil shipments began to dwindle after the late 1970s (Hauk, 1981).

Between 1957 and 2000, the number of hard coal mines in West Germany decreased from 153 to 12; by 2007, the number had been cut to eight. From a post-Second World War peak of 151 Mt in 1956, production has fallen steadily to just 24 Mt in 2006. In East Germany, lignite production rose steadily during the 1970s and 1980s, reaching a peak of 312 Mt in 1985. Following German reunification in 1990, demand for lignite slumped to 64 Mt in 1998, with a slight recovery to 78 Mt in 2006. This contrasts with the situation in the west of the country, where the demand for lignite from the low-cost, highly mechanised mines operated by RWE has been maintained to supply state-of-the-art power plants (Figure 7.8).

The impact on employment in both the hard coal and lignite mining industries was significant, although over different timescales. Employment in German hard coal mining fell from 600 000 in the mid-1950s to 58 000 in 2000 and to 35 400 by 2006. In former East Germany, employment in the lignite industry fell from 139 000 in the late 1980s to around 10 000 at the end of the 1990s and 8 100 in 2006.

Hard coal played a critical role in supplying West Germany's energy needs in the years immediately following the Second World War. However, by 1957 government policy had changed in favour of increased reliance on cheaper imported oil, and the hard coal industry entered a crisis period since production capacity was significantly greater than falling demand. Between 1958 and 1970, two-thirds of the hard coal industry's employment was lost, the total falling from 600 000 to 200 000 – a rapid reduction in workforce that was often uncontrolled and uncoordinated (Hessling, 1995). In response to the crisis in the Ruhr coalfield, amalgamation of 26 out of the existing 32 coal companies resulted in the formation of Ruhrkohle AG (now RAG) in 1968 – a private-sector company jointly owned by the leading German steelmakers and electricity generators. Mine closures coincided with a period of strong economic growth in Germany, such that there was a shortage of labour in many other sectors, including construction. New local employment opportunities included an Opel car plant and there was considerable migration of former miners to find alternative work in other parts of the country. There was minimal assistance at this stage from the federal and state governments. Special arrangements between Ruhrkohle and its principal customers – the steel and electricity supply industries – provided long-term market guarantees for its output, albeit at a significant cost in terms of state subsidies.

The high cost of German hard coal production resulted in part from the social burden and legacy costs of environmental clean-up borne by RAG; in other countries, these are often the responsibility of the state. Although the average cost of production in the company's Ruhr coalfield mines had fallen to around USD 120/t by the end of the 1990s, this was around four times world market prices. Under an agreement that ran from 1997 to 2005, the German government paid a reducing annual subsidy to the coal industry, falling from EUR 6.7 billion (USD 8.7 billion) in 1996 to EUR 2.7 billion (USD 3.4 billion) in 2005 (Frondel *et al.*, 2006).

In 2007, the German federal government, the state governments of Nordrhein-Westfalia and the Saarland, RAG and mine worker unions reached an agreement to end subsidised hard coal production by 2018, subject to a review in 2012 when Germany's energy supply requirements would be re-assessed. Despite the remarkable rise in international coal prices (Figure 3.10), German hard coal production for power generation remains uncompetitive, with costs averaging around USD 250/t in early 2008. Some coking coal production may be profitable. Assuming that subsidised production does come to an end in 2018, this will mean the loss of the remaining 35 400 direct jobs in the hard coal industry, plus additional employment losses in supplier companies.

Since the 1970s, a deterioration in the general economic situation in Germany resulted in fewer alternative job opportunities, so the governments and industry developed "socially acceptable" schemes in which miners made redundant were guaranteed alternative work, offered "pre-early" retirement, or were financially compensated. In the period from 1972 to 2000, an estimated 120 000 former mine workers took early retirement. Under the schemes, the first five years of early retirement benefits are paid by the federal and state governments in the form of adaption assistance, with monthly payments in the order of USD 1 600 for former underground miners and USD 1 330 for surface workers. After five years, a normal miners' pension is paid, which provides a slightly higher income, and about one-third more than a general workers' pension. The aim is for retired ex-miners to receive between 80% and 90% of their former net salary. Those taking early retirement also receive a lump-sum payment of up to USD 40 000, the exact amount depending on their length of service.

By 2006, the average age of the remaining workforce was 43 years, with over 63% aged between 40 and 50. Thus, early retirement will continue to be the main tool for compensating job losses in the future, with underground workers eligible from the age of 50 and surface workers from 58. RAG continues to provide apprenticeships to young trainees, but with an emphasis on skills that will be transferable to other industries, and younger members of RAG's workforce understand that they will have to find other employment as and when the remaining mines are closed.

The concept of the "employment company" has been used in both the hard-coal and eastern lignite-mining districts of Germany as a means of helping those left unemployed. Established as autonomous bodies, and generally operated by both industrial companies and local authorities, these organisations have provided training and employment on a temporary basis for individuals seeking employment or awaiting retirement. In the hard coal sector, RAG has provided this function.

One of RAG's strategies for handling the continuing decline in the coal sector has been to refocus its activities into other business areas that assist Nordrhein-Westfalia to reduce its dependence on coal. Thus, RAG subsidiaries operate in the fields of energy provision, specialist chemicals, environmental services, engineering and technology, coal trading, real estate and training. In some cases, mine sites have been converted to new uses (Figure 7.10). In the Saarland, Saarbergwerke AG has followed a similar strategy. The agreement in 2007 over the future of hard coal production in Germany will see RAG split into two, with its subsidised coal-mining operations separated from its other activities.

Vocational training is one of RAG's major contributions to creating employment opportunities in the former coal-mining communities. The company trains around 12 000 mainly young people annually, both in Germany and in other countries. Although RAG's own training requirements for its mining operations are now limited, its training facilities provide the wider business community with a level of pre-apprenticeship training that few small- or medium-sized enterprises would otherwise be able to sustain. Training in management and administration is available in addition to technical or trade skills. RAG also provides retraining to ex-miners and helps to find suitable new employment after re-skilling has been completed (RAG, 2000). The costs of retraining are borne in part by government agencies and in part by the company itself, through internal cross-subsidy from its profitable non-coal activities. In some cases, training is tailored to the specific requirements of new employers, such as German railway operators. There is also a partly government-funded "crafts" training programme under which ex-miners can obtain "hands-on" vocational training; after six months in what is effectively a "later-life apprenticeship", trainees can either move to their new trade or return to RAG for alternative retraining. Trainees are eligible for a lump-sum payment of between USD 20 000 and USD 25 000, dependent on their length of service. Around 60% of those choosing this scheme are successful in finding alternative employment, with the government providing a salary "top-up" for the first two years to compensate where wages are lower than in the mines.

Figure 7.10 .... Zollverein coal mine industrial complex, Essen, Germany

Once one of Europe's largest mines (3.6 Mtpa) and coking works, the mine closed in 1986 and is now a UNESCO world heritage site housing a museum, cultural centre, restaurant and exhibition halls.

Despite the success achieved by the various schemes in offsetting the impact of mine closures, the situation in some towns in Germany's hard coalfields remains problematic. Unemployment remains higher than national and regional averages in some parts of the Ruhr coalfield, especially in areas where the most recent round of mine amalgamations and closures has taken place. In 2006, a report from the Ruhr Regional Association confirmed that many of the jobs that had been lost in the coal and steel industries in the region between 1999 and 2004 had not been replaced, leading to long-term, high unemployment rates (RVR, 2006).

France

Since 1974, following the first oil price shock, France's energy policy has centred on the use of nuclear power. Demand for coal fell from the 1960s onwards and the last deep coal mine closed in 2004. Management of this long-term decline was undertaken by the state-owned Charbonnages de France (CdF), a company established in 1946 when coal production was nationalised. Production fell from 56 Mt in 1960 to 18 Mt in 1980 and to 5 Mt in 1999, and finally ceased five years later – a year ahead of the planned closure schedule – partly because too few skilled people remained to ensure safe operations. Employment reduced steadily from around 350 000 at the end of the Second World War to 216 800 in 1960, 60 900 in 1980 and 9 000 in 1999. In the 1950s and early 1960s, jobs were lost largely through natural wastage as miners retired or chose to leave the industry. By the early 1960s, CdF entered into the first of a series of agreements with its workforce on the terms and conditions for redundancy or retraining, followed later by *Plans Sociaux* (Social Plans) for the Nord-Pas de Calais coalfield.

At the same time, the French government set about diversifying the mining regions by creating regional development agencies and providing finance for new ventures – complementing the measures being adopted by CdF. These moves culminated in 1967 with the creation of the special industrial development organisation, *Société Financière pour favoriser l'Industrialisation des Régions Minières* (SOFIREM), a wholly owned CdF subsidiary with responsibility for promoting and supporting the development of alternative industrial employment in the mining regions.

The strategy adopted by CdF was to end production in individual coalfields across France, such that the remaining capacity became concentrated in the north of the country. The impact of mine closures on local communities varied from location to location. Some in the south of France, where the economy was primarily based on agriculture and where coal-sector employment was less significant in overall terms, were much less affected than the industrial areas of northern France. For example, the number of mineworkers in the Nord-Pas de Calais coalfield halved between 1960 and 1970, then reduced by a further two-thirds over the next ten years, reaching just 26 000 in 1980. Mining ended in 1993 and unemployment remains higher in the former mining areas than elsewhere in the region. The period of heavy job losses, in the 1960s and early 1970s, coincided with major restructuring in the area's steel and textile industries. It was against this background that CdF introduced its first programmes, aimed at finding new employment for younger ex-miners. Few major employers came into the area, the state-owned Renault car-maker being one exception, and job opportunities were mainly at small- and medium-sized enterprises. During the 1990s, Toyota also opened a new car plant, against the general trend of limited industrial development in the area. In the last coalfield to be worked, Lorraine (a continuation of the Saar coalfield in Germany), coal industry employment shrank from 43 300 in 1960 to 24 000 in 1980 and 7 250 in 1999, before the last mine closed in 2004.

Introduced in 1986, *Plans Sociaux* represented a new concept: a comprehensive package of measures to offset the social and economic effects of industrial decline. Whereas previous measures had helped younger miners to leave the industry and provided incentives for older workers to take early retirement, the *Plans Sociaux* provided support for those directly affected by mine closure (Drouard, 2000). In each case, the scheme was introduced two years before mine closure was scheduled, giving those involved time to make choices about their futures, as well as providing time for retraining. A major change was the introduction of the *congé charbonnier de fin de carrière* (CCFC), allowing older miners aged over 45 and with 25 years' service the right to remain CdF employees, but effectively on an "end-of-career holiday" until reaching retirement age. Early retirement was available for workers aged 50-60, and normal retirement under the state scheme from age 60. Benefits under the CCFC included 80% of net salary plus the continued provision of subsidised housing and heating. Under the *Plans Sociaux*, CdF had an obligation to find new employment for its younger miners, who could return to the company within two years if the new job did not prove satisfactory. This obligation ended in the 1990s, once the decision had been made to cease coal mining in France.

A key principle adopted by CdF during its closure programmes was the concept that it is better to spend money on training for new jobs rather than on lump-sump redundancy payments. Individuals who preferred to seek new work on their own, or to undertake longer training courses, were entitled to "adaption leave" for up to one year, with the company paying 65% of their previous net salary plus normal benefits (Franz, 1994). CdF operated its own training centres and funded training for its former employees from outside agencies. The success rate achieved in terms of training for long-term future employment depended to a certain extent on the prevailing economic conditions. For instance, during the general economic "boom" years of the late 1980s, there was high demand for workers for transport, construction and similar occupations. In the 26 years from 1973 to 1999, CdF spent a total of over USD 800 million on training for its employees. The proportion of its workforce that underwent some form of training varied from 25-30% at the start of the period to around 70% at the start of the 1990s. Between 1984 and 1999, 13 200 underwent some form of training, 19 400 took retirement and 12 200 left the industry voluntarily.

Aside from providing training, CdF had a major input into attracting alternative employment opportunities to coalfield areas. In Lorraine, SOFIREM's initial promotion activities during the 1970s helped create around 7 000 new jobs, with a further 5 000 during the 1980s. New, large-scale employers included companies manufacturing heating equipment (Viessmann), vehicle tyres (Continental), glass (Interpane-Pilkington), pumps (Grundfoss), automotive parts and Smart cars (Daimler-Chrysler). Overall, SOFIREM and another organisation, FINORPA (in the Nord-Pas de Calais), assisted in the creation of around 100 000 jobs in the French coalfields with an estimated investment of USD 500 million – more than sufficient to absorb the number of former miners who remained in the job market. Typical costs in the late 1990s were about USD 8 800 per job created.

Poland

With the largest hard coal mining industry in Europe, Poland provides a valuable guide to industrial restructuring during a period of change from a command to a market economy. One of the principal features of heavy industry in centrally planned economies was the integrated nature of the operations, with a single company having much greater social and community responsibilities than would be found elsewhere in the world. Thus, a coal company's remit might include housing, medical, social and educational provision for its employees, whereas in Western European countries, for instance, housing was often provided but other services were the responsibility of the local authorities. In Poland, this provision of services has had to be separated from business as an integral part of the restructuring process, thereby placing additional administrative and funding demands on local, regional and national governments.

With unemployment still unacceptably high within some mining communities, the focus in Poland has been on the development of small and medium enterprises (SMEs) and on improving the educational and skills base of the workforce. In comparison with Western Europe, Poland still offers the advantage of low labour costs, and this has attracted some larger employers. Mining regions have also benefited strongly from European Union development funding.

The current restructuring of Poland's coal industry provides some relevant experience and guidelines for China; these are discussed fully in Annex V. However, Poland is dealing with the geographically concentrated impacts of closing a small number of large coal mines, whereas China is faced with the broader impacts of closing a large number of small mines.

United Kingdom

When coal production in the United Kingdom reached a peak in 1913, there were 1.1 million workers employed (DTI, 2001). When the industry was nationalised in 1947, 707 000 workers were employed in 1 542 mines of which 958 came under the newly established National Coal Board. The coal industry became strongly unionised and was able to bring down an elected government during the 1970s through strike action, ostensibly for better pay. A year-long, national strike in 1984-85 saw much civil unrest among miners who believed they were fighting for their jobs as pit closures accelerated. Eventually, the government was able to better manage the industry's contraction through various assistance schemes and regional regeneration projects, but the pain and bitterness of the earlier period lives on. By 1992, just 50 mines were left in operation, employing 28 000 people. In 2007, with the industry back in private ownership, six deep mines and 28 opencast sites remained with a workforce of around 3 600 (BERR, 2007).

The first phase of restructuring, from the late 1940s until the late 1970s, mainly reflected the falling domestic market for coal. It resulted in a steady reduction in the number of deep mines and while huge numbers of people left the industry during this period, there was little provision for socio-economic support. Job losses were handled through a combination of voluntary redundancy, early retirement and internal transfers within the National Coal Board (later British Coal). Typically, 60-70% of workers took redundancy, with the remainder preferring to transfer.

Of much greater significance, in terms of the need to provide alternative employment, was the second phase of mine closures during the 1980s and early 1990s. Over 275 000 jobs were lost, leading to widespread, long-term unemployment within many former coal-mining communities, despite assistance schemes and regional regeneration projects. British Coal's voluntary early retirement scheme, that operated up to 1990, offered miners aged 60 or over with at least two years' service a lump-sum payment and enhanced unemployment benefit until the normal state retirement age of 65. The threshold was subsequently reduced to age 50. The total cost of early retirement payments between 1967 and 1987 was USD 4.0 billion, with on-going costs of around USD 100 million annually to support the scheme. These costs and redundancy payments rose steeply, such that between 1985 and 1995 almost USD 10 billion was dispersed. During the final round of mine closures, transfer opportunities were virtually non-existent, and the potential for early retirement had been exhausted. British Coal then introduced a series of increasingly attractive lump-sum payment schemes aimed at inducing progressively younger miners to leave the industry voluntarily. Based on the individual miner's age and length of service, payments ranged up to USD 60 000 for older, experienced men in the late 1980s.

While mine workers who took early retirement had some level of financial security, those that chose to seek alternative work often experienced difficulties. Indeed, social research undertaken in the early 2000s indicates that, while a significant number of new jobs were created in many of the coalfield areas since the mid-1990s, there is still a shortfall that leads to high, localised unemployment. Even more significantly, it is estimated that the unemployment rate was around five times the official rate in the coalfield areas, since many people had been transferred off unemployment benefits onto other state security systems such as health-related incapacity benefits (Beatty *et al.*, 2005). These authors estimated that 60% of coal-industry jobs had then been replaced, leaving a shortfall of around 90 000 jobs for which new employment opportunities were still needed.

Many of those who lost their jobs during the restructuring of the UK coal industry had poor educational standards. This made it harder to acquire new skills, while existing skills were often rejected by prospective employers as being inappropriate or inadequate. British Coal's principal training and counselling service, the Job and Career Change Scheme (JACCS), helped ex-miners identify alternative employment opportunities, in addition to those available from the normal state employment service. JACCS offered a structured approach:

- establishment of a "job shop" at the mine to be closed;
- an interview for each employee with a trained counsellor;
- identification of the employee's skills;
- search within the local area for employers seeking such skills; and
- matching available skills to existing job vacancies.

Under the JACCS programme, retraining was only offered if this would improve the chances of an ex-miner finding a new job. During the period from the mid-1980s to the early 1990s, about 50% of redundant miners used the service, with an average cost per placing, including retraining where necessary, of USD 2 800 (Pickering, 1995).

British Coal Enterprise (BCE) was established in 1985 with the aim of helping – through work with other agencies – to create jobs and stimulate economic regeneration in the coalfield areas. BCE was organised on a regional basis, each covering a major coalfield. Its activities centred on:

- business funding;
- support for enterprise agencies;
- managed workspace and small business development;
- outplacement services; and
- marketing its expertise in the UK and Europe on a consultancy basis.

By early 1994, BCE claimed to have, "helped to create over 106 000 job opportunities" at a cost per job ranging from USD 700 for the provision of workspace to USD 1 900 for each outplacement and USD 2 500 for its loan services. However, BCE was not acting alone in attempting to attract new employment to the coalfields, and "job opportunities" do not necessarily convert into well-paying alternative work. Other estimates of the organisation's impact suggest 7 000 to 24 000 new jobs, with a cost per job of between USD 6 900 for workspace provision and USD 26 550 for outplacement and retraining, comparing favourably with other job-creation agencies (Fothergill and Guy, 1994).

Box 7.2 .... Responsibilities of the UK Coal Authority

The Coal Authority, established under the Coal Industry Act 1994, is responsible for the proper exploitation and licensing of the UK's coal and CBM/CMM resources, while providing information and addressing coal mining liabilities. It is under a statutory duty to provide public access to all the nation's coal mining abandonment plans, dating back to the 18th century, so that subsidence and other risks can be assessed, and potential new coal-mining areas identified (Figure 7.11). To facilitate this, all mining plans have been digitised. The costs of repairing subsidence damage, managing tips and treating pollution from mining activities before the industry's privatisation in 1994 is now a significant part of the Authority's GBP 41 million (USD 81 million) annual budget; it has been able to make good all accepted subsidence-damage claims and has a programme to treat water discharges from abandoned coal mines. The Coal Authority is accountable to Parliament. Its annual reports and full details of every aspect of its work and performance can be found on its website.[1]

---

1. www.coal.gov.uk

Figure 7.11 .... Managing the UK's coal mining legacy

Mine-water discharge treatment at Pool Farm to remove ochre (left) and plans of mine workings dating back to before 1800 (right) are the responsibility of The Coal Authority in the UK.

Following its rationalisation, the UK coal industry was privatised in 1994, with regulatory functions vested in The Coal Authority (Box 7.2), and the country no longer attempts to meet domestic coal demand solely from domestic sources. Under

the same free-market policies, all major coal users, including the power generation and steel industries, have been privatised and are now almost entirely foreign owned. Domestic coal production in 2006 was 18 Mt, such that most of the 67 Mt demand was met by imports, mainly from Russia. Remaining coal production has been forced to meet increasingly stringent fuel specifications, particularly for sulphur (Figure 7.12). The United Kingdom's leading private coal-mining company, UK Coal, has substantial land assets totalling over 19 300 ha, in addition to its operating mines which produced 9.5 Mt in 2006. The company aims to develop these mainly brown-field sites as business parks, housing estates and for other uses, thereby attracting new investment into former coal-mining areas as well as increasing the value of its property portfolio threefold to around USD 2.0 billion by 2012 (UK Coal, 2007). At the end of March 2008, there were a further nine companies with opencast coal mining operations in the UK and twelve companies operating deep mines (BERR, 2008). Some of these operations are very small, around 200 tpa in the case of Cannop drift mine in the Royal Forest of Dean.[4] However, they are regularly inspected by HM Inspectorate of Mines to ensure compliance with current health and safety law (Box 7.8).

Figure 7.12 .... UK Coal's Thoresby colliery

The coal preparation plant at Thoresby colliery uses two Baum jigs dating from the 1950s together with centrifuges and a froth flotation plant to produce a range of products, including power station fuel that meets stringent specifications for sulphur, ash, moisture and calorific value.

In 1998, the UK government committed substantial resources to the coalfield areas, including the allocation of over USD 1.5 billion annually for regeneration and to establish regional development agencies. As with other economically deprived coalfield regions in Western Europe, much of the subsequent job creation has come from SMEs, with few major employers being attracted, despite substantial incentives. In addition, the employment structure has changed, with an increase in the service sector where mostly women rather than men have found employment. Beatty *et al.* (2005) noted, however, that while the older generation of former miners often rejected this type of work on gender grounds, recent job seekers have been more willing to undertake what was previously seen as "women's work", such as jobs in call centres, supermarkets and food-preparation factories.

Other European countries

Coal-industry restructuring also took place in three other Western European countries, *i.e.* Belgium, the Netherlands and Spain, during the second half of the 20th century. The approaches adopted to economic regeneration in the coalfield areas differed widely in these countries, with varying results.

---

4.   www.forestofdeancoal.co.uk

In Belgium, regeneration efforts focused on the Limburg coalfield, where the last mine closed in 1992. Since the 1960s, the coal industry here had been heavily reliant on state subsidies to cover its costs. By the early 1980s, it provided work for around 19 000 people in a region that was economically depressed, with unemployment rates around 25%. In the mid-1980s, the Belgian government committed USD 2.5 billion to cover the costs of closing the industry and for future regional economic regeneration. Strong incentives were provided for a rapid mine closure programme, with more funds then being available for regeneration. Options for redundant miners included early retirement, accepted by around 40% of those affected, or a lump-sum redundancy payment and the opportunity for retraining. Between 1988 and 2000, the Limburg Mining Region Counselling Service (*Begeleidingsdienst Limburgs Mijngebied* or BLM vzw) and other training centres assisted a total of around 25 000 people, including around 8 000 ex-miners, with advice, specialised training services and job placement.[5] Those of Belgian and Italian extraction were most successful in finding new work, compared to those of Turkish or Moroccan origin who lacked language skills and technical training.

In the Netherlands, closure of the coal industry was completed in the 1970s, and was followed by a regional economic regeneration programme that included diversification of the state mining company, Dutch State Mines, into a chemical producer, the relocation of government administration offices to the South Limburg region, and the establishment of several technical education centres. Around 49 600 people were directly affected by the mine-closure programme. Half of these left the workforce during the period 1965-75 through retirement, early retirement and migration, or through social work-provision schemes that provided subsidised employment, mainly for invalids. The remaining 25 200 people required new jobs (Vermeulen, 2001). Despite the creation of over 17 000 new jobs in South Limburg between 1965 and 1990, unemployment quickly rose above the national average, and only began to recover in the 1990s when the region capitalised on its geographic location as a logistics centre for Western Europe.s Government funding for the South Limburg economic regeneration programme totalled USD 4.35 billion (in 2000 dollars) between 1965 and 1977, and a further USD 3.25 billion between 1978 and 1990.

Hard coal mining in Spain is concentrated in the north of the country, with both state-owned and private producers. The state mining company, Hulleras del Norte SA (HUNOSA), operates in the province of Asturias, in economically deprived areas with few alternative sources of employment and relatively high unemployment. The company's workforce shrank from over 26 000 in the late 1960s to 7 500 by 2000, with many of the job losses occurring during the 1990s. More recently, direct employment at HUNOSA has fallen from 6 150 at the start of 2002 to around 3 400 at the end of 2005 (HUNOSA, 2007). In 1998, the company, the national and provincial governments, and unions agreed an eight-year support plan under which production and employment were to be cut further and subsidies reduced. The USD 6.6 billion cost of the plan included USD 3.5 billion targeted investment to generate new jobs in the coalfields. Early retirement has been the principal tool used to reduce the workforce; minimum age and length of employment requirements were relaxed to include more people. Vocational training was also provided, mainly to younger

---

5. www.blmgenk.be

people who would have previously sought mining work. Local organisations, such as the Coalfields Development Agency (SODECO), provide support for regeneration projects, including financial and business services, and the provision of subsidised premises for start-up companies.

Coal industry restructuring in the United States

Restructuring of the coal industry in the eastern United States provides some valuable lessons in terms of the social impact of the widespread closure of small coal mines. Following a surge in demand for coal during the 1970s, resulting in many small mines being opened, the industry in the eastern coalfields suffered recession in the 1980s in the face of new competition from the huge, low-cost surface mines of the Midwest (Figure 7.13). In consequence, large numbers of miners lost their jobs in a region that was already home to above-average levels of poverty.

Figure 7.13 ....Map showing the principal coalfields of the United States

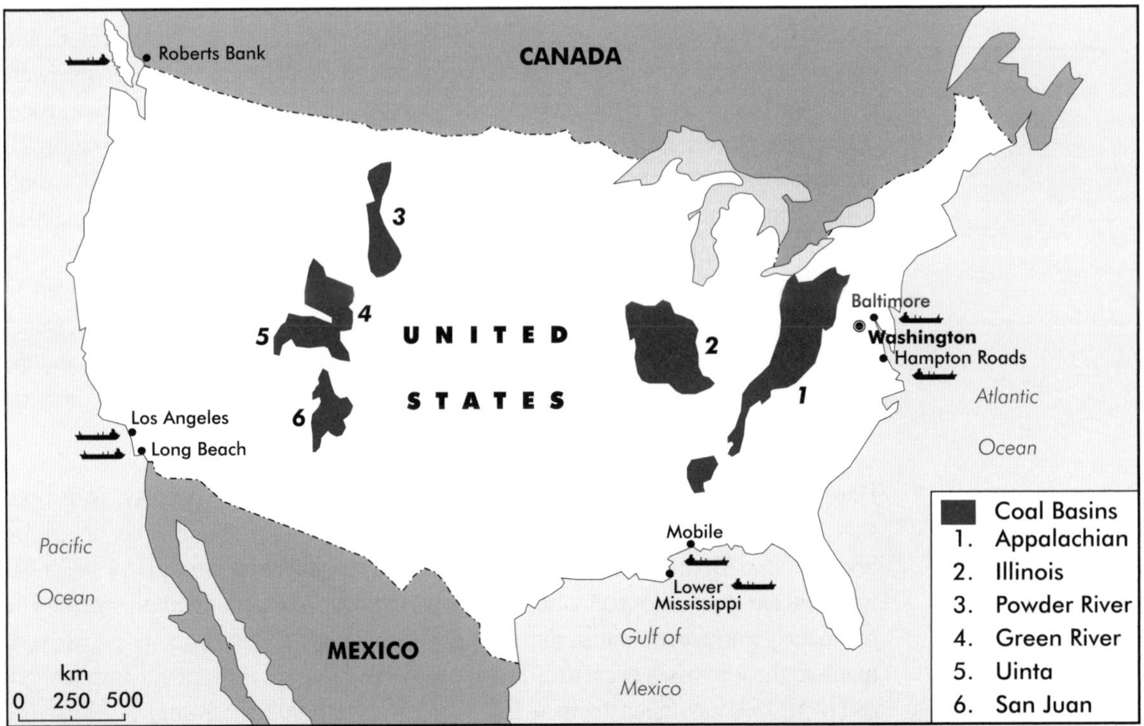

The boundaries and names shown and the designations used on maps included in this publication do not imply official endorsement or acceptance by the IEA.

Source: IEA (2003).

Employment trends in the US coal industry are similar to those of other countries (Figure 7.3). The widespread adoption of mechanised mining had a major impact on employment during the mid-20th century in what had previously been a labour-intensive industry. Since the 1980s, the development of huge open-pit coal mines in Wyoming and Montana has shifted the emphasis away from the traditional coalfields of the eastern US. Powder River Basin producers are able to sell their coal profitably to electricity utilities who, 50 years ago, would have been captive to mining companies based in the Appalachian states. This competition has meant that only the most efficient eastern producers have been able to retain market share and remain in business, sometimes through consolidation.

Increasingly stringent environmental control of coal mining operations has also influenced industry structure. The Surface Mining Control and Reclamation Act of 1977 (SMCRA), regulations under this act and other federal laws (*e.g.* laws requiring water quality and endangered species protection) have all led to the rigorous enforcement of measures to protect the environment. The need for specialised environmental personnel has served as another factor driving economies of scale and coal industry consolidation.

The US coal industry differs from those in Western and Eastern Europe, in that it has always been subject to free-market forces. In the period from the end of the Second World War until the 1990s, European coal production was largely state-owned with guaranteed markets. In contrast, US coal production has always been sold on the basis of contracts negotiated between mine owners, traders and consumers. Thus, market downturns have had a much greater effect on US coal producers, and hence on employment, than was the case in Europe. The principal victims of weak markets have often been small-scale operations that do not have sufficient financial resources to survive until the next upturn in the economic cycle. While such "swing" producers are a feature in the production of any commodity, the US coal industry was unusual in terms of the sheer number of mines that could be started and stopped at short notice. This was nothing new: in the 1920s and 1930s, coal-mining communities in the eastern US had a precarious existence, with mines frequently closed or "idled" as demand for coal fell, or more competitive suppliers took their business (Caudill, 1962). Conversely, small producers were also able to resume production very quickly when coal demand was strong. This cyclical opening and closing of mines remains a feature of the eastern US coal industry even now. The major difference between the industry today, compared to 40 or 50 years ago, is that there are far fewer mines that can fulfil this swing function. During the 1980s and into the 1990s, half of the coal mines operating in the US were permanently closed, mainly small mines.

This section looks at the impact of coal-sector restructuring on the Appalachian coalfields of the eastern US.[6] Although economically important to the regional economy, the coalfields occupy only a small proportion of Appalachia, a region that covers 13 states from New York in the north to Mississippi in the south. The coalfields lie mainly within the states of Ohio, Kentucky, Pennsylvania and West Virginia, but also extend into smaller portions of Virginia, Tennessee and Alabama. Communities range in size from major cities such as Pittsburgh to isolated valley-bottom townships. Because of the terrain, most of the Appalachian coalfield remains rural, in contrast to

---

6. As Black and Sanders (2004) point out, relatively few studies have been undertaken into regional poverty and income inequality in Appalachia in general, or the coalfields in particular. In consequence, it is difficult to obtain a detailed picture of the situation facing many small communities in Appalachia, or to differentiate between general economic recession and the effects of coal-mine closures. This contrasts with the situation in some European countries, where organisations such as the Centre for Regional Economic and Social Research (CRESR) at Sheffield Hallam University in the UK have carried out detailed investigations of conditions in former coal-mining communities.

other areas of eastern US. The topography of the region, much of which consists of deeply incised valleys lying between long ridges, helped in the development of the coal industry by providing easy access to rich seams, but has also left many communities isolated from their neighbours.

Appalachia, including large areas of the four main coal-producing states, is economically deprived compared to other areas of the US. In addition to coal mining, traditional industries have included logging, agriculture, textile production and tobacco growing, all of which have been affected by long-term economic decline. In response to this underlying poverty, the US government established the Appalachian Regional Commission in the 1960s, charged with economic regeneration of an area that extends to 520 000 km², and has a population of around 23 million. For comparison, Shanxi province has a population of 34 million living in an area of 156 300 km² (NBS, 2007). Plishker *et al.* (2002) described Appalachia as, "an area that is undergoing significant changes in its social and economic well-being, yet it continues to lag behind the rest of the nation in education and income. Decades ago its economy depended on industry, agriculture, and mining; today, human capital and the service sector are growing more critical to economic growth. While some areas within the region have made substantial strides, others have shown only limited progress. Measures such as the number of persons living in poverty, high school completion rates, employment rates, and job growth rates are but a few of the indicators that illustrate the gaps that exist between the citizens of Appalachia and the overall population of the US. With poverty rates continuing to decrease and educational attainment and employment rates continuing to grow, the gap is narrowing". A fundamental role for the Appalachian Regional Commission has been to establish public infrastructure, including transportation, to support economic diversification as the coal industry has become more productive and its employment base has declined.

The "boom and bust" of eastern US coal mining

The period of greatest relevance to the current situation in China is the early 1970s to the mid-1990s. Initially, a dramatic increase in coal prices led to increased output from the Appalachian coalfields. While employment in the coal industry rose steeply, a key feature of this boom was the high wages that were available to a largely low-skilled, male workforce – higher than paid elsewhere in Appalachia. This boom period for eastern US coal lasted from the first oil-price shocks of the early 1970s to the early 1980s. Kuby and Xie (2001) noted that, "from 1975 until 1980, higher prices due to the 1973-74 oil embargo encouraged the entry of small inefficient operators, allowed for profitable mining of geologically marginal deposits, and created a less competitive market that tolerated inefficiency". However, the strong market for coal brought increased competition from other sources, both domestic and foreign. In addition, deregulation of US railway freight rates, increased competition from natural gas and improved industrial relations within the US coal industry led to the domestic market for thermal coal becoming chronically over-supplied in the early 1980s. On top of this, the introduction of the Clean Air Act Amendments in 1990 effectively pushed users away from higher-sulphur Appalachian coal in favour of lower-sulphur coal from the new Midwestern mines.

The subsequent slump in demand had a huge effect on both the number of mines in operation, and the number of miners employed. Kuby and Xie (2001) reported that the number of mines across the US producing more than 10 000 short tons per year (9.1 ktpa) fell from 3 969 in 1980 to 2 104 in 1995. The average mine production rose from around 150 000 short tons per year (136 ktpa) to 491 000 short tons per year (445 ktpa) during this period. By 1995, 198 mines producing over one million short tons per year (907 ktpa) accounted for 73% of national coal production. Looked at another way, in 1980, mines producing less than 2 short tons per worker-hour (1.8 tonnes) accounted for 40% of US production, but had fallen to 3.5% by 1995 (*ibid.*). The reduction in mine numbers has continued since then; only 1 415 mines were operating in the US in 2005, with an average size of 725 ktpa (EIA, 2006b).

A further illustration of the disproportionate impact on the small-mine sector is provided by a breakdown of the number of mines operating in the Appalachia coalfields within different production capacity ranges (Table 7.1). The effect on the labour force was severe, with the majority of job losses falling on already-deprived Appalachian communities. That new jobs were being created in the Midwestern states of Montana and Wyoming was of little comfort to redundant miners here, since their experience was largely limited to small-scale underground production with few skills that could be transferred to the distant high-capacity surface mines that were gaining market share.

Table 7.1 ....Appalachian annual coal production (thousand tonnes), number of mines and employment by mine size, 1980-2006

	1980	1985	1990	1995	2000	2005	2006
Total production (thousand short tons)	438 604	422 315	488 993	434 861	419 419	396 666	391 159
Total production (ktpa)	397 895	383 118	443 607	394 499	380 491	359 849	354 853
Total number of mines	3 498*	2 962*	2 377*	1 848	1 280	1 230	1 254
number of mines producing <91 ktpa	2 444*	1 788*	1 389*	1 122	680	n/a	n/a
number of mines producing <9.1 ktpa	n/a	n/a	n/a	367	219	n/a	n/a
Total employment	170 746	122 102	95 421	64 153	48 021	53 509	55 632
employment at mines producing <91 ktpa	n/a	n/a	n/a	9 713*	5 616*	6 831	7 199

* These figures exclude mines producing <9.1 ktpa (*i.e.* <10 000 short tons per year).

Sources: EIA (1981, 1986, 1991, 1996, 2001, 2006b and 2007).

Direct employment in the Appalachian coalfields fell from 171 000 in 1980 to 54 000 in 2005, having risen above the 2000 figure of 48 000 as new mines were developed or old mines were re-opened at short notice to meet buoyant demand (Table 7.1). In terms of the share of overall employment provided by mining, Bradley *et al.* (2001) noted that between 1993 and 1998, this fell from 1.2% to 0.9% in Appalachia, with a proportionately similar fall from 0.7% to 0.5% across the nation. These figures also illustrate Appalachia's greater reliance on the coal mining industry – nearly twice as reliant on mining for employment as the country overall. In some counties, the contribution of coal mining is even greater. Thompson *et al.* (2001) noted that in some counties in eastern Kentucky, coal production accounts for over half of the gross county product. These authors added that in 1997, "the total economic impact [of regional coal mining] accounted for 3.1% of employment, 3.4% of earnings and 2.7% of value-added in the Northern Appalachian region. In the Central Appalachian region, the total impact accounted for 29.9% of employment, 27.6% of earnings and 29.8% of value-added". At that time, coal was produced in 118 out of 399 counties within the Appalachian Regional Commission's remit.

One unusual feature of the eastern US coal industry has been the relative ease with which companies can open and close mines, as well as suspend and restart production at short notice. Dunne and Merrell (2001) noted that, for example, 34% of all the mines closed in 1975 would subsequently reopen and 41% of mines opened in 1985 were re-openings of mines that had previously been suspended. By 1990, however, barely 20% of mines closed that year would reopen later, illustrating the trend towards permanent closure of small mines. Stronger environmental requirements have made coal mine permitting more difficult over the past decade, particularly the large area mines ("mountaintop removal") that have sustained Appalachian production in the face of reserve depletion at underground mines. This has contributed to rising coal mining costs in the US over the past few years.

Miner characteristics While it is impossible to describe an "average" Appalachian coal miner, some demographic pointers can be established. During the 1970s and 1980s, the workforce had a high proportion of poorly educated people with few transferrable skills to other industries. According to Black *et al.* (2005), "coal employment [in 1970] was predominantly low-skilled. In 1970, 45.2% of coal workers had less than eight years of education and a full 68.7% had less than a high school education. By comparison, in 1970, only 18.0% of all workers had less than eight years of education and 41.3% had less than a high school education. Even though education levels had increased overall by 1990 [11% of miners had less than eight years of education, compared to 45% in 1970], coal workers in 1990 were still more likely to have low education than were workers in general". Age distribution is central to this fourfold reduction in the proportion of the mining workforce with minimal schooling. In 1970, many of those with little education would have been older men who subsequently retired, with their place being taken by the younger and better educated. Nonetheless, there remained a substantial shortfall, with the 11% in the mining industry having less than eight years of education contrasting with just 4% in the wider US workforce. Poor educational achievement and a shortage of skilled labour continue to hamper coal mining regions in eastern US. The industry also needs to attract a very different type of employee than it did 30 years ago. Today's mine workers need a higher level of technical education to operate the advanced equipment that has allowed productivity gains.

In terms of age distribution, Black *et al.* (2002) noted that in 1970, 18.5% of the workforce was aged 55 or over. By 1990, this proportion had fallen to 9.6% while the proportion in the 25-44 age range had risen from 22% to 40%. The coal mining industry had shed many of its older workers, but had not provided many employment opportunities for younger people – the proportion of workers aged under 25 more than halved between 1970 and 1990.

Declining employment in the coal industry added to the wider economic hardship experienced by many of the communities in Appalachia. The incidence of "economically distressed" counties – those that the Appalachian Regional Commission has identified as having high rates of both poverty and unemployment (over 150% of the national average) as well as low per-capita income (less than 67% of the national average) – correlates with declining coal production in these counties (Thompson *et al.*, 2001).

Welfare and incapacity benefits

There has been significant investment by the federal and state governments since the 1960s, as well as by organisations such as the Appalachian Regional Commission, in poverty-alleviation programmes. Overall, the effect has been positive, with falling unemployment in the economically distressed counties. In central Appalachia, home to the highest concentration of coal mining in the region, Black and Sanders (2004) reported a 28% fall in unemployment between 1990 and 2000 as economic regeneration became more effective. They also noted that Appalachia's farming and mining counties received higher than average payments from federal assistance programmes, while the 20% growth in disability insurance payments across the Appalachian region was larger than the 12% national average during the 1990s.[7]

As in some Western European countries, of which the United Kingdom and the Netherlands are good examples, this increase in the uptake of disability benefit often masks the true level of unemployment within former coal-mining communities. There can be strong political and social incentives for former miners to seek incapacity status rather than merely registering as being unemployed, especially if there are few alternative job opportunities within their immediate locality. This in turn introduces a grey area of "under-employment", where people who would normally be considered to be of working age have effectively withdrawn from the official job market, although they may well have unofficial income to supplement their social security payments.

Targeted economic regeneration

Although by no means unique in the US in terms of poverty, Appalachia has been the target for sustained investment in economic regeneration for over 40 years. These programmes have not concentrated solely on the coalfield areas, but have attempted to create a new economic environment across the whole region. For this reason, it is difficult to single out specific initiatives aimed at coalfield regeneration, since poverty ran deep. According to Black (2007), economic regeneration programmes within Appalachia have been funded at both federal and state level, with state governments often receiving federal assistance for their own projects. The Appalachian Regional Commission (ARC) has also been active, although the scale of its budgets, and hence its operations, have been limited. The ARC is a federal-state partnership that creates opportunities for self-sustaining economic development and improved quality of life. The Commission awards grants and contracts for projects, from funds appropriated to it annually by the US Congress, through its Area Development Program and its Highway Program with the following strategic goals:

■ to increase job opportunities and per-capita income in Appalachia to reach parity with the nation;
■ to strengthen the capacity of the people of Appalachia to compete in the global economy;
■ to develop and improve Appalachia's infrastructure to make the region economically competitive; and

7. The Food Stamp Program enables low-income families to buy food with coupons or electronic cards. Social Security Disability Insurance benefits are paid to disabled individuals who have worked five out of the previous ten years. Supplemental Security Income benefits are paid to individuals who are poor and disabled, regardless of whether or not the individual has worked in the past. Temporary Assistance for Needy Families provides income supplements to poor families with children, with work requirements that vary by state.

■ to build the Appalachian Development Highway System (ADHS) to reduce Appalachia's isolation.

Construction of the ADHS has been under way since 1965, with over 4 200 km completed or under construction by the end of 2006. Completion of the 4 975 km authorised remains a top priority for the ARC (ARC, 2007). Under the scheme, West Virginia's highway network has been expanded substantially over the past 40 years, with new roads often cutting across natural geographical features that would have hindered construction in the past. A new road was built in Kentucky to link the major centre of Lexington with the impoverished coal-mining communities in the east of the state. While the scheme has required massive investment, it is probably the most visible evidence of the effort to make Appalachia a more attractive location for new businesses.

At state level, funding has been targeted at vocational training, often with federal assistance. Heavily subsidised community colleges have been established, while the federal government has also had its own schemes aimed at providing "displaced workers" with better skills and qualifications.

For businesses, incentives in the form of tax concessions and other instruments are available for those wishing to either set up or expand their operations within Appalachia. However, such incentives often cover the entire region, so apply in major cities with established communications links as well as in isolated rural communities – a lack of targeting that can miss those areas with most to gain.

Despite these programmes, and the success that has been achieved in reducing unemployment, Appalachia's economy is not stable. The drain of young people with better educational achievements and transferrable skills continues, leaving behind a residual population that is less able to cope with engrained poverty. Black (2007) noted that between 1980 and 1990, six eastern Kentucky counties each lost 25% of their population by migration to other parts of the US, and while the level of population decline has slowed, the trend remains.

In terms of the costs involved in poverty-alleviation programmes in Appalachia, little academic research appears to have been undertaken. Costly social security mechanisms are certainly more widely used in this region than in many other parts of the US. However, measures such as these are essentially palliative rather than addressing the root causes of poverty. An example of the costs of ARC's vocational training programme was provided by Plishker *et al.* (2002). ARC grants examined in their study ranged from USD 15 000 to 900 000 for total project values ranging from USD 27 000 to over USD 1 million. Although many agencies and funding bodies have participated in regeneration projects across Appalachia over a long period, no overall funding total has ever been estimated, but it is likely to total billions of dollars.

Another factor in Appalachia's economic diversification has been local economic development programmes, including those funded by the coal industry. In Appalachia, most coal mining states have employed a "severance tax" (*i.e.* a tax imposed by a state or local government on the extraction of a natural resource, generally as a fixed percentage of gross revenues). A common use for the severance tax is investment in local infrastructure, especially for maintenance of roads that are used heavily by the coal industry. However, some localities have decided to invest a portion of severance tax funds in local economic development. An example of such an organisation is the Virginia Coalfield Economic Development Authority[8] which has brought at least 10 000 jobs to Virginia's coalfield region, many in higher-paying industries. Coal industry leaders participate in the organisation's management along with local government officials, thus helping to assure that the organisation's economic development investments complement the coal industry's operations.

Miner-specific programmes

The United Mine Workers of America (UMWA) has provided targeted assistance to redundant miners. Established by the union in 1996, and funded by grants from the US Department of Labor and other state agencies, UMWA Career Centers (UMWACCs) continue to provide a "one-stop shop" for support, retraining and job identification. Since its establishment, the UMWACC programme has received over USD 40 million in grant funding and has helped around 7 000 redundant miners who remained eligible for assistance for five years after being laid off (Adams, 2007). Of these, 4 000 have received training and have been placed in jobs that pay above the national average, although below a miner's average wage. Typical annual running costs for administration and the provision of support services are USD 1.5 million, plus income maintenance that can amount to USD 8-9 million per programme.

As an illustration of how the service works, the Career Centers provided retraining and living expenses for 751 out of about 1 200 former miners who lost their jobs when three coal mines closed during 2000. Funding came from a special USD 6.5 million grant from the US Department of Labor. In early 2001, the UMWA reported that of these, 348 had been placed in new jobs and a further 340 had been enrolled in various retraining programmes. Skills acquired during retraining included operation of heavy construction equipment, with job opportunities in Appalachia's highway construction project (Murtha, 2001).

The Appalachian coalfield includes both unionised and non-union coal mines. Unionised mines operate under national agreements negotiated between the UMWA and the coal companies, agreements that include benefits such as healthcare provision and retirement pensions. Since the 1980s, coal miners in unionised operations that have closed have had opportunities to take early retirement at age 50 and over, with their healthcare and pension benefits guaranteed (Adams, 2007). Conversely, those in non-union mines have to rely on general social support programmes until they can find alternative work.

Relevance to China

Despite clear differences between Appalachia of the 1980s and China today, there are striking similarities. Setting aside the different national political systems in both countries, miners face essentially the same challenges in terms of securing alternative sources of future income following the closure of small mines. Small mines have been the hardest hit during restructuring in the US coal industry since the early 1980s. Many opened in response to rising coal prices in the 1970s and small mines supplied a substantial proportion of US coal by 1980. A subsequent refocusing on larger, more competitive production units operated by a few large companies, plus the impact of new health, safety and environmental legislation, left small-scale mining less viable, and led to widespread closures and job losses. China's policy to close small mines and to create large coal bases will have a similar impact.

---

8.   www.vaceda.org

Summary of experience in IEA member countries and relevance to China

In some respects, there are significant differences between the experience in coal-sector restructuring gained in IEA member countries and current needs of the Chinese coal industry. The most obvious of these relate to the scale of restructuring and the socio-economic conditions of the people most affected. In China today, the emphasis is on closing small coal mines, which employ substantial numbers of migrant miners as well as people from the local communities. In IEA member countries, by contrast, restructuring has often affected the whole coal industry, with massive job losses among the entire workforce.

Whatever the strategic background, communities faced with job losses need extensive social support (Box 7.3). This applies equally to a peripatetic workforce as to people within established communities who have lost their principal employment. Social deprivation knows no boundaries and the potential for social unrest exists wherever those affected by mine closures have no alternative income sources. Experience from IEA member countries has shown repeatedly that workers with readily transferable skills, such as welders, electricians and mechanics, have much greater success in finding alternative employment than unskilled workers. Improved education and the provision of vocational training are widely seen as vital components of social and economic regeneration in areas affected by the closure of heavy industries, enabling redundant workers into new jobs. Given the age structure of the workforce in China, redundancies are more likely to affect younger miners; hence, opportunities for early retirement would be limited. Of much greater importance will be the provision of education and training to give redundant miners new skills, allowing them to find jobs with comparable incomes to those in the mining sector.

Box 7.3 .... Positive experience of coal industry restructuring in IEA member countries

Economic regeneration of former mining areas is more likely to be successful if the stakeholders (national and local governments, mining companies, workforce, unions and communities) work together. The establishment of good lines of communication between all of the interested parties is essential, and where this has happened, regeneration has achieved more and has cost less. Early retirement has been a major tool used to achieve workforce reductions during coal industry restructuring. In France, this was coupled with a strongly paternalistic approach from the state, through the state-owned mining company, Charbonnages de France, which guaranteed alternative employment or training to miners who had lost their jobs but did not qualify for retirement. Across Western Europe, coal industry restructuring has been painful, but with a positive outcome – many of the regions affected are now prospering and governments have avoided the worst consequences of economic deprivation. In the US, targeted support has provided a safety net for those unable to find alternative employment, while economic regeneration projects in the Appalachian region have assisted communities faced with depopulation and failure to create new economic activity and to find new purpose.

Financial support mechanisms used in IEA member countries have ranged from lump-sum payments to long-term commitments for pre-retirement payments, pensions, housing, fuel allowances and other benefits (Box 7.4). While the former requires the payment of significant sums on a one-off basis, the latter route can entail substantial expenditure over long periods; costs can run into billions of dollars where thousands or hundreds of thousands of people are involved. Experience has shown that people with poor financial skills must receive appropriate advice, in order to optimise the use of any payments until they have found alternative work.

Communities affected by mine closure need support in terms of maintaining their basic infrastructure, such as schools, medical clinics, public transport and communal facilities. Where these have been paid for by mining companies, other funding arrangements need to be made. The more deprived a community becomes in relation to its basic services, the harder it is to attract new business. In addition, funding has to be provided to encourage new businesses to start up. This may include assistance in providing facilities such as workshops or office space, venture capital to cover initial outlays and administrative support.

Box 7.4 .... Economic regeneration incentives used in IEA member countries

Governments have often offered incentives for inward investment into economically deprived areas, typically:

■ financial assistance to new businesses through grants or soft loans;
■ business advice and support at low or no cost;
■ deferral of corporation tax ("tax holidays") for a specific period after business start-up;
■ provision of land for constructing low-cost workshops and small industrial units;
■ training programmes for potential workers;
■ improved infrastructure (transport and communications);
■ environmental improvements, such as landscaping and pollution control;
■ relocating government activities to these areas, in order to encourage spin-off businesses;
■ relocating education and research organisations that provide direct employment and offer the potential for spin-off businesses;
■ promoting alternative uses for industrial sites, including recreation and tourism; and
■ maintaining local services until the economy has recovered sufficiently to support them again.

None of these is sufficient in isolation – attracting new businesses, often in competition with other potential locations, is dependent on factors such as transport links, access to markets and the availability of a motivated, skilled workforce. In economically depressed former coalfield areas, some or all of these attributes are often missing.

The worst possible option is to do nothing. Having a population of former mine workers with low educational achievements, no transferable skills and no hope of finding alternative work is a recipe for social discontent (Box 7.5). Migration from rural to urban areas will result, with implications both for service provision in the cities and for the long-term survival of the rural economy. Experience in IEA countries has demonstrated that former coal mining regions can recover from major job losses and re-establish a viable economic base. However, this requires a long-term commitment on the part of government at all levels, and incurs substantial costs.

In China, government policy supported the widespread development of township and village enterprise coal mines during the 1980s in response to growing demand for energy. As in the Appalachian coalfields of the US, this had the side benefit of helping to reduce levels of poverty in rural areas (Andrews-Speed, 2007). Today, however, concerns over safety, environmental impacts and resource utilisation favour the development of high-capacity mining units, with small mines either being closed or consolidated into larger units. The analogy with Appalachia is apparent, especially in relation to future income sources for redundant miners.

Box 7.5 .... Negative experience of coal industry restructuring in an IEA member country

In some IEA member countries, the economic burden of subsidised coal production has been the main driver for industry restructuring. Where restructuring plans have been negotiated and agreed by all stakeholders, restructuring has been effective but slow, as in France and Germany. In confrontational situations, industrial restructuring becomes a much more daunting challenge. The restructuring of the UK coal industry provides a lesson in poor management, with a lack of understanding, communication and common purpose between the major stakeholders. Industrial disputes, coupled with extreme personal antipathy between the government and mine workers trade union, led to bitter confrontation in the British coal industry during the early 1980s. On the government's agenda was the forthcoming privatisation of the electricity industry, which had long-term coal off-take commitments with the state-owned coal industry. The closure of individual mines was often rushed, with little time for the workforce to come to terms with events as they occurred. Large redundancy payments proved to be of little benefit to workers who had only limited experience in handling financial affairs. The government also underestimated the close social ties that exist within coalfield communities that made it harder for redundant miners to leave for alternative work. In addition, their often poor educational achievements and low level of transferrable skills meant that former miners found new work hard to come by. Government policies of the time, particularly in respect of the relationships between central and local administrations, made it difficult for local communities to access European Union support funds. Subsequent investment in regeneration projects in the late 1990s helped to offset much of the economic deprivation in the former coalfields, but the legacy of unemployment, under-employment, worker incapacity and bitterness remains.

Often with little to offer in the way of transferrable skills, such people have few options other than to return to farming, to seek menial, low-wage labouring jobs in the locality, if there are any such jobs available, or to leave to find work in other parts of the country. The latter option, has serious implications both for the demographics of the communities that remain, and for the cities which already face a large influx of low-skilled workers. There is also the problem of migrant contract miners, who may have been drawn from neighbouring counties or provinces to a particular area by job opportunities, but who have no job security once that work has ended. Migration from the countryside to cities is nothing new; in many industrialised countries it has been underway for centuries, and it is a common feature of developing countries today. While the incentive of higher-paid work is completely understandable, the risk remains that the communities left behind fall into a long-term spiral of poverty and neglect. Another similarity is the way in which small coal mines can be rapidly opened and closed in response to changing economic or regulatory conditions. In Appalachia, both of these acted as drivers at different times; in China, inadequate regulatory enforcement is widely reported as being an incentive for small mines to reopen if their operators perceive even short-term economic opportunities.

In most IEA member countries where coal-industry restructuring has occurred, governments have tried to create an economic framework within which private-sector enterprise can sow the seeds of regional regeneration. The creation of the Appalachian Regional Commission in the US in the late 1960s was a direct result of the recognition of regional poverty, and although it has had a much wider remit than purely coalfield communities, the Commission's achievements have resulted in real improvements in the economic well-being of Appalachia as a whole. Reductions in unemployment and in the number of people living in poverty testify to that. Poverty reduction is a key to achieving social stability. Conversely, high levels of poverty lead to social deprivation, petty crime, increased alcoholism and drug-use, family breakdown and a host of other personal and community-related problems. Poor communities in Appalachia have had more than their fair share of these. While the task for government is to put in place sufficient support mechanisms for individuals and communities, and infrastructure development to enable new employment opportunities, some communities will simply be too remote and isolated ever to attract business investment. In the coalfields of Appalachia, topography and a community's distance from the nearest regional centre play an important role in determining the success of regeneration efforts. This may also be the case in some parts of China, so economic regeneration has to be undertaken with local conditions in mind – blanket programmes are unlikely to be appropriate for all situations.

DEVELOPMENT OF COAL-RELATED LAW IN IEA MEMBER COUNTRIES

Legislation applicable to coal production covers exploration and mining permits, health and safety, operating standards, worker welfare and environmental protection. Establishing legislative and regulatory frameworks in IEA member countries was a long process, beginning with worker safety and employment conditions, followed by operating standards to ensure that safety became an inherent feature of all mining operations. With rare exceptions, environmental protection legislation came into force relatively recently, in response to heightened environmental awareness during the second half of the 20th century.

In most cases, rights to coal and other mineral resources are owned by the state, which then issues licences to companies who can prospect, evaluate and produce the resources. Even where resources are privately owned, such as in the eastern United States, regulatory permits are still required for exploration and production. Resource exploitation usually involves payment of a production royalty to the mineral-rights owner, be it the state or a private entity. Technical competence and the ability of a company to optimise resource recovery are often major factors when awarding production licences. In practice, resource recovery depends on the deposit geology, surface topography and land-use priorities, as well as the proposed mining method, but the permitting process is key to ensuring that resources are not wasted by inappropriate mining practices.[9] Tonnage-based production royalties, paid to the state or private owner of mineral rights, do not achieve this aim, although they can be considered a resource-depletion tax.

This section provides an overview of how coal mining legislation has developed in IEA member countries, then offers a view on how certain aspects might inform and influence future measures in China.

The development of rights acquisition and tenure law

The security of tenure over mineral resources is of fundamental concern to mining companies. Without protected resource rights, or in situations where long-term tenure is unclear, they are unlikely to invest in new production capacity, given the risk of expropriation or seizure. For large, multinational mining companies, sovereign risk is a major factor that influences investment decisions.

Mineral resource ownership now lies with the state in most countries. In the UK, for example, coal resources were nationalised in the 1930s, removing ownership from holders of private rights. In the US, coal resources to the west of the Mississippi River are largely under federal, state or native American ownership, while resources in the eastern coalfields remain mostly in private hands. In both Queensland and New South Wales in Australia, coal resources are state-owned with minor, long-standing exceptions, although the issue of Aboriginal rights is still under discussion (Box 7.6). In countries that have a shorter history of coal production, state resource ownership is almost universal. However, there remain significant differences in administration between countries that have a legal system based on common law, and those with civil law codes. In a common law country, such as Australia, India, the UK and the US, a mining agreement is equivalent to holding title over a property. In a civil law jurisdiction, such as Germany, Russia and China, a mining agreement has to be supplemented by a separate mining title. The role of the state differs under the two

---

9. Poor mining practices and land-use planning can "sterilise" coal resources by making them inaccessible to future mining operations. For example, it is generally not safe to re-work a mine district (the longwall mining of coal pillars after bord-and-pillar mining being an exception); coal left behind after mining is sterilised. Opencast coal extraction from a greenfield development site is possible prior to the site's development, but not after. CBM projects can sterilise coal resources because it would not be safe to use mechanised mining equipment in coal seams littered with steel well casings. Enhanced CBM projects, which store CO_2 as methane is displaced, sterilise coal because the CO_2 would be released to atmosphere if the coal were ever mined.

legal systems: under common law, the state acts as a contracting party while, under civil law, it has a sovereign role (Walker, 1998). A further complication might be the separation of land ownership from mineral rights. Thus, the award of tenure over mineral rights may only be part of the process of acquiring access to a property, with additional agreements required with the land owner, either through purchase, renting or wayleave.

Box 7.6 .... Coal exploration and mining in Australia: licences and royalties

Each year Australia produces more than 300 Mt of saleable hard coal and more than 70 Mt of brown coal or lignite from more than 100 privately owned mines. Over 80% of production is from opencast operations, with the remainder from underground mines. Australia exported 231 Mt of hard coal in 2006, making it the world's largest coal exporter. The coal industry is dominated by BHP Billiton (Australia-UK), Anglo American (UK), Rio Tinto (UK-Australia), Xstrata (Switzerland) and Peabody (US).

The ownership of Australia's coal resources is vested in state governments, each of which has its own mining legislation. Exploration and mining licences are granted to mining companies on a reasonably consistent basis across Australia and production royalties are collected. Each state publishes guides, summarising mining law and how it is administered during exploration, retention and mining. Other state and federal government legislation, such as native title and environmental legislation, may also apply to mining approvals.

An application for an exploration licence must include an access arrangement with the landowner, a work programme, evidence that the applicant has adequate financial and technical resources, and an application fee. The applicant is required to pay compensation to the landowner for any surface or property damage sustained. In some states, a security deposit or bond is held to ensure that there is no default on any due payments. Exploration licences are generally granted for periods from two to six years, with renewals being subject to "relinquishments" or reductions in area. A retention licence allows a company to maintain title to an identified coal deposit until opportunities for development arise. During this time mine plans and layouts might be developed and environmental impact assessments undertaken.

Holders of exploration or retention licences have priority when applying for a mining licence. Applicants must provide plans of their mining proposals and assessments of the potential impacts of those plans. Public notification of the application is required, with provision for objections and public hearings, before a decision is made by a minister who might reject the proposal or impose particular conditions when granting approval. Consent of the owner or occupier of private land is required before mining operations can take place near to private dwellings, and a negotiated compensation is generally payable to landowners for mining activity, although many mining companies prefer to buy the land on which they are developing major projects.

A royalty is a payment made to the Crown for the right to use the state's coal resources and is generally paid when the resource is sold. The landowner has no right to royalties or any portion of royalties. Three types of royalty exist: a quantum royalty, charged at a flat rate per tonne; an *ad valorem* royalty, being a percentage of the coal's value; and a profit-based royalty. In Australia, *ad valorem* royalties are now favoured by government and coal producers; these are typically levied at between 5% and 7% of the coal's sale value.

Procedures for obtaining the various permits needed for exploration, evaluation and production vary, but security of tenure is of major importance in any jurisdiction, with investors preferring sequential processes in which the discovery of a viable deposit automatically gives precedence to the holder of the exploration permit when a production licence is sought. Countries where this right does not exist are at a distinct disadvantage in terms of perceived investment risk. The type and number of permits required before a coal mine can be brought into production is highly variable. In addition to the actual mining licence from the relevant administrative body, permits may be needed from many others, ranging from local planning authorities to wildlife protection agencies, covering water resource management, waste disposal, effluent control, land use and protection of historical monuments. As an example, in the US 20 or more federal agencies can have interests and input into the permitting process for a new coal mine. Overall, there is a similarity between the licensing processes found in many of the world's major coal-producing countries. Typically, exploration involves a separate licence that costs less than a production licence, because of the greater risk involved. Exploration licences often include a relinquishment requirement on each renewal, thereby releasing prospective blocks of land for other companies to explore, as well as reducing the potential for speculative land holding.

In summary, those countries that have a clearly established legal framework to administer large-scale minerals exploration and production, together with fiscal stability, are more likely to attract investment for coal production than those where there is less transparency and less certainty over tenure rights.

The development of health and safety law

Much of today's health and safety legislation was enacted in response to mining accidents rather than through enlightened foresight. For example, the Hartley disaster of 1862 in England resulted in the requirement for each mine to have at least two means of access, while the Courières explosion in France in 1906 prompted the coal mining industry to investigate the role of coal dust in explosion propagation. Restrictions on the employment of women and children underground in the United Kingdom followed public inquiries into labour practices during the 1840s, while the licensing of supervisors and mine management demanded recognised levels of qualification and expertise (Bryan, 1975). Health and safety legislation continues to evolve. For example, in response to the Sago mine accident in the United States in 2006, the Mine Improvement and New Emergency Response (MINER) Act of 2006 requires operators of underground coal mines to improve accident preparedness (MSHA, 2007b).

Box 7.7 .... European Union law concerning health and safety in coal mining

Through its Commission, Council and Parliament, the European Union agrees and enacts directives which must be transposed into the national legislation of the 27 EU member states. The earliest European input to improving health and safety in coal mining came with the establishment of the European Coal and Steel Community (ECSC) in the early 1950s, under which various research and information dissemination programmes were set up. This was followed in 1957 by the creation of the Safety and Health Commission for the Mining and Other Extractive Industries, set up to assist in the preparation of laws aimed at preventing the occurrence of major accidents and health impacts such as pneumoconiosis in the coal sector. Major areas of research included:

- dust suppression in underground mines;
- monitoring underground conditions in mines;
- understanding the links between dust exposure and pulmonary diseases caused by dust inhalation; and
- improving safety consciousness and individual safety procedures.

Subsequent major directives and legislation that affect health and safety in the coal industry include:

1980 Directive 80/1107/EEC on the protection of workers against risks related to exposure to chemical, physical and biological agents at work;

1982 Directive 82/605/EEC on the protection of workers from the risks related to exposure to metallic lead and its ionic compounds at work;

1983 Directive 83/477/EEC on the protection of workers from the risks related to exposure to asbestos at work;

1986 Directive 86/188/EEC on the protection of workers from the risks related to exposure to noise at work;

1987 Directives adopted under Article 118A of the Single European Act lay down minimum requirements concerning health and safety at work;

1990 Directive 90/269/EEC on the minimum safety and health requirements for the manual handling of loads where there is a particular risk of back injury to workers;

1992 Directive 92/91/EEC concerning the minimum requirements for improving the safety and health protection of workers in the mineral-extracting industries through drilling;

1992 Directive 92/104 on the minimum requirements for improving the safety and health protection of workers in surface and underground mineral-extracting industries;

1997 Directive 97/42/EC on the protection of workers from the risks related to exposure to carcinogens at work; and

2002 Directive 2002/44/EC on the minimum health and safety requirements regarding the exposure of workers to the risks arising from physical agents (vibration).

It should be noted that European Union directives establish the required end result and the time frame in which it must be achieved; individual member states are autonomous to decide the means of implementation.

Source: EC (2008a).

While individual countries have enacted their own health and safety laws on coal mining, there are many common threads. Maximum allowable methane concentrations are based on internationally recognised criteria, with comparable limits prescribed in law worldwide. Maximum working hours are typically stipulated under general employment law, with groups of countries, such as members of the European Union, often sharing comparable regulations (Box 7.7). Exposure to dust and vibration is also governed by legislation, with provisions made for invalidity resulting from past exposure. In some cases, this is through state compensation, while in the US, the Black Lung programme is funded through a production levy. In the UK, the government expects to pay a total of GBP 6.4 billion (approximately USD 13 billion) in compensation to mineworkers who continue to suffer health problems as a result of working underground after 1954 in the case of respiratory disease and after 1975 in the case of hand-arm vibration syndrome (or vibration white finger), these being the dates when the National Coal Board should have been aware of the risks. This is the biggest single personal injury compensation scheme in British legal history and possibly in the world.

The existence of comprehensive health and safety legislation is meaningless unless there is adequate inspection and enforcement by a competent body in which all stakeholders have confidence. The concept of an independent mines inspectorate dates from the mid-1800s, with stringent qualifications now needed (Box 7.8). In the UK, for example, an inspector must be accredited to the level of coal mine manager, while US Mine Safety and Health Administration (MSHA) inspectors are trained in the organisation's own college (Box 7.9). In both countries, compliance with inspectorate directives is mandatory and inspectors have the authority to close an unsafe mine immediately.

Box 7.8 .... Coal mining health and safety legislation in the UK

Legislation aimed at enhancing the health and safety of the individual worker has been developed progressively since the early 1800s, with ever-greater levels of individual protection and responsibility. Key pieces of legislation (mainly acts of Parliament) are listed here. In addition, numerous regulations and approved codes of practice provide specific details so that the laws can be properly understood and enforced.

1850 Inspection of Coal Mines in Great Britain Act: established the role of independent safety inspectors and the requirement for regular inspection of coal mines.

1855 Inspection of Coal Mines Act: set out required safety standards for coal mines.

1860 Mines Regulation and Inspection Act: updated and reinforced the previous legislation.

1862 Amending Act: required every mine to have two means of egress.

1872 Coal Mines (Regulation) Act: updated earlier legislation, authorised an increase in the number of mine inspectors and set out the concept of "contraband" items (such as matches and open flames) that could not be taken underground.

1886 Amending Act: gave the government powers to hold formal investigations into mine accidents.

1887 Coal Mines Regulation Act: updated and extended previous legislation.

1896 Coal Mines Regulation Act: introduced the concept of "permitted" explosives for use in coal mines, and set out requirements for dust control by water sprinkling.

1906 Notice of Accidents Act: tightened accident reporting requirements.

1908 Coal Mines Regulation Act: limited daily work to eight hours.

1910 Mines Accidents (Rescue and Aid) Act: required mines to maintain rescue equipment and to have trained rescue brigades.

1911 Coal Mines Act: established requirements for stone-dusting to reduce the risk of underground explosion.

1914 Coal Mines Act: updated previous legislation.

1920 Mining Industry Act: established a separate Mines Department within the Civil Service, and created the Miners' Welfare Fund, financed by production levies, that invested in social and welfare facilities at mines, and in research into mine safety through the Safety in Mines Research Institute.

1954 Mines and Quarries Act: defined the broad principles of good mining practice in relation to changes that had occurred in technology and management structures since 1911. Also introduced the concept of "workmen's inspections" to carry out independent inspections and accident investigations on behalf of mineworkers.

1969 Mines and Quarries (Tips) Act: enacted in response to the Aberfan coal mine tip disaster to regulate the construction and maintenance of waste-disposal facilities.

1971 Mines Management Act: permitted the appointment of assistants to the manager of large-scale coal mines.

1974 Health and Safety at Work Act: consolidated health and safety legislation covering all industries, and introduced the "duty of care" concept that applies to everyone involved. Also introduced statutory compensation for industrial diseases such as pneumoconiosis, and the establishment of the Health and Safety Executive, of which the Mines Inspectorate became a part. Regulations introduced in 1999 generally make more explicit what employers are required to do to manage heath and safety under the Act, notably to perform risk assessments.

1993 Management and Administration of Safety and Health at Mines Regulations (MASHAM): recognised the importance of formal supervision, professional training, quality reporting and adequate inspection of all districts (zones).

Source: updated from Bryan (1975).

To give an example of the thoroughness with which MSHA operates, each year it is obliged by law to undertake four complete inspections of every underground coal mine and two inspections of every surface operation. In 2006, the organisation achieved a 95% inspection rate, down from over 99% the previous year as several major incidents took up inspectors' time. An average of 161 hours per mine was spent on inspections during the year. In the coal sector, MSHA operates on a district basis, with 45 field offices covering 11 mining districts across the US. The agency employed 759 coal-sector inspectors at the end of 2007, an increase of over 200 in the preceding two years, and estimates that it costs USD 20 150 to train each entry-level inspector (Fontaine, 2008). Violations of safety law can lead to penalties of up to USD 220 000 per citation, although the actual amount will depend on the severity of the non-compliance, the actions of the mining company to remediate the problem promptly and the impact that the penalty would have on the company's financial standing. Penalties charged for coal industry safety violations during 2006 totalled USD 22.5 million (MSHA, 2008a). For perspective, the total value of US coal production in 2006 was estimated to be USD 26.8 billion (NMA, 2007).

Box 7.9 .... Federal legislation affecting coal mining health and safety in the United States

Legislation has developed over the last 100 years such that coal mining in the US is a relatively safe industrial activity, with 33 fatalities in 2007, a rate of 0.03 fatalities per million tonnes mined.

1910 Organic Act: established the US Bureau of Mines

1966 Federal Metal and Non-metallic Mine Safety Act: demanded annual inspections of underground mines.

1969 Federal Coal Mine Health and Safety Act: established the Mining Enforcement and Safety Administration (MESA) which later became the Mine Safety and Health Administration (MSHA).

1970 Federal Occupational Safety and Health Act: created the National Institute for Occupational Safety and Health (NIOSH) within the US Department of Health and Human Services, and the Occupational Safety and Health Administration (OSHA) under the US Department of Labor. OSHA is responsible for developing and enforcing workplace health and safety regulations, while NIOSH is tasked with research, information, education and training in occupational health and safety.

1977 Federal Mine Safety and Health Act: combined coal, metal and non-metallic mining under one legislative instrument. Established the National Mine Health and Safety Academy which operates under the US Department of Labor.

2007 Mine Improvement and New Emergency Response (MINER) Act: requires operators of underground coal mines to improve accident preparedness.

Sources: MSHA (2007b, 2008a and 2008b).

The critical factors for a successful inspection and enforcement regime are trust and incorruptibility. Based on their practical experience, inspectors play an advisory as well as a regulatory role and, as such, are a major asset in the introduction and maintenance of safe working practices in all countries. For much of the past 200 years, the development of health and safety legislation focused on improvements on working practices and protection for miners. More recently, increasing emphasis has been placed on the responsibility of the individual to ensure safe working conditions and practices within these regulatory requirements; in other words, individuals now bear more of the responsibility for their own safety and that of their fellow workers.

Pollution control and other environmental legislation

Concern over damage to the environment has grown since the mid-20th century, once the impact of human industrial activity became clearer. Environmental protection legislation is generally more recent than that relating to mining and to health and safety (Box 7.10). However, it has evolved quickly under both national and international frameworks, reflecting the transboundary impacts of pollution.

Box 7.10............. Environmental law

"Environmental law is a comparatively new branch of law and has evolved mainly over the last thirty years. It is therefore as yet in a formative stage and is undergoing a process of rapid development inspired also by a quantum leap in our understanding of the environmental challenge." (UNEP, 2005).

Environmental problems are related mainly to two areas of human activity:
- use of resources at unsustainable levels; and
- contamination of the environment through pollution and waste at levels beyond the capacity of the environment to absorb them or render them harmless.

Resulting ecological damage seen around the world includes (*ibid.*):
- biodiversity loss;
- pollution of water and consequent public health problems;
- air pollution and resulting increases in respiratory diseases and the deterioration of buildings and monuments;
- loss of soil fertility, desertification and famine;
- depletion of fishing resources;
- increase in skin cancers and eye diseases in certain areas as a result of ozone depletion;
- new diseases and more widespread disease vectors; and
- damage to future generations.

To be effective, environmental protection legislation has to be unambiguous, conform with international standards, and be enforceable in practice. Complexity, in terms of both regulatory requirements and administrative implementation, leads to opportunity for misinterpretation and can be a deterrent to investment. Table 7.2 provides a template for the structure of effective environmental legislation, identifying the targets, suggesting mechanisms for implementation and demonstrating the need for enforceability.

To enable the sharing of knowledge and experience in the field of environmental law, the OECD has co-operated with the European Environment Agency to develop a database of instruments used for environmental policy and management of natural resources.[1] It is a rich source of information, drawn from OECD member countries and some non-member countries, on environment-related taxes, fees and charges, tradable permit systems, deposit refund systems, environmentally motivated subsidies and voluntary approaches.

---

1. www2.oecd.org/ecoinst/queries

Table 7.2 ....Template for elements of environmental management law and regulations

General objective of the legislation	Pollution prevention and control (reduce risk; improve, maintain and restore environmental quality; prevent and control pollution; sustain environmental uses; clean up past pollution).
Scope or relevant areas of regulation	Framework environmental laws provide the general principles of environmental regulation, including providing for cross-cutting issues and mechanisms. Specific topics include: Air pollution Noise pollution Freshwater pollution Pollution of the coastal and marine environment Land degradation and soil pollution Sustainable use of environmental resources
Selection of environmental management approaches	Command and control, the use of economic and market-based instruments, risk-based instruments, pollution prevention (regulatory, voluntary, liability), standard setting (ambient, technology, performance, economic and voluntary standards), permits/authorisation, inspection and monitoring compliance, use of economic instruments such as the "polluter pays" principle to internalise the cost, use of market-based mechanisms to discourage or encourage behaviour with its incentives or disincentives, self-regulation, clean development mechanism, land-use planning and zoning, international co-operation, environmental impact assessments, integrated resource management, training, education and public awareness.
Types of national actions and laws	Legislation, regulations, permits and licences, court cases/precedents, taking administrative action (e.g. fines and other penalties), setting up compliance programmes.
Ensuring compliance and enforcement of laws	Promoting and monitoring compliance, and reviewing and evaluating the effectiveness of national legislation to ensure adequate enforcement mechanisms are in place, institutions have capacity, laws act as a deterrent and are enforceable.

Source: UNEP (2006).

In the past, coal mining often had a poor environmental record. While working practices are now vastly improved, past mining activity has left a substantial legacy of environmental damage that will require huge investment to remediate. To avoid future damage, major coal-producing countries have enacted environmental protection legislation, although some developing countries still lack the institutional capacity to ensure that it is fully implemented, monitored and enforced (Foster, 2001). Efforts to minimise environmental impacts begin during the permitting process for a new mine. For example, UK planning authorities define times when blasting is permitted, allowable truck movements, noise and dust limits, operating hours, standards and schedules of restoration, levels of compensation paid to landowners, project duration and many other details that must be adhered to before mining can commence and throughout the project (CLG, 1999). Regulatory authorities and international financial institutions require the preparation of environmental impact assessments and statements, along with remediation plans, covering wastewater treatment, airborne emissions, noise, visual impact, land restoration, spoil tip landscaping and long-term, post-mining monitoring and maintenance. Most countries now require that financial bonds be posted before permits can be issued. These are released once site restoration has reached an approved state, often several years after a mine has ceased production, and the sustainability of restoration has been demonstrated. Bond requirements vary from country to country, but must be sufficient to incentivise proper land restoration; bonds can be as high as USD 10 000/ha of affected land (Foster, 2001).

Solid waste disposal has been a problem for the coal mining industry for centuries, with the potential for long-term environmental impacts from dust, acid water runoff and landslides caused by slope instability. While current legislation requires that waste-disposal sites are well designed with prompt site restoration, many countries have a legacy of abandoned mine workings and spoil tips. Costs of their remediation can be borne in a number of ways, ranging from inclusion in wider economic regeneration programmes to site-specific projects that are part-funded by levies on coal production, as in the US (Box 7.11). Subsidence damage has also been a major problem, especially where water resources have been disrupted. Repair of third-party property damage must be managed – ideally by the mining company responsible for the damage, but ultimately by governments who must therefore ensure that the mining industry finances these costs. In the US, an Abandoned Mine Land fee is levied on a per-ton basis to fund further remediation beyond what is achieved through best practice.[10]

Box 7.11.............. Environmental law in the United States

The current legislative regime that regulates the environmental impact of coal mining and utilisation dates mainly from the 1970s, with some subsequent amendments. Within a body of around a dozen laws, three are predominant in this respect:

- Clean Air Act of 1970 and the Clean Air Act Amendments of 1990;
- Clean Water Act of 1977; and
- Surface Mining Control and Reclamation Act of 1977.

The National Environmental Policy Act (NEPA) of 1969 is also of importance (Foster, 2001). Overall implementation at the federal level is the responsibility of the US Environmental Protection Agency (US EPA), although a host of other agencies are involved, at both federal and state level.

The Clean Air Act of 1970 (and its subsequent amendment in 1990) was one of the most influential pieces of legislation ever to affect the coal industry. Its impact on coal utilisation in particular has been enormous, and was mainly responsible for the growth in low-sulphur coal production from the US Midwestern coalfields, notably the Powder River Basin. In addition to placing limits on atmospheric sulphur dioxide (SO_2) emissions, the 1970 legislation also required dust-emission controls to be implemented. Its success in this respect is shown by the fact that between 1970 to 1993, PM_{10} emissions declined from coal-fired power stations while coal consumption to produce electricity more than doubled.

Permissible emissions from coal utilisation have become increasingly restricted since the laws were first enacted, with tighter controls on SO_2 and particulates and, more recently, nitrogen oxides (NOx). The process is continuing, with pressure growing for the removal of mercury and other trace elements from coal-utilisation emissions.

The Surface Mining Control and Reclamation Act (SMCRA) of 1977 sought to bring environmental responsibility to the surface-mining sector of the US coal industry and to repair environmental damage that had been caused by earlier, unregulated mining activity. It established the Office of Surface Mining Reclamation and Enforcement (OSM) as the principal regulatory agency for the legislation. While implementation

---

10. Fees of USD 0.35 per ton of surface-mined coal, USD 0.15 per ton of coal mined underground and USD 0.10 per ton of lignite mined are deposited in an interest-bearing Abandoned Mine Reclamation Fund, which is used to pay reclamation costs of approved projects (OSM, 2008). States and Indian tribes have priority access to 50% of the fees collected in their regions, but face the hard choice of which projects to fund – revenues are insufficient to reclaim all eligible sites.

is predominantly at state level, with the OSM overseeing compliance, the OSM is also the controlling agency in states and tribal territories that do not have appropriate laws of their own.

Key features of the SMCRA include:

■ the need for mining companies to apply for and receive a mining permit before mining can begin – the permit application must describe plans for operations, including environmental compliance;

■ the need for mining companies to post bonds (financial guarantees) to cover the cost of final reclamation and environmental closure of a mine site;

■ the establishment of responsibility by company directors for environmental compliance – substantial failure by a mining company to satisfy environmental obligations results in a prohibition against parties who were in an "ownership and control" role from taking out new mining permits, either as individuals or as owning or controlling parties of other companies;

■ the establishment of an inspection system to ensure compliance with the law;

■ banning mining in some areas, such as national parks; and

■ the establishment of the Abandoned Mine Reclamation Fund, financed by a per-ton levy on coal production, and used to reclaim derelict mined land from before 1977. The law was later amended to allow the fund to be used on post-1977 projects as well.

Sources: EPA (2008a and 2008b) and OSM (2008).

The impact of coal mining is by no means all negative, with major economic advantages to local communities that have to be balanced against the potential for short- or longer-term environmental disruption. Thus, the economic value of coal production at both national and local levels is one of the considerations during permitting.

Transport choices largely reflect the relative cost of transport fuels and the investment that has been made in transport infrastructure. In China, low levels of tax on diesel fuel have encouraged widespread coal transport by truck – higher taxes would discourage long-distance haulage in favour of rail or water-borne transport. Uncontrolled road transport of coal has major impacts on the road network in terms of increased maintenance costs, and on the surrounding corridor in terms of dust, water runoff, noise and road safety. In some cases, conveyor systems have been used to eliminate the environmental impact of road transport on local communities (Walker, 2006).

In terms of coal utilisation, measures such as the UK Clean Air Acts of 1956 and 1968, and the US Clean Air Act of 1970 and its subsequent amendments brought power plant emissions under public scrutiny. The first UK Act introduced zones in which only authorised smokeless fuels may be burnt, including hard coke, anthracite, petcoke and blends. Precise details of the zones and all authorised fuels are published by the government.[11] More recently, in Europe, the EC Large Combustion Plants Directive of 1988 and its revision in 2001 limit allowable sulphur dioxide, NOx and particulates emissions from plants rated above 50 MWth (see Box 7.12 for an example

---

11. www.uksmokecontrolareas.co.uk

of one such plant), with little leeway for plants that are unable to meet strict emission limits – reduced operation and the closure of older plants being one intended outcome, and the installation of pollution control equipment being another (Nalbandian, 2007).

Box 7.12.............. Drax power station, UK

The 4 000 MWe Drax power station in the UK is owned by Drax Power plc, a private company listed on the London stock exchange (Figure 7.14). The last of six 660 MW units at the station was commissioned in 1986, but it was 1993 before a flue gas desulphurisation (FGD) plant was constructed and operational. The performance of this FGD plant is crucial in meeting the UK's national SO_2 emission target of just 585 ktpa in 2010 under EU and international law. In 1998, the FGD system suffered from a major design fault that necessitated its fans being taken out of service for repair. Special and exceptional permission was granted for Drax to emit SO_2 above its permitted level until all of the fans had been repaired, allowing the plant to remain in operation. Since that incident, FGD performance has been improved and the proportion of low-sulphur, imported coal increased such that SO_2 emissions are now below 1 g/kWh. Emissions from the plant are continuously monitored and publicly reported by the Environment Agency of England and Wales on its website. As required under UK company law, Drax Power publishes an annual environmental report with full information on fuel use and emissions to air, land and water. Reducing emissions of CO_2 are now the focus of attention – Drax Power is a major player in the EU Emission Trading Scheme. It has a target to replace 10% of its fuel input with biofuels by 2009 and is investing up to GBP 100 million (USD 200 million) to upgrade the performance of its six steam turbines with more efficient, re-profiled blading. To reduce fuel costs, Drax Power also burns a small quantity of petcoke; this was authorised after more than ten years of debate and public scrutiny of tests demanded by the Environment Agency.

Sources: Drax Power (2008) and Environment Agency (2008).

Figure 7.14 ....Drax power station, UK

At 3 945 MW, Drax power station is one of the largest thermal power plants in Europe (the 4 440 MW Bełchatów lignite-fired power plant in Poland is the largest).

Emissions from cokemaking are controlled in IEA member countries, with non-compliance having resulted in plant closure or renewal in some cases (Reeve, 2000). Atmospheric emissions from cokemaking include carcinogenic organic compounds as well as dust and noxious fumes, all of which are subject to regulatory control. Personal exposure to potential carcinogens is an area of particular importance, with several countries having enacted regulations that provide time limits to exposure and requirements for respiratory protection.

Box 7.13.............. Environmental law and water management in Australia

Australia's environmental legislation was consolidated in 1999 when the Protection of the Environment Operations (POEO) Act 1997 came into force. This replaced a number of earlier laws relating to water, air, noise and pollution, and was itself amended in 2006 in the light of experience. Implementation of "scheduled" activities (of which coal mining is one) is through individual state Environmental Protection Agencies.

Key features of the POEO Act include:
- permitting coal mines in relation to dust, wastewater, solid waste and noise emissions;
- requiring financial bonds to cover post-mining reclamation; and
- establishing the "polluter pays" principle, with companies that cause pollution liable to make good any damage at their own cost.

The extraction of surface and groundwater at levels close to or exceeding sustainable limits in Australia has seen water bodies deteriorate and salinity increase. The Murray-Darling Basin is particularly affected by the competing demands of agriculture and the mining industry. In response, the government's National Water Initiative aims to establish a nationally co-ordinated water market that reflects the true value of water. This will improve water management and increase investment in water recycling, desalination and infrastructure, but will require careful design to balance the legitimate needs of residential consumers and farming, against the economic value of water to industry. In this respect, water quality is as important as water volume. Thus, the market must be structured to ensure that industry values the use of poorer-quality water while farmers are frugal in their use of good-quality water.

Source: New South Wales Department of Environment and Climate Change (2008), NWC (2007) and COAG (2004).

Coupled with well-crafted laws and regulations that are effectively enforced, financial frameworks offer an additional means of directing industry towards better working practices (*e.g.* Box 7.13). The widespread adoption of the "polluter pays" principle has been an effective way to ensure greater compliance with environmental regulation. While pollution charges or fines for non-compliance are fundamental to the principle, in practice their size must reflect the polluter's ability to pay, the level of non-compliance and the degree of environmental damage caused and consequent socio-economic costs (Sloss, 2003). In any event, they should encourage pollution prevention, since this is preferable to pollution control or remediation of damage caused by pollution. Pollution control legislation must reflect international, national and local concerns. For example, the environmental impacts of power stations can cross national boundaries and has led to treaties such as the UNECE LRTAP (initially applied to North-West Europe, and later extended; Box 7.14), whereas the permitting process for a new coal mine is more concerned with local issues such as dust emissions or any potential effects on a community's water resources. In all situations, transparency in the application of environmental legislation, adequacy of inspection and enforcement, and public information on compliance and any shortfalls are all vital to achieving the goal of a clean environment.

Box 7.14.............. Environmental law in the European Union

Key areas covered by European Union environmental legislation in relation to coal production and utilisation include air pollution and waste disposal.

Overall, European Union environmental law can be categorised as covering:

■ Pollution and nuisances, under which fall:
nuclear safety and radioactive waste;
water protection and management;
monitoring of atmospheric pollution;
prevention of noise pollution; and
chemicals, industrial risk and biotechnology.
■ Space, environment and natural resources, covering:
management and efficient use of space, the environment and natural resources;
conservation of wild fauna and flora; and
waste management and clean technology.

European law is also influenced by the UNECE Convention on Long-Range Transboundary Air Pollution (LRTAP) under which signatory countries within Europe (and elsewhere) committed themselves to limiting, and to gradually preventing and reducing, their discharges of air pollutants. Coming into force in 1983, the convention was supplemented by eight protocols:

1987 reduction of sulphur emissions by at least 30%;

1988 long-term financing of the co-operative programme for the monitoring and evaluation of the long-range transmission of air pollutants in Europe;

1991 nitrogen oxides;

1997 volatile organic compounds (VOC);

1998 additional reduction of sulphur emissions;

2003 persistent organic pollutants (POPs);

2003 heavy metals; and

2005 acidification, eutrophication and ground-level ozone.

The EC Large Combustion Plants Directive (LCPD) of 1988 (88/609/EEC) established emission standards for new combustion plants with a capacity of more than 50 MWth, including power plants fuelled by coal. The LCPD was revised in 2001 (2001/80/EC), imposing more stringent emission limits for SO_2, NOx and particulates. Operators of existing plants have the choice of complying with these new emission limit values (ELVs), or effectively derating their plants in order to achieve equivalent reductions before mandatory closure. In addition, plants with a capacity larger then 20 MWth fall under the EC Integrated Pollution Prevention and Control (IPPC) Directive of 1996 (Box 7.15) and are now obliged to participate in the European Union Emissions Trading Scheme for CO_2 emissions.

Sources: EC (2008b); EUR-Lex (2008); Nalbandian (2007); and Sloss (2003).

Box 7.15 .... State-of-the-art clean coal technologies: assessments by the European Integrated Pollution Prevention and Control Bureau

The EC Integrated Pollution Prevention and Control Directive (96/61/EC) is a key piece of legislation to reduce industrial pollution and improve the quality of air, water and land within the European Union. When granting permits, pollution control inspectors have to judge whether industrial plant operators are using "best available techniques", as demanded by the directive. To help in this task, the European Integrated Pollution Prevention and Control Bureau was established in Seville, Spain to exchange and collate technical information in the form of Reference Documents on Best Available Techniques (BREFs). The 580-page BREF covering large combustion plants (>50 MWth) for power and heat generation was issued in July 2006 and provides a detailed account of the technical characteristics, costs and performance that can be achieve from all types of combustion plants with the pollution control technologies that are currently available. The document also covers upstream and downstream activities such as fuel preparation and ash treatment. BREFs covering large industrial sectors, including iron and steel, are freely available to download from the Bureau.[1]

---

1. http://eippcb.jrc.es/pages/FActivities.htm

Table 7.3 .... Best available techniques for reducing NOx from coal- and lignite-fired combustion plants: an example of plant performance data contained in BAT (best available techniques) Reference Document (BREF)

Capacity, MWth	Combustion technique	NOx emission level associated with BAT, mg/Nm³			BAT options to reach these level
		New plants	Existing plans	Fuel	
50 – 100	Grate-firing	200 – 300*	200 – 300*	Coal and lignite	Pm and/or SNCR
	PC	90 – 300*	90 – 300*	Coal	Combination of Pm and SNCR or SCR
	CFBC and PFBC	200 – 300	200 – 300	Coal and lignite	Combination of Pm
	PC	200 – 450	200 – 450*	Lignite	
100 – 300	PC	90* – 200	90 – 200*	Coal	Combination of Pm in combination with SCR or combined techniques
	PC	100 – 200	100 – 200*	Lignite	Combination of Pm
	BFBC, CFBC and PFBC	100 – 200	100 – 200*	Coal and lignite	Combination of Pm together with SNCR
>300	PC	90 – 150	90 – 200	Coal	Combination of Pm in combination with SCR or combined techniques
	PC	50 – 200*	50 – 200*	Lignite	Combination of Pm
	BFBC, CFBC and PFBC	50 – 150	50 – 200	Coal and lignite	Combination of Pm

* Some split views appeared in these values.

Notes: PC: pulverised combustion; BFBC: bubbling fluidised bed combustion; CFBC: circulating fluidised bed combustion; PFBC: pressurised fluidised bed combustion; Pm: primary measures to reduce NOx; SCR: selective catalytic reduction of NOx; SNCR: selective non-catalytic reduction of NOx. The use of anthracite hard coal may lead to higher emission levels of NOx because of the high combustion temperatures.

Source: EIPPCB (2006).

Elements from legislation in IEA member countries relevant to China

China does not lack legislation covering coal-mine administration, health and safety and environmental issues. Yet, as Andrews-Speed (2007) notes, in many cases laws, "have been developed and promoted by institutions with noticeably different agendas. This, combined with the reactive nature of these regulatory initiatives ... has led to duplication, inconsistency and ambiguity." There have also been significant shortfalls in enforcement by the regulatory authorities for a variety of reasons, the sheer number of mines that require inspection being an obvious one. Conflicts of interests have also been apparent, especially within the small-mine sector, since local governments derive income from these mines. China is not unique in facing such issues; all the countries reviewed above have had to deal with similar challenges.

Clearly there are opportunities for China to clarify the various aspects of its legislative framework covering coal production, transport, utilisation and environmental issues, perhaps by adopting a structure that is easier to understand, interpret, administer and enforce. By doing this, opportunities for misinterpretation and non-compliance would be reduced significantly. Table 7.4 collects together key elements that have proven to be important in establishing effective legislation in IEA member countries.

In addition, there appears to be a pressing need to strengthen the mines inspectorate in terms of resources and capabilities, such that a truly independent inspectorate can be established in which all parties involved in coal production can have confidence. China's largely uneducated, often migrant workforce can be open to exploitation in terms of workplace conditions and safety. In the US, each MSHA inspector has responsibility for regulatory enforcement on behalf of an average of around 160 coal miners. By contrast, the huge number of miners working in the coal industry in China, likely to be over 4 million men,[12] presents the existing inspectorate body with an impossible challenge when considered on a comparable basis.

Table 7.4 ....Key elements of licensing, health and safety, and environmental legislation in IEA member countries

Licensing	Health and safety	Environmental
• security of tenure • continuity of title • administrative clarity • recognition of the economic value of coal production within planning processes • relinquishment of a proportion of land rights upon exploration licence renewal to reduce speculation and to encourage resource evaluation	• technical competence of the mine operator • strict adherence to health and safety legislative requirements • strict adherence to approved mining method statements • responsibility of each individual for personal safety, and that of fellow workers • an independent mines inspectorate to monitor, advise and enforce	• statutory control of emissions to air, land and water • safe long-term storage of solid wastes • mine-site rehabilitation to original state or better • minimisation of transport system impacts on local communities • "polluter pays" principle • pollution prevention is better than control or remediation

---

12. Table I.E in Appendix I reports 2.7 million employees in state-owned mines, to which must be added miners working in TVE and private coal mines, estimated here to be over 1.5 million.

From this brief review of the development of coal mining legislation and regulation in IEA member countries, the following recommendations emerge as ones that might help address the challenges that China faces today:

■ ensure technical competence of mine operators;

■ strengthen the mines inspectorate and guarantee its independence;

■ ensure a clear and consistent regulatory framework for coal-mining operations;

■ enhance enforcement of existing pollution control requirements;

■ enhance training of underground workers;

■ ensure that existing penalty structures for regulatory non-compliance are enforced;

■ extend the policy of consolidating small mines into larger, more efficient units; and

■ extend social support provisions for ex-mine workers and those injured in accidents.

International co-operation and collaboration at government and industry levels could assist in the implementation of these and other recommendations. However, current Chinese resource management legislation provides for state ownership of coal resources and while majority foreign investment in coal production is allowed, it is not favoured, creating a barrier to industrial participation. A more open policy towards foreign investment in the energy sector could allow experience from other countries to spread more quickly into China than government-to-government co-operation can achieve alone. Despite this, there has been significant foreign interest in areas such as coal-to-liquids and coal mine methane, with joint ventures between local and overseas partners. Chapter 9 examines international co-operation from the perspective of deploying clean coal technologies.

CONCLUSIONS FROM COAL-POLICY EXPERIENCE IN IEA MEMBER COUNTRIES AND THE GLOBAL TRANSITION TO CLEANER COAL

Given the abundance of coal resources in comparison to other energy resources, there is widespread consensus that coal will remain a vital component of world primary energy supply (CIAB, 2008). In China, coal resources outweigh other domestic energy resources, so coal exploitation will grow as the national economy expands. Neither the 11th Five-Year Plan for Energy or the IEA *World Energy Outlook 2007* envisage a decline in coal's share of primary energy supply. Thus, coal supply will increase more or less in line with overall economic growth. However, concerns over the environmental and social impacts of markedly higher production, greenhouse gas emissions from coal use and climate change are all creating pressure to improve the technologies for coal production and utilisation. The rapid deployment of clean coal technologies is essential if coal is to retain its position in world energy markets without adding to the risk of climate change. The Chinese government's target of reducing primary energy intensity by 20% between 2005 and 2010 is one response and its National Climate Change Programme, launched in 2007, is another. One potential constraint to achieving the goals set out is the relative complexity of the tiered administration systems (national, provincial and local). In this section, coal-related policies that have been effective in IEA member countries are considered against China's needs.

Applying coal-policy experience in China

With its fast-growing economy, China has the financial resources to invest in clean coal technologies along the whole supply chain, from production to end use. The country's established construction and technology industries provide it with the capability to install and operate state-of-the-art combustion and other utilisation systems, thereby gaining the benefits of higher efficiencies and reduced environmental impacts. However, in order for China to become a world leader in the adoption of clean coal technologies, major amendments need to be made to the investment and regulatory structures governing the coal cycle from production to end use. This study finds that recommendations made by the World Bank (2004) remain relevant, *i.e.* the licensing and administration systems for resource acquisition and tenure have to be strengthened, mine safety monitoring and regulation needs unequivocal support from all stakeholders, environmental licensing, monitoring and rehabilitation requires greater commitment from national government downwards, and the investment climate for constructing and operating state-of-the-art power plants, coking plants and industrial furnaces must be improved. While the progress already made by the Chinese authorities in a number of these areas is to be commended, many of the World Bank's recommendations remain as valid now as they did in 2004. China's current focus is on improving efficiency through the better use of resources, but leaves many other issues to be addressed. These include the social impact of coal-mine closures on rural areas, and the environmental legacy of under-regulated mining activities and coal transport. That said, China does not have to work in isolation, since much experience has been gained in IEA member countries on all aspects of clean coal. The potential to transfer this experience means that China can benefit by avoiding mistakes that others have made in the past. With this in mind, the remainder of this section comprises a selection of relevant experiences that, if acted upon, could help in that direction.

Industry restructuring Successful coal industry restructuring programmes always involve all stakeholders. Imposing new structures without adequate consultation or investment in alternative employment opportunities leads to high, long-term support costs for the affected communities. The alternative is community breakdown, worker migration and rural depopulation. Experience from IEA member countries in implementing well-planned coal industry restructuring programmes can provide transferrable measures that will assist China as it goes through the same process.

Rural regeneration China's promotion of small-scale coal mining in the 1980s and 1990s was in part aimed at alleviating rural poverty, in which it has been successful (Andrews-Speed, 2007). More recently, the widespread closure of small mines has led to a loss of employment opportunities, mainly within the same rural economies. It is a waste of human resources if redundant mine workers have no alternative means of earning a living within their own communities. Increased investment will be needed in rural areas affected by coal-mine closures – in training programmes to enhance workforce skills, and in attracting replacement job opportunities.

Stakeholder involvement

Establishing strong community relations and civic pride is vital to the success of coal-mine and power-plant projects during development, operation and closure. Greater community participation should be sought as a means of increasing awareness of the benefits of clean coal technologies and techniques throughout the coal supply chain.

Technology development

Today's broad sustainable development targets mean that state-directed projects can be less effective than multi-participant approaches to developing new technologies (see Chapter 9). China should encourage multi-partner input to the future development of coal production, environmental protection and coal utilisation technologies.

Coal transport

Coal transport in China, especially at a local level, has a huge environmental impact on the surrounding countryside and communities, notably dust pollution and damage to the road infrastructure. Further investment is needed in alternatives to road transport for coal movements to major users, a task that should be made easier if the number of mines in operation is reduced.

Landscape restoration

Uncontrolled coal mining in the past has left a legacy of environmental degradation in many countries. China also faces long-standing problems of deforestation and desertification. In IEA member countries, governments have responded with policies and many cost-effective remediation techniques that China could adapt (Figure 4.15). There will be a long-term need for investment in landscape restoration in China, with particular emphasis on revegetation and slope stability in an effort to combat further erosion and loss of soil fertility.

Figure 7.15 .... UK Coal's Waverley opencast coal mine

After producing 5 million tonnes of coal from a former, heavily-contaminated industrial area, restoration of the 295 ha Waverley opencast site is nearing completion with a housing development, business areas and a country park.

Social demands

Clean coal encompasses more than just state-of-the-art mining and power-station technology. Social and environmental aspects have to be taken into consideration as well, especially now that large sections of China's population are becoming more environmentally aware. Planning of the coal sector should take a holistic approach that includes social and environmental aspects to help justify the use of clean coal technologies.

The global transition to cleaner coal

To date, the introduction of clean coal technologies in IEA member countries has been mixed. Certainly, initial steps have been taken, with the adoption of supercritical steam cycles for pulverised coal-fired power stations. However, the higher costs of building, operating and maintaining plants that make use of state-of-the-art technologies, from supercritical to ultra-supercritical steam conditions and integrated gasification combined cycle (IGCC), have constrained their implementation. In the case of IGCC, only demonstration plants have been built and operated to date, and hence there has been no experience of cost reductions over time at commercial units. Of even greater concern, Kessels *et al.* (2007) suggest that, unless mandatory implementation is brought in, "CCTs are not likely to be implemented on a large scale in the short or mid-term".

The same authors commented that, "[t]wo different market imperfections currently prevent uptake of CCTs: insufficient internalisation of environmental externalities and those imperfections associated with technological change". They suggested that accelerating deployment will require changes at national and international level. If commercial deployment of CCTs is to occur, companies will need investment certainty with a policy framework that recognises the costs and economic risks of long-term capital investment in supercritical technologies, IGCC and CO_2 capture and storage (CCS). Kessels *et al.* go on to suggest that existing emissions trading schemes have had only limited impact on efforts to cut CO_2 emissions under the Kyoto Protocol, with structures that have failed to encourage CCTs or CCS. Indeed, successful CCS has yet to be demonstrated at any coal-fired power plant. Although moves to change this are already underway, the over-riding risk is that there are still too few positive incentives for countries with developing economies, including China, to adopt clean coal technologies as a matter of course when proven, subcritical power generation systems are readily available. Clearly, construction cost is a major factor here, yet this is a short-term perspective on what is a long-term issue, since a subcritical power plant built in 2007 is likely to be operating in 2050.

In practice, the development and introduction into commercial service of clean coal technologies and techniques has largely come about through a multi-participant approach. Government provides the framework, and in some cases carries out early-stage research, within which utilities and equipment suppliers can then feel sufficiently confident to move to further stages of demonstration plants and deployment. Support for projects such as coal-based IGCC units comes from consortia of interested parties; project risks are too great for individual participants to carry on their own, but become more manageable when shared.

In summary, the adoption of supercritical and ultra-supercritical equipment in power stations marks a stepping stone on the path towards greater deployment of clean coal technologies. However, a milestone will be reached when commercial CCS becomes a reality; unless there is greater near-term commitment to developing and demonstrating these systems through the use of grants and investment incentives from national governments and international bodies alike, this will not happen until the cost of carbon universally reaches much higher levels than seen anywhere recently. With supercritical coal-fired power stations already in operation in China, the country has a great opportunity to bypass existing subcritical technologies for new plants and to focus on achieving higher conversion efficiencies for its new electricity generating capacity.

REFERENCES

Adams, L. (2007), personal communication, UMWA Career Centers Inc., Washington, PA, September.

Andrews-Speed, P. (2007), "Marginalisation in the Energy Sector: the Case of Township and Village Coal Mines" in H. X. Zhang, B. Wu and R. Saunders (eds.), *Marginalisation In China*, Ashgate Publishing Ltd., Aldershot, UK, pp. 55-80, August.

ARC (Appalachian Regional Commission) (2007), "About ARC – Highway Program", Appalachian Regional Commission, Washington, DC, www.arc.gov/index.do?nodeId=1006, accessed 15 July 2008.

Beatty, C., S. Fothergill and R. Powell (2005), *Twenty Years on: Has the Economy of the Coalfields Recovered?*, Sheffield Hallam University, Centre for Regional Economic and Social Research, Sheffield, UK.

BERR (Department for Business Enterprise and Regulatory Reform) (2007), *Digest of United Kingdom Energy Statistics 2007*, The Stationery Office, London.

BERR (2008), *Digest of United Kingdom Energy Statistics 2008*, The Stationery Office, London.

Black, D. A. (2007), personal communication, University of Chicago, IL, September.

Black, D., K. Daniel and S. Sanders (2002), "The Impact of Economic Conditions on Participation in Disability Programs: Evidence from the Coal Boom and Bust", *The American Economic Review*, 92 (1), pp. 27-50. March.

Black, D. A., T. G. McKinnish and S. G. Sanders (2005), "Tight Labor Markets and the Demand for Education: Evidence from the Coal Boom and Bust", *Industrial and Labor Relations Review*, 59 (1), pp. 3-15.

Black, D. A. and S. G. Sanders (2004), *Labor Market Performance, Poverty, and Income Inequality in Appalachia*, Appalachian Regional Commission, Washington, DC.

Bogalla, B. (2008), personal communication, Gesamtverband Steinkohle (GVSt – German Hard Coal Association), Essen, Germany, June.

Bradley, D. H., S. A. Herzenberg and H. Wial (2001), *An Assessment of Labor Force Participation Rates and Underemployment in Appalachia*, Appalachian Regional Commission, Washington, DC.

Bryan, A. (1975), *The Evolution of Health and Safety in Mines*, Ashire Publishing, Letchworth, UK.

Caudill, H. (1962), *Night Comes to the Cumberlands*, Little, Brown and Co., New York.

CIAB (Coal Industry Advisory Board) (2008), *Clean Coal Technologies: Accelerating Commercial and Policy Drivers for Deployment*, CIAB, International Energy Agency, Paris.

CLG (Department for Communities and Local Government) (1999), *Minerals Planning Guidance 3: Coal Mining and Colliery Spoil Disposal*, CLG, The Stationery Office, London, March, www.communities.gov.uk/documents/planningandbuilding/pdf/154812.pdf.

COAG (Council of Australian Governments) (2004), *Intergovernmental Agreement on a National Water Initiative between the Commonwealth of Australia and the Governments of New South Wales, Victoria, Queensland, South Australia, the Australian Capital Territory and the Northern Territory*, COAG, Canberra, ACT, Australia, 25 June, www.nwc.gov.au/resources/documents/Intergovernmental-Agreement-on-a-national-water-initiative.pdf.

Drax Power (2008), "Environmental Performance", Drax Power plc, www.drax-group.co.uk/corporate_responsibility/environment/envperformance, accessed 15 July 2008.

Drouard, J. (2000), personal communication, Charbonnages de France, Rueil-Malmaison, Paris, October.

DTI (Department of Trade and Industry) (2001), "Trends in Coal Production and Consumption", *Energy Trends*, URN 01/79B, DTI (now Department of Energy and Climate Change), London, September, www.berr.gov.uk/files/file40592.xls.

Dunne, T. and D. R. Merrell (2001), "Gross Employment Flows in US Coal Mining", *Economics Letters*, 71, pp. 217-224.

EC (European Commission) (2008a), "Health and Safety at Work", EC Directorate-General for Employment, Social Affairs and Equal Opportunities, Brussels, Belgium, http://ec.europa.eu/social/main.jsp?catId=148&langId=en, accessed 15 July 2008.

EC (2008b), "Summaries of Legislation – Air Pollution", SCADPlus website, European Commission, Brussels, Belgium, http://europa.eu/scadplus/leg/en/s15004.htm, accessed 5 August 2008.

EIA (Energy Information Administration) (1981), *Coal Production 1980*, EIA, Office of Coal, Nuclear, Electric, and Alternate Fuels, US Department of Energy, Washington, DC.

EIA (1986), *Coal Production 1985*, EIA, Office of Coal, Nuclear, Electric, and Alternate Fuels, US Department of Energy, Washington, DC.

EIA (1991), *Coal Production 1990*, EIA, Office of Coal, Nuclear, Electric, and Alternate Fuels, US Department of Energy, Washington, DC.

EIA (1996), *Coal Industry Annual 1995*, DOE/EIA-0584(95), EIA, US Department of Energy, Washington, DC.

EIA (2001), *Coal Industry Annual 2000*, DOE/EIA-0584(2000), EIA, US Department of Energy, Washington, DC.

EIA (2006a), *Coal Production in the United States – an historical overview*, EIA, US Department of Energy, Washington, DC, www.eia.doe.gov/cneaf/coal/page/coal_production_review.pdf.

EIA (2006b), *Annual Coal Report 2005,* DOE/EIA-0584(2005), EIA, US Department of Energy, Washington, DC.

EIA (2007), *Annual Coal Report 2006.* DOE/EIA-0584(2006), EIA, US Department of Energy, Washington, DC.

EIA (2008), *Natural Gas Navigator*, EIA, US Department of Energy, Washington, DC, http://tonto.eia.doe.gov/dnav/ng/ng_sum_top.asp, accessed 5 August 2008.

EIPPCB (European Integrated Pollution Prevention and Control Bureau) (2006), *Reference Document on Best Available Techniques for Large Combustion Plants*, EIPPCB, European Commission, Seville, Spain, http://eippcb.jrc.ec.europa.eu.

Environment Agency (2008), "Pollution Inventory", Environment Agency for England and Wales, Bristol, UK, www.environment-agency.gov.uk/business/444255/446867/255244, accessed 15 July 2008.

EPA (Environmental Protection Agency) (2008a), *Clean Air Act*, US Environmental Protection Agency, Washington, DC, www.epa.gov/air/caa, accessed 15 July 2008.

EPA (2008b), *Summary of the Clean Water Act*, US Environmental Protection Agency, Washington, DC, www.epa.gov/lawsregs/laws/cwa.html, accessed 15 July 2008.

EUR-Lex (2008), "Legislation in Force – Environment", Office for Official Publications of the European Communities, Luxembourg, http://eur-lex.europa.eu/en/legis/20080701/chap1510.htm, accessed 5 August 2008.

Fontaine, R. B. (2008) personal communication, Mine Safety and Health Administration, US Department of Labor, Washington, DC, January.

Foster, S. M. (2001), *Comparative Environmental Standards - Deep Mine and Opencast*, CCC/44, IEA Coal Research, London.

Fothergill, S. and N. Guy (1994), *An Evaluation of British Coal Enterprise*, Coalfield Communities Campaign, Barnsley, UK.

Franz, H-W. (1994), *Manual: Social Crisis Management in the Coal and Steel Industries*, Office for Official Publications of the European Communities, Luxembourg.

Frondel, M., R. Kambeck and C. M. Schmidt (2006), *Hard Coal Subsidies: A Never-Ending Story?*, Discussion Paper No. 53, Rheinisch-Westfälisches Institut für Wirtschaftsforschung (RWI), Essen, Germany, November.

Hauk, P. (1981), "German Democratic Republic", *Mining Annual Review 1981*, The Mining Journal Ltd., London.

Hessling, M. (1995), "The Role of Ruhrkohle AG in Germany's Energy and Coal Policy" in C. Critcher, K. Schubert and D. Waddington (eds.), *Regeneration of the Coalfield Areas: Anglo-German Perspectives*, Pinter, London, pp. 55-68.

HUNOSA (Hulleras del Norte SA) (2007), "About Us – History – Present and Future", Hulleras del Norte SA, Oviedo, Spain, www.hunosa.es/en/framework.hps, accessed 15 July 2008.

IEA (International Energy Agency) (2003), *Coal Information 2003*, IEA, Paris.

Kasztelewicz, Z. (2006), "Węgiel brunatny: optymalna oferta energetyczna dla Polski" ("Lignite: Optimal Energy for Poland"), PPWB w Bogatynia, redakcja *Górnictwa Odkrywkowego*, Bogatynia, Wrocław, Poland.

Kessels, J., S. Bakker and A. Clemens (2007), *Clean Coal Technologies for a Carbon-Constrained World*. CCC/123, IEA Clean Coal Centre, London.

Kuby, M. and Z. Xie (2001), "The Effect of Restructuring on US Coal Mining Labor Productivity, 1980-1995", *Energy*, 26, pp. 1015-1030.

Minerals Bureau (1988), *South Africa's Minerals Industry 1988*, Minerals Bureau, Pretoria, South Africa.

Mining Annual Review (2000-2007), *Mining Annual Review*, various years, The Mining Journal Ltd. / Mining Communications Ltd., London.

Mitchell, B. R. and P. Deane (1962), *Abstract of British Historical Statistics*, Department of Applied Economics Monographs, No. 17, Cambridge University Press, Cambridge, UK.

MSHA (Mine Safety and Health Administration) (2007a), *Coal Fatalities for 1900 through 2007*, MSHA, Arlington, VA, www.msha.gov/stats/centurystats/coalstats.asp.

MSHA (2007b), *Mine Improvement and New Emergency Response Act of 2006*, MSHA, Arlington, VA, www.msha.gov/MinerAct/2006mineract.pdf.

MSHA (2008a), "Mine Safety and Health at a Glance", MSHA, Arlington, VA, www.msha.gov/MSHAINFO/FactSheets/MSHAFCT10.HTM, accessed 15 July 2008.

MSHA (2008b), "Legislative History", MSHA, Arlington, VA, www.msha.gov, accessed 15 July 2008.

Murtha, J. (2001), "Miners Complete Heavy Equipment Training Program", www.house.gov/murtha/news/nw010223.htm, February, accessed 1 October 2007.

Nalbandian, H. (2007), *European Legislation (revised LCPD and EU ETS) and Coal*, CCC/121, IEA Clean Coal Centre, London.

NBS (National Bureau of Statistics of China) (2007), *China Statistical Yearbook 2007*, China Statistics Press, Beijing.

NCB (National Coal Board) (1958), *Coal Mining in Poland: The Report of the Technical Mission of the National Coal Board*, National Coal Board, London.

New South Wales Department of Environment and Climate Change (2008), "About POEO Legislation", www.environment.nsw.gov.au/legislation/aboutpoeo.htm, accessed 15 July 2008.

NMA (National Mining Association) (2007), *Most Requested Statistics – US Coal Industry*, NMA, Washington, DC, March.

NWC (National Water Commission) (2007), *Water Reform in Australia*, NWC, Canberra, ACT, Australia, January, www.nwc.gov.au/resources/documents/water-reform-Australia-PUB-0107.pdf.

OSM (Office of Surface Mining Reclamation and Enforcement) (2008), "Regulation of Active Coal Mining and Reclamation", OSM, US Department of the Interior, Washington, DC, www.osmre.gov/osmreg.htm, accessed 5 August 2008.

Pickering, D. (1995), "British Coal Enterprise" in C. Critcher, K. Schubert and D. Waddington (eds.), *Regeneration of the Coalfield Areas: Anglo-German Perspectives*, Pinter, London, pp. 104-110.

Piekorz, J. (2004), "Restructuring of the Polish Hard Coal Industry", paper presented to Ad-hoc Group of Experts on Coal in Sustainable Development, UNECE, Geneva, Switzerland, 7-8 December, www.unece.org/ie/se/pp/csddec.html, accessed 15 July 2008.

Plishker, L., G. Silverstein and J. Frechtling (2002), *Evaluation of The Appalachian Regional Commission's Vocational Education and Workforce Training Projects*, Appalachian Regional Commission, Washington, DC.

RAG (2000), *RAG - mit den Regionen im Wandel (RAG – in the Transition Regions)*, RAG Aktiengesellschaft, Essen, Germany.

Reeve, D. A. (2000), *Coke Production and the Impact of Environmental Legislation*, CCC/35, IEA Coal Research, London.

RVR (Regionalverband Ruhr – Ruhr Regional Association) (2006), *Struktur und Entwicklung der sozialversicherungspflichtig Beschaeftigten im Ruhrgebeit (Employment Structure and Social Development in the Ruhr Area)*, RVR, Essen, Germany.

Sloss, L. L. (2003), *Trends in Emission Standards*, CCC/77, IEA Clean Coal Centre, London.

Statistik der Kohlenwirtschaft (2006), *Der Kohlenbergau in der Energiewirtschaft der Bundesrepublik Deutscheland im Jahre 2005 (Coal Mining in the Energy Sector of the Federal Republic of Germany in 2005)*, Statistik der Kohlenwirtschaft eV (Coal Industry Statistics Association), Essen, Germany.

Statistik der Kohlenwirtschaft (2008), data from Statistik der Kohlenwirtschaft eV (Coal Industry Statistics Association), Essen, Germany, www.kohlenstatistik.de, accessed 4 August 2008.

Thompson, E. C., M. C. Berger, S. N. Allen and J. M. Roenker (2001), *A Study on the Current Economic Impacts of the Appalachian Coal Industry and its Future in the Region*, Appalachian Regional Commission, Washington, DC.

UK Coal (2007), "Land and Property – The Strategy", UK Coal plc, Doncaster, UK, www.ukcoal.com/lp-the-strategy1, accessed 15 July 2008.

UNEP (United Nations Environment Programme) (2005), *Judicial Handbook on Environment Law*, UNEP, Nairobi, Kenya.

UNEP (2006), *Training Manual on International Environmental Law*, UNEP, Nairobi, Kenya, www.unep.org/law/PDF/law_training_Manual.pdf.

Vaux, G. (2004), "The Peak in US Coal Production", www.fromthewilderness.com/free/ww3/052504_coal_peak.html, accessed 15 July 2008.

Vermeulen, W. C. M. (2001), personal communication, ETIL bv, Maastricht, The Netherlands, January.

Walker, S. (1998), *Coal Licensing and Production Tax Regimes*, CCC/07, IEA Coal Research, London.

Walker, S. (2000), *Major Coalfields of the World*, 2nd edition, CCC/32, IEA Coal Research, London.

Walker, S. (2001), *Experience from Coal Industry Restructuring*, CCC/48, IEA Coal Research, London.

Walker, S. (2006), "ATH Re-equips for Growth: ATH Resources Invests in New Equipment for its Flagship Scottish Sites", *Coal International*, 254 (5), pp. 12-19, Sep-Oct, Tradelink Publications Ltd., Tuxford, UK.

World Bank (2004), *Toward a Sustainable Coal Sector in China*, World Bank, Washington, DC.

VIII. DOMESTIC AND INTERNATIONAL STUDIES RELEVANT TO CLEANER COAL IN CHINA

OVERVIEW OF DOMESTIC STUDIES

China's coal resource is rich and has a low development and utilisation cost. For these reasons, coal will continue to be the country's dominant energy source for a considerable period to come – a view expressed by Pu Hongjiu, Vice President of the China National Coal Association (Pu, 2005). This section reviews studies undertaken by Chinese organisations since 2002 in response to China's dependence on coal and the challenges this poses. The aim here is to give an insight into current thinking in China on the coal industry – an insight that is often missing from foreign-language reports.

Coal industry development

In response to the State Council's *Opinions on Promoting the Sound Development of the Coal Industry* (State Council, 2005), the National Development and Reform Commission (NDRC), in collaboration with coal enterprises and experts, drafted the 11th Five-Year Plan for the Coal Industry. The main tasks identified in the plan are: to optimise the industry's structure; to strengthen its management; to ensure coal production is sufficient to meet demand; to establish large-scale regional coal mining bases that exploit coal resources efficiently and in an environmentally responsible manner; to create large coal mining enterprises; to consolidate some small mines and eliminate those with low resource recovery rates; to improve coal mine safety with better equipment; to accelerate technological innovation; to improve worker safety through training; to improve the living standards and social security of miners; and to address legacy issues such as subsidence damage (NDRC, 2007).

The development philosophy for the coal industry in China is changing from "new coal mines as the main driving force, with consolidation of mines as a supplementary force" to the current thinking of "consolidation of mines as the main driving force, with new coal mines as the supplementary force". Hence, coal industry consolidation will become the main focus in the future and production from large-scale coal mines will displace that from small-scale mines. According to the 11th Five-Year Plan, the share of production from large coal mines will increase from 48% in 2005 to 65% in 2010, and that from small mines will decrease from 38% to 21%. China will close 4 000 small mines over the plan period. It is predicted that coal production capacity lost due to the closure of these small mines will amount to 60 Mtpa. During the 11th Five-Year Plan period, the Chinese government will moderate investment in the coal industry. Nevertheless, the planned additional production capacity from new coal mines opened between 2005 and 2010 is around 300 Mtpa, concentrated in the north and east of Shanxi, eastern Inner Mongolia, northeast China, north Shaanxi, Huainan, Huaibei, Yunnan and Guizhou.

The northwest region[1] will be the main coal supply base in the future, but conventional coal mining and utilisation techniques would put additional pressure on its fragile ecosystem. To understand this further, a public-benefit project was carried out by MOST during 2005/06: Study on Ecological and Environmental Issues and Measures for Coal Development in Northwest China (MOST, 2006). Water resource conservation and other ecological issues, such as desertification and salinisation of arable land and grasslands, were surveyed during site visits, and techno-economic evaluations made to select environmentally friendly coal mining and utilisation technologies suitable for the conditions found in the northwest. The MOST report concluded that the use of coal, coal products and coal wastes should be optimised to realise the benefits of a "circular economy",[2] and that the location of coal mining areas should be planned on a rational basis. It goes on to recommend that "green" coal mining technologies should be adopted to control ecological impacts at source; water-saving and environment-friendly processes, such as coke dry quenching (CDQ) coke production and air-cooled power generation, should be widely deployed to reduce environmental damage caused by coal processing and utilisation; and other advanced technologies developed at an economic scale. Based on the project's recommendations for technological development, a sustainable strategy for energy development in northwest China can be contemplated.

Some coal-related studies have been conducted by non-governmental organisations. A study by the China Energy Research Society puts an emphasis on the strategic and rational development of coal resources through macro-economic adjustments and optimisation of China's industrial structure in general (CERS, 2005). The study makes various recommendations, including the need to strengthen coal supply security,[3] to provide further support for mine consolidation, to raise technical ability, to improve production safety and environmental protection, to reform the management cost-accounting systems used by industry and to adopt a rational system for coal pricing. A study from the China Industry Research Network investigated the status and competitiveness of the coal industry, the drivers that will influence its future development, markets for coking coal, steam coal and anthracite, the key coal-producing regions and the main coal enterprises; and analysed the risk and value of investing in the coal industry (CIEDR, 2006). The study concluded that coal industry investment would depend primarily on government policies, but anticipated the growth of state- and privately owned enterprises, supplemented by foreign direct investment, to create much larger and stronger coal corporations. The share of production from small mines would thus dwindle. Like the government, the China Industry Research

1. Inner Mongolia Autonomous Region, Ningxia Hui Autonomous Region, Xinjiang Uygur Autonomous Region, Shaanxi Province and Gansu Province.
2. See footnote 6 in Chapter 5.
3. By planning new and expanded mines to ensure future coal supply, ensuring long-term investment in coal prospecting and stricter monitoring of resource utilisation.

Network suggests that the development and investment priorities at older coal bases should be the refurbishment and consolidation of existing mines to expand production capacity, while, at the proposed large coal bases, the focus should be on planning the rational distribution of productive new mines across the coalfields.

Prospects for coalbed methane and coal mine methane

The Chinese government is paying more attention to coalbed methane (CBM) and coal mine methane (CMM) resources due to the scarcity of oil and gas resources (Guo Banfa [2005], No. 35). According to the National 11th Five-Year Plan on CBM and CMM Development and Utilisation, NDRC predicts that the total production of CBM and CMM will be 10 bcm in 2010 and additional CBM reserves of 300 bcm will be proven (NDRC, 2006a). An industrial base for CBM and CMM development and utilisation will be set up. Development principles include a combination of surface and sub-surface drainage, independent development by Chinese enterprises with government support and overseas co-operation, and local gas utilisation with export of any remaining gas for household use and industrial applications. By 2015, the production of CBM and CMM is planned to be 20 bcm, with a 100% utilisation rate for CMM. By 2020, the production of CBM and CMM is expected to be 30 bcm, with the CMM utilisation rate held at 100%. It is predicted that by 2020, an additional 500 bcm of CBM reserves will be proven.

From the point of view of reducing greenhouse gas (GHG) emissions, a two-year study on emissions and control of low-concentration CMM showed that emissions of methane in 2003 were in the range of 9.4-11.4 bcm, of which less than 10% was utilised (MOST, 2004). The study found that the most promising technologies include power generation from CMM with low to medium methane concentration. In addition, ventilation air methane, which accounts for 80% of methane emissions, can also be used to generate power by means of a reverse-circulation reactor (Section 5.1). However, this technology is unproven at scale and expensive, although benefits through the Kyoto Protocol's Clean Development Mechanism (CDM) may improve its prospects. The study concluded that further policy support was needed to encourage the capture and use of CMM.

Development and commercialisation of clean coal technologies

Against a backdrop of China's rapid economic growth and soaring energy demand, the Medium- and Long-Term Energy Development Plan Outline (2004-2020) lays out an energy strategy for the comprehensive development of oil, natural gas and new energy sources, with coal as the base for power generation (State Council, 2004). The important position of coal in the national energy strategy is further reinforced. However, coal utilisation can cause serious environmental problems due to low efficiency of plant and insufficient pollution control, leading to unnecessary air pollution from emissions of sulphur dioxide (SO_2), NOx, dust and GHGs. The State Council observes that currently available clean coal technologies (CCTs) can increase efficiency and reduce pollution, ensure energy security, respond to new challenges and opportunities in the world, and achieve the sustainable development of China's national economy.

In order to solve problems during coal mining and coal utilisation, CCTs and near-zero emission coal technologies have become important subjects for study in China, drawing on overseas experience. The current status and development of coal preparation, processing and combustion, power generation, coal conversion, pollution control and other CCTs have been reviewed in a number of reports, all sponsored by NDRC. The development obstacles, market demand and trends, and policy recommendations are covered in these reports which include: Commercial Development and Policy Study of Major Clean Coal Technologies, carried out in 2006/07 (ONELG, 2007); and a national key R&D project, Policy Research on Clean Coal Technology, carried out during the 10th Five-Year Plan period (BRICC, 2004). Perhaps not surprisingly, further R&D features in BRICC's recommendations, but so does enforcement of environmental standards and the use of market-based mechanisms to ensure that only low-sulphur, low-ash coal is used in industrial boilers. More technically based studies include a Policy Study on Promotion of Steam Coal Washing and Coal-Water Mixtures in China (BRICC, 2005), carried out over the period 2002-04.

Power generation, steelmaking, cement production, industrial boilers and kilns, and chemical production are the main uses of coal in China, accounting for more than 90% of total coal consumption. For power generation, the average coal consumption per unit of generation is higher than in some other countries and pollutant emissions have had a serious impact on the environment. Use of advanced technologies and optimisation of existing coal-fired power generation are both necessary. Measures suggested in the NDRC-sponsored reports include closure of small units, installation of flue gas pollution control equipment at existing units and development of energy-efficient, environment-friendly larger units, such as supercritical and ultra-supercritical units, circulating fluidised bed combustion (CFBC) and integrated gasification combined cycle (IGCC). Pollutant emissions from industrial boilers and kilns exceed those from coal-fired power stations in some cities due to low combustion efficiency and outdated pollution control equipment. This is one of the difficulties hindering attempts to improve urban air quality and has become the focal point of regional emission reduction strategies.

A joint US-China study on energy futures and urban air pollution was carried out by the National Academy of Engineering and National Research Council of the US in collaboration with the Chinese Academy of Engineering and the Chinese Academy of Sciences (National Academies, 2008). Based around case studies in four cities – Pittsburgh, PA and Los Angeles, CA in the US and Huainan, Anhui and Dalian, Liaoning in China – it examines the sources of urban air pollution and its impacts on human health, ecosystems and the built environment. A chapter on coal combustion and pollution control gives an overview of the current status of CCTs in China and the US. Conclusions and recommendations from the study, aimed at policy makers in the US and China, focus mainly on regulatory matters. In China, enforcement of existing air quality standards is found lacking and strengthening national and local environmental protection authorities is seen as a priority, coupled with greater public participation and easier access to data. In both countries, higher charges on polluters are recommended to encourage greater efficiency, better pollution control and the adoption of cleaner energy sources and technologies.

Prospects for coal-to-liquids

In 2006, total oil supply in China was 352 Mt, and net imports were 184 Mt – a 52% import dependence on foreign oil (Section 4.2). The industrial and transport sectors consumed about two-thirds of the total oil consumed. It is predicted that oil demand will grow to 450-500 Mt and that dependence on foreign oil imports will be more than 60% in 2020 (NDRC, 2006b). Transport oil consumption will then account for more than 50%; incremental consumption in the transport sector is expected to account for more than 85% of the total incremental increase in oil consumption. How to ensure China's continued energy security and sustainable development under such circumstances have been the subjects of various reports which have pointed to coal-based liquid fuels as a strategic option for oil substitution.

These include a Study on Alternative Energy Development in China (NDRC, 2006c), a national key project which studied seven sectors and involved over thirty organisations and nearly one hundred experts, initially reported to the State Council at the end of 2006. NDRC itself has published the National 11th Five-Year Plan on Coal Chemical Projects (NDRC, 2006b), having conducted an earlier Investigation of Coal-Based Multi-Product Technologies (NDRC, 2005). These reports conclude that by using its abundant coal resources, China has the means to solve the problem of scarce indigenous oil supply, and could realistically limit its dependence on imported oil to around 50%. From its solid development base, coal-to-liquids is identified as a field where Chinese-developed and -owned technology could lead the world. It is seen as a feasible strategic option, prior to large-scale utilisation of renewable energy. However, for the orderly and sound development of coal-based liquid fuels production, the reports highlight that the issues of energy efficiency, environmental pollution and water consumption will need to be solved through further follow-up studies and technical progress.

Coal chemical products are of strategic importance for ensuring China's supply of chemicals; production is currently growing strongly and is likely to continue to do so in the near and medium term. Application of coal-based methanol fuels can be seen in some regions, but further studies on the impacts on human health and the environment are needed. The studies cited above recommend substitution of dimethyl ether (DME) for liquid propane gas (LPG) and demonstration of DME as a vehicle fuel. R&D and demonstration of coal-based olefins is recommended to reduce the oil demand for chemical production. The advanced processes and management experience found in other countries are seen as necessary to promote the commercial demonstration of coal liquefaction projects in China, but only on the basis of accelerating development and demonstration of technologies with Chinese-owned intellectual property. During the period of transition to a sustainable energy system, CCT development and utilisation will play a crucial role. Development of coal-based liquid fuels is seen by some as a response not only to China's own energy issues, but to global energy issues as well.

From a longer-term perspective, NDRC identifies coal-based polygeneration as offering the prospect of optimal coal utilisation. It can co-produce chemicals, liquid fuels and power according to market demand and so maximise returns. It has the desirable features of a flexible product mix, low production costs and high energy conversion efficiency. Pollutants can be treated and used as resources. For example, recovery and utilisation of sulphur, and utilisation and sequestration of high-concentration carbon dioxide (CO_2) can allow near-zero emissions, including of GHGs. The economics of polygeneration will require further careful study (Carpenter, 2008).

Prospects for local coal utilisation

Chinese cities make use of a variety of energy sources. Some use mainly coal while others have access to alternative energy sources such as natural gas and renewable energy. Environmental performance is also different between cities. SO_2 is the main air pollutant in some cities and particulates in others, but both cause serious air pollution in very many cities. Levels of technical and administrative development also differ, thus affecting the choice of CCTs.

Reports on these issues include: one supported by Beijing Municipal Science and Technology Commission[4] titled, Study on the Application of Clean Coal Technology in Beijing (2003-04); the 11th Five-Year Plan on Coke and Coal Chemical Projects for Shanxi Province (2005-06) (NDRC, 2006d); and Guidelines for Urban Clean Energy Action Plan (MOST, 2005), a key project under the 10th Five-Year Plan. A common conclusion from these reports is that coal occupies an important position in the local energy supply mix, but that its utilisation has led to serious environmental issues. Some of the reports deal with the development of local regions, analysing energy resources, economic growth, transport needs and demand for commercial energy. Cities, in particular, are identified as needing CCTs, often coupled with other energy conservation and emission reduction measures, such as the use of cleaner energy (*e.g.* natural gas and renewables), use of low-sulphur coal, development of district heating, restructuring of urban industries and elimination of processes and equipment with low energy efficiency and excessive pollutant emissions.

Beijing is the centre of Chinese politics, commerce and culture as well as one of biggest energy-consuming cities in China. Coal is expected to continue to occupy an important place in Beijing's energy supply, though with a downward trend. The large quantity of coal consumed, 30 Mt in 2006, has an environmental impact that has put Beijing under huge pressure to make urgent improvements, partly driven by its hosting of the Olympics in 2008.[5] The study for the Beijing Municipal Science and Technology Commission recommends that, for existing coal-fired power plants, advanced flue gas desulphurisation (FGD), de-dusting and de-NOx technologies should be used

---

4. A department of Beijing municipal government.
5. For example, coal combustion accounts for 80% of SO_2 emissions in Beijing and a large proportion of NOx, dust and soot emissions.

(Beijing Municipal Science and Technology Commission, 2005). Coal-fired industrial boilers should use low-sulphur, low-ash coal or briquettes with wet desulphurisation plus high-efficiency de-dusting technologies. CFBC with in-bed desulphurisation plus high-efficiency dust filtration is also seen as having a role. For coal-fired kilns, the study says that outdated technologies should be eliminated and advanced technologies employed at large scale, including foreign technologies. In the meantime, effective environmental management methods should be introduced and the emissions from coal users continuously monitored and controlled.

OVERVIEW OF INTERNATIONAL STUDIES AND PROJECTS

This section and those that follow complement Section 8.1 by providing an assessment of international studies and projects that have addressed coal-related issues in China, with an emphasis on those either undertaken or completed since 2002. The studies and projects are grouped into five categories and the review focuses on the drivers, objectives and motivation of each organisation engaged in preparing them.

■ **Coal industry restructuring studies and projects.** These are studies and projects supported by the World Bank and Asian Development Bank, with strong co-operation from Chinese government or nominated organisations. In general, work on coal production and use has focused on infrastructure and industrial restructuring issues, including economic, environmental and social impacts, reflecting objectives found in China's Five-Year Plans. In many cases, recommendations have been made to the Chinese government for changes to policies and procedures in order to achieve the declared objectives.

■ **Market-related technology assessments.** These include studies supported by organisations such as the European Commission, in co-operation with Chinese government departments, to provide market-related technology assessments. Findings have been disseminated to both Chinese and overseas stakeholders to further industrial co-operation, with support from both the international organisation and the Chinese government.

■ **Strategic and commercial studies.** These studies are typically supported by national organisations from countries outside China and examine the likely impact of Chinese policies on various strategic and commercial activities in those countries. They aim to assist industrial organisations to make future plans for their dealings with China. The findings have generally been made public, but there is no obligation to agree such findings with the Chinese government prior to publication.

■ **Topical reviews.** These are studies that review topics of interest in China to national governments and industrial organisations from outside China. They are often undertaken with Chinese assistance, but are not in any way endorsed by the Chinese government prior to publication.

■ **Other studies.** Such studies include information on China but are not specific to China.

MOTIVATION OF ORGANISATIONS ENGAGED IN STUDIES AND PROJECTS

The rationale for undertaking the studies and projects, categorised above, is described here and the motivation of those funding such work is explored.

Coal industry restructuring studies and projects

Organisations supporting coal industry restructuring studies and related projects have worked closely with the Chinese government and state organisations over the last few years to shape and implement changes to the national industrial structure in line with the aims of the current Five-Year Plan. The studies themselves may be designed to substantiate recommendations or may be precursors to the sanctioning of large financial loans. Consequently, the coal sector may be just one aspect of a broader study. In China, the Asian Development Bank (ADB) works closely with the government, NGOs, civil society groups, the private sector and other development agencies to prepare an agreed Country Strategy and Program (ADB, 2006a). Current strategic priorities in the energy sector encompass (ADB, 2008a):

■ improvements in sector governance, including restructuring;
■ environmental improvements through the increased use of clean fuels, modern production technologies and market-based instruments;
■ tariff reform;
■ an enabling framework for private sector involvement;
■ improvements in the efficiency of energy production and utilisation;
■ improving access for the poor to reliable supplies of electricity and addressing biases against poverty in the structure and levels of tariffs; and
■ regional co-operation in the energy sector.

Similarly, the World Bank's Country Partnership for China aims to help further integrate China into the world economy, reduce poverty, manage resource scarcity and environmental challenges, expand access to capital and build institutional capacity (World Bank, 2006a). As with the ADB, this is a very wide-ranging remit where coal is one sector among many. Nevertheless, demonstrating more efficient ways of using coal, creating a more competitive electric power market and addressing climate change are key aspects. The World Bank's draft Strategic Framework on Development and Climate Change aims to: integrate climate actions in development strategy; mobilise concessional and innovative finance; facilitate the development of innovative market mechanisms; leverage private sector finance; increase support to technology acceleration; and step up policy research, knowledge and capacity building (World Bank, 2008a).

Market-related technology assessments

The European Commission has developed close links with China on energy-related issues (Chapter 9) and has encouraged activities to support the competitiveness of EU industry in its dealings with the Chinese coal sector. Projects have mostly been established through competitive bidding processes via the Framework R&D Programmes, with consortia submitting proposals in line with the quite broad terms of reference of the various calls issued by the Directorate General for Energy and Transport.[6] Recent studies have aimed to improve the understanding of market prospects in the coal-fired power generation sector. Results are made available to all EU industrial organisations that have the potential to undertake coal-related business in China and various initiatives have assisted commercial co-operation between such organisations and their Chinese counterparts.

In addition, the Organisations for the Promotion of Energy Technologies (OPET Network) is an initiative of the European Commission that aims to promote public awareness of current energy research, further the deployment of innovative technologies and increase the pace of market uptake in respect of research that supports EU energy policy priorities. The extension of this activity to suggest technologies for China was one aspect of a broad-ranging study, although its scope was primarily the new and candidate countries of the European Union (CERTH/ISFTA, 2003).

Strategic and commercial studies

The most active organisations in this area have been those in Australia and Japan, reflecting the growing economic dominance of China in the Asian-Pacific region as well as the major environmental concerns arising from China's ever-growing use of coal, most of which is still burned directly with very limited pollution control.

The Australian Bureau of Agricultural and Resource Economics (ABARE) is a government research agency that provides independent analysis and forecasting to contribute to the competitiveness of Australia's agricultural, fishing, forestry, energy and minerals industries, and to the quality of the Australian environment.[7] It has carried out economic research and analysis on the issues that will impact on the opportunities for Australia to export coal to the Asian region, in particular to China. Such information, together with regular quarterly forecasts for the full range of export commodities, assists the Australian coal industry to better plan its future operations. At the same time, ABARE research contributes to some of the most important items on the Australian and international policy agenda, including plans for minerals exploration and responses to rising GHG emissions and climate change.

The Institute of Energy Economics, Japan (IEEJ) was established by the Ministry of International Trade and Industry to undertake energy-based research activities.[8] It is tasked to support the sound development of the Japanese energy-supplying and energy-consuming industries, covering technical, economic, environmental and social issues. IEEJ offers independent advice by analysing energy problems and providing the basic data, information and reports necessary for policy formulation. A key research area is the impact on energy markets of the structural changes that have been brought

---

6.   http://ec.europa.eu/dgs/energy
7.   www.abareconomics.com
8.   http://eneken.ieej.or.jp

about by a combination of demand-side fluctuations, including the globalisation of the world economy and the economic development of emerging countries, together with the political, economic and social changes occurring within supply countries.

Topical reviews

The current focus of work at the IEA is on climate change policies, market reform, energy technology collaboration and outreach to major producers and consumers of energy, like China, India, Russia and OPEC member countries. As such, it conducts a broad programme of energy research, data compilation, publications and public dissemination of the latest energy policy analysis and recommendations on good practices. This has included a study specific to the Chinese power sector (DTI, 2004).

The IEA Clean Coal Centre (IEA CCC), which is an implementing agreement under the auspices of the IEA, provides unbiased information on the sustainable use of coal worldwide.[9] Services are delivered to the governments of member countries and industrial sponsors through direct advice, review reports, facilitation of R&D and provision of networks. It provides technical assessments, economic reports and market studies on specific topics throughout the coal chain. It also provides databases on coal characterisation, coal-fired power plants and emissions standards. The Centre has produced two studies on China in recent years (Minchener, 2004 and 2007a).

Capgemini provides consulting and related technology services to its commercial clients in the energy, utilities and chemical sectors. It has provided a public-domain review of opportunities within the Chinese power sector, in part as a means to publicise its own involvement in the fast-growing Chinese economy.

In addition, various other institutions and universities prepare and publish reports on aspects of coal use, some of which have focused on China (see also Bibliography).

Other studies

These include statistical energy reports and projections of energy use worldwide. Providers include the IEA and organisations such as the US Energy Information Administration. While they are mentioned for completeness, they are not considered further in this chapter.

KEY RECOMMENDATIONS FROM STUDIES

Asian Development Bank studies and projects

Studies financed under the ADB technical assistance programme for the energy sector cover: energy efficiency and conservation; clean energy promotion; power sector and market restructuring; tariff reform; and private sector participation. Within this

---

9.   www.iea-coal.org.uk

framework, the ADB focuses on the use of cleaner fuels and the development of renewable energy. Coal technology studies are currently very limited, with the ADB only supporting the use of CMM, improved coal mine safety, new opportunities such as the CDM, energy conservation and resource management. In the late 1990s, ADB funded an important study for what would have been China's first commercial-scale IGCC demonstration project to be located at Yantai in Shangdong (ADB, 1999). However, there have been only five coal-related technical assistance projects proposed, started or completed since 2002, as outlined below.

ADB is also considering financing for a 250 MW IGCC power plant at Tianjin (the GreenGen project described in Section 9.1). An environmental impact assessment has been prepared by Huaneng Greengen Co. to support a loan application for USD 200 million, this being a substantial part of the total USD 300 million project cost (ADB, 2008b).

Waste Coal Utilization Study

The goal of Project 36471, completed in 2003, was to improve environmental quality and conservation of energy in Shanxi by promoting waste coal utilisation (ADB, 2004a). Waste coal arising from coal production accounts for approximately 25% of all industrial solid waste in China. Large stockpiles of waste coal damage the ecology of mining areas and occupy valuable land. However, it can be used as fuel for power generation, as a raw material for construction products, as a source of useful minerals and for land reclamation. While waste coal utilisation has been increasing due to encouraging government policies, there are still many barriers (ADB, 2007a). The specific project objectives were to:

■ study the technical options, financial and economic viability, environmental issues and impact on social development;

■ encourage private sector participation in waste coal utilisation and create a policy and regulatory framework; and

■ draw up a framework for medium- and long-term co-operation between the provincial government and ADB.

The 11th Five-Year Plan places a high priority on resource conservation and includes the mandatory target of a 20% reduction in energy intensity, thus encouraging the greater use of waste coal for power generation. As such, local ownership of the project in China's major coal-producing region was strong. The project identified several policy-related obstacles to power generation from waste coal and stakeholders undertook an international field study to gain information on the latest trends and international best practices in waste coal treatment and use. It is expected that the ADB will continue its dialogue with the Shanxi provincial government to promote waste coal-based power generation. There is also the likelihood of a further study to determine the techno-economic feasibility, the required unit size that would eventually allow all the accumulated waste coal to be consumed in power plants, and the design parameters of such power plants to minimise emissions (*ibid.*).

Poverty Reduction in Coal Mine Areas in Shanxi Province

Project 37616 was designed to assist the Shanxi provincial government with the design of a socio-economic and environmental programme to focus on retraining redundant miners, developing job creation schemes and mitigating environmental impacts for poor communities following the closure of large numbers of small coal mines in accordance with national and provincial policies (ADB, 2005a). A study was undertaken in the area around Jincheng to identify and evaluate specific options, such as training for re-employment in the larger mechanised coal mines or in other sectors of the regional economy. The implications for wider application of this model training and job creation programme in coal mining areas throughout China were also assessed. This led to practical policy recommendations for local government and the design of specific assistance programmes, including an action plan for training and job creation. To create jobs and reduce environmental damage in Shanxi and elsewhere in China, the project report recommended that the coal industry should diversify into CBM and waste coal utilisation (ADB, 2006b). It was noted that the ADB would most likely continue the dialogue with the provincial government to monitor the impact in terms of number of miners re-employed and poverty reduction (ADB, 2008c).

Coal Mine Safety Study

Project 39657 is concerned with assisting the Chinese government to contain worsening environmental pollution and improve coal mine safety to demonstrate that rapid economic growth and a cleaner environment are compatible (ADB, 2006c). It will consider how to improve safety through restructuring of small coal mines and various other policy reforms (*e.g.* consolidation into larger units and strengthening safety codes and standards for mine design and operation). The expected outcome will be the adoption and implementation of a strategy for enhancing safety in coal mines. For the Zhengzhou mining area, a coal mine safety action plan to 2020 is proposed that will include:

- an assessment of the technologies and management practices for accident prevention and control, particularly gas drainage technologies, and for improving CMM utilisation, monitoring and control;
- a strategy for public dissemination of the project's findings and recommendations, including comprehensive training courses; and
- the identification and prioritisation of candidate coal mine safety projects for possible ADB financing and evaluation of a pilot project.

Alternative Livelihood Options to Facilitate Coal Sector Restructuring

Project 37618 examined alternative employment opportunities for miners made redundant by the closure of coal mines as a result of the national restructuring plan (ADB, 2005b). A case study was undertaken at Jiaozuo, Henan which examined the economic and social structure, developed a methodology for identifying alternative economic activities and formulated a job creation and training programme to assist redundant miners to find alternative livelihoods. The study showed that almost 60% of the 9 460 miners expected to lose their jobs by 2010 were below the age of 40 and that about 70% of them had only 9 years of formal education, while their average annual wage of around RMB 10 000 contributed 70-75% of their household incomes (ADB, 2007b). While the coal mining industry has declined, other industries in Jiaozuo are growing: non-ferrous metals, food processing, paper making, energy, chemicals and auto parts suppliers. Based on these observations, it was recommended that the local government should provide vocational training programmes to displaced workers in welding, fitting, basic mechanical and electrical skills, and small business development, including farming. It was also expected that the ADB would continue its dialogue with the China National Coal Association to determine how best to establish a comparable scheme on a national basis (ADB, 2007c).

Coalbed Methane II Project

Finally, Project 37613 proposes investments that will increase the production, capture, and use of methane in association with underground coal mining and so improve mine safety, reduce GHG emissions and reduce the use of coal (ADB, 2006d). The driver is the poor safety record at China's coal mines, particularly at small mines. Underground methane explosions are largely responsible for the high number of fatalities, so the Chinese government is promoting more efficient removal of underground methane and its capture for clean energy production. Reducing the emissions of methane to atmosphere is an added benefit. Separately, ADB is providing a USD 117 million loan to finance almost half the capital cost of a 120 MW CMM project at Sihe mine, owned by Jincheng Anthracite Mining Group Co. Ltd., where 100 vertical CBM wells will supplement the CMM produced from Sihe and other nearby mines (ADB, 2004b).

World Bank studies and projects

The World Bank has a number of completed and ongoing coal-related investment projects, some of which are described below, including (Takahashi, 2006):

- CBM projects in various provinces;
- FGD projects in Shandong Province;
- an IGCC project at Yantai in Shandong Province, including options for CO_2 capture and storage;
- coal-fired power plant construction and rehabilitation projects in various provinces;
- coal preparation in Hunan Province; and
- a multi-product gasification project in Inner Mongolia Autonomous Region.

For example, the World Bank has part-financed the construction of ten coal-fired power plants in China since 1985, with a total capacity in excess of 10 GW and total investment of around USD 9.4 billion (Table 8.1). The purpose has been to assist China in its programme of electrification to satisfy the growing demand for electric power, while improving coal utilisation efficiency and reducing adverse environmental impacts, the latter mainly through the use of low-sulphur coal (<1%). Alongside technical developments, the World Bank has supported aspects of power sector reform through improvements to management and financing systems.

In addition, the World Bank has supported improvements to China's coal mining sector, such as the Changcun coal project which established the first of a series of large-scale, fully mechanised coal mining operations in the country in the early 1990s. The Bank financed: the transfer of technology in mine design, engineering and construction (especially the mechanised, highly productive longwall mining technology); the introduction of modern coal washing technology; and the fostering of improved management techniques for project scheduling, monitoring and cost control (World Bank, 1985).

Table 8.1 ....World Bank lending for coal-fired power plants in China

Name	Location	Date commissioned	Capacity (MW)	Total investment, million USD	World Bank investment, million USD
Beilungang I	Zhejiang	May 1986	600	1 044.9	225.0
Beilungang II	Zhejiang	May 1987	600	289.7	165.0
Wujing	Shanghai	Feb 1988	2 × 300	354.1	190.0
Yanshi	Henan	Dec 1991	2 × 300	459.6	180.0
Zouxian	Shandong	Mar 1992	2 × 600	957.4	310.0
Beilungang II	Zhejiang	Mar 1993	2 × 600	1 350.0	400.0
Yangzhou	Jiangsu	Feb 1994	2 × 600	1 081.4	350.0
Tuoketuo	Inner Mongolia	Apr 1997	2 × 600	1 300.5	400.0
Waigaoqiao	Shanghai	May 1997	2 × (900 to 1 000)	1 898.0	400.0
Leiyang	Hunan	Mar 1998	2 × 600	678.0	300.0
Total			**10 200 to 10 400**	**9 413.6**	**2 920.0**

Source: Fritz (2004).

The World Bank continues to assist China in resolving policy and technical issues associated with improving coal-fired power generation efficiency as part of a strategy to achieve an environmentally sustainable energy sector. This will be based on a new investment framework, which will promote clean energy and energy efficiency by combining carbon financing with funds from the Global Environment Facility (GEF) and the newly established Climate Investment Funds. It is proposing to support a multi-task project within three provinces. This will include (World Bank, 2007 and 2008b):

■ Mechanisms to support the closure of small inefficient coal-fired generation units in Shandong Province (4 300 MW) and Shanxi Province (2 870 MW) by 2010. This will assist county and municipal power companies in the two provinces to address the social and financial impacts of closures.

■ Demonstration of thermal power plant efficiency improvement and GHG emission reduction through three different types of investment activities: conversion of mid-sized power generation units into combined heat and power (CHP) units at Huangtai power plant in Shandong; waste heat recovery for district heating at Jinan Beijiao power plant in Shandong; and improvement of power generation efficiency following energy audit recommendations at Yangguang power plant in Shanxi.

■ Efficient generation dispatch to reduce system-wide coal consumption and GHG emissions from power generation by supporting a transition from current dispatch practices to ones that maximises coal savings through a pilot scheme in Guangdong.

Shandong currently has the highest emissions of SO_2 of all the Chinese provinces. During the 11th Five-Year Plan period, it is gradually closing small power generation units (<100 MW), replacing heating and industrial boilers with CHP facilities, and installing SO_2 control facilities at all CHP units. The World Bank is supporting the development and implementation of this province-wide SO_2 emission reduction programme in the province's CHP sector. This includes the installation of various types of FGD equipment, together with on-line monitoring equipment, at a number of existing and new-build coal-fired CHP plants. The current list include 40 units, ranging from simple systems on small (35 tph) boilers up to wet lime/limestone scrubbers for 300 MW units at thermal power plants (World Bank, 2008c).

Policy Recommendations to SDPC Supporting the Development of CCTs in China

Alongside these major investment initiatives over the last two decades, there have been some significant CCT assessment and policy advice studies. The World Bank undertook a study to provide the State Development Planning Commission (SDPC) with policy recommendations regarding the development and deployment of CCTs for China's 10th Five-Year Plan covering the period 2001-05. Although issued in 2001, it is still a useful document to review (World Bank, 2001). The rationale was that China could only establish a sustainable use of coal if it addressed the negative environmental impacts that had reached critical levels, affecting human health and agricultural production. It was assumed that CCTs were the critical link to balance continued coal use and environmental protection. The study's aim was to review all issues related to the development of CCTs in China and to recommend the policy initiatives needed to ensure their deployment in preference to conventional technologies. The recommendations were regarded as necessary short-term initiatives to attain the longer-term objectives of a more competitive energy market with adequate consideration of environmental concerns. They included:

■ Pursue the rapid acquisition of FGD and CFBC technologies. A key challenge was to make it possible to finance FGD investments at all plants required to have such pollution control equipment, as defined by state environmental regulations. For CFBC, the key challenge was to scale up the technology to a unit size of 300-500 MWe, while keeping costs competitive.

■ Deploy supercritical technology wherever it would be cost-effective and increase domestic manufacture of all major components. The experience initially had been encouraging and steps had been taken to acquire more know-how to increase the percentage of plant components manufactured domestically. However, further actions would need to be taken to adopt this technology more widely and to move to higher steam temperatures and pressures, and hence higher plant efficiencies.

■ Develop and demonstrate pressurised fluidised bed combustion (PFBC) and IGCC technologies as long-term options. These technologies would require demonstration and adaptation to local requirements before they could be deployed widely, but their higher costs and greater technical risks than conventional technologies would make it difficult to finance demonstration projects. Furthermore, the limited domestic expertise in manufacturing key components would not provide any opportunity to reduce costs, at least not in the short term.

While the technical issues associated with the development and deployment of CCTs were seen as important, the more critical issues related to the regulatory framework, decision-making process and the transition to a market economy. As such, the report also made recommendations on the need to create an enabling environment that, in the long term, would foster rational decision making when selecting technologies and pollution control techniques within a more market-driven energy sector. Accordingly, the report suggested that improvements in the following areas were needed:

■ environmental laws and regulations, including market-based instruments such as pollution fees, emission allowances and trading;

■ environmental monitoring and enforcement;

- pricing and distribution of coal; and

- retirement of small power plants.

Options and Opportunities for CCTs in China's Non-Power Sectors

In a follow-on to the above study for SDPC, the World Bank sponsored work to assess the options and opportunities for CCTs in China's non-power sectors (World Bank, 2003a). The study comprised five tasks:

- overview of current status and key issues;

- environmental regulations, economic incentives, monitoring and enforcement;

- coal gasification and liquefaction;

- biomass co-firing, combustion, gasification and utilisation; and

- use of refinery residues for gasification.

Several recommendations were made to the Chinese government which the World Bank suggested could be actioned through technical assistance programmes, carbon financing and the Global Environment Facility:

- To enhance existing and create new policies that encourage more efficient, economic and cleaner coal-based industrial applications. For example, the quality, sulphur content and graded size of coal should be appropriate for use with the specific technologies found in different sectors. High-sulphur coal should be distributed to users operating large plant with FGD, while low-sulphur coal should be provided to smaller users where FGD would not be economic. Such an initiative would help to address problems of inefficient, highly polluting coal use in industrial boilers and for the production of metals, ammonia, cement, glass and chemicals.

- To modify its command-and-control environmental policies, often with inconsistent enforcement, and develop a market-based approach for pollution control within the coal-fired industrial sector. In addition, policies should encourage the use of CMM for cost-effective commercial applications, in parallel with programmes to test advanced technologies that can use low methane content mine gas.

- To conduct a feasibility study for a cost-effective demonstration of coal gasification technology for the production of electricity and a variety of chemical and gaseous by-products.

- To conduct a biomass-coal co-firing demonstration project to prove the economic use of biomass in reducing emissions from coal-fired boilers. The demonstration should be coupled with analysis of the economic incentive options that could best be used to rapidly expand utilisation of this promising, low-cost and clean technology option. Together with financial incentives, programmes should be developed to disseminate information on co-firing technology and build capacity in China to use it.

Toward a Sustainable Coal Sector in China

The World Bank with the UNDP also supported a project, Capacity Building for National and Provincial Socially and Environmentally Sustainable Management of Coal Resources in China, because the coal sector's reform process was seen as lagging behind China's accelerating economic reforms and the restructuring of its power sector (World Bank, 2004a). The coal sector institutional and regulatory framework was examined at central, provincial and local levels, culminating in case studies at two mine sites. Possible policy solutions and capacity-building needs were identified at national level (some of which have since been addressed) and at provincial and lower levels. The report concluded that, although the government had already taken some positive actions, in order to reap the full benefits of improvements in the sector, laws and regulations needed to be consistently enforced, the sector rationalised and the investment climate improved. Specific regulatory and institutional changes were recommended:

■ Safety inspection and enforcement departments should be further strengthened and inspectors given authority to impose substantial fines and immediately terminate operations at any mine where there is imminent risk to worker safety. Preventive safety policies and practices that enable mining companies to achieve above-average safety performances should be institutionalised for all mines.

■ Local officials and mine management should be held accountable for environmental management failures. The capacity and resources of the local environmental protection bureaus (EPBs) should be strengthened to ensure enforcement. All mines should be required to post sufficient bonds to cover the cost of reclaiming mining areas when production has ceased. The magnitude of environmental liabilities that could arise from small mine closures should be researched to enable formulation of a suitable reclamation policy. Regulations should be imposed and enforced on the coal supply side to prevent low-quality coal from being used, especially for domestic purposes. Incentives are needed to encourage use of cleaner coal by all consumers.

■ To help meet China's energy needs in an environmentally and socially acceptable way, and to improve regional development prospects, small-scale mining should be consolidated and transformed into large-scale mining. Strict field monitoring should ensure that closed mines do not reopen. Funds should be made available by central and provincial governments for supervision of small mine closures, post-closure policing, site rehabilitation, mitigation of social problems, promotion of economic opportunities and compensation for mine owners.

■ An annual "holding fee" should be established for mining rights, sufficiently high to discourage speculative holding of coal resources. The mining licence period should be limited to 25 years, automatically renewable when all terms and conditions of mining and reclamation have been met. The government should retain the option to issue licences based on exploration expenditure commitments for areas where the coal potential is largely unknown and there is no investor interest for auctioned rights. A review should be conducted, by government and industry, of geological data to identify areas where coal exploration has been inadequate with the aim of increasing the availability of information and stimulating active exploration that leads to the award of new mining rights.

■ The provincial government's licensing role should be strengthened and the production capacity criterion for referral of projects to national level removed. Provincial government intervention in the operation of the large, state-owned mining groups should be reduced and all mining enterprises should be allowed to manage their activities in accordance with commercial criteria. Mine design and equipment regulations should be revised and aligned with current international mining practices and standards.

Hunan Urban Development Project

More recently, the World Bank is supporting the Hunan Urban Development Project, which aims to foster greater integration in the Changsha-Zhuzhou-Xiangtan (CZT) region of Hunan Province through support of a selected set of priority investments that address specific regional needs in a sustainable manner (World Bank, 2004b). The Zhuzhou clean coal component of the project aims to reduce air pollution in the CZT region through the production and sale of low-sulphur coal. Coal is imported into the province due to the lack of significant local sources. As such, there is a better recognition of the potential benefits of using prepared and processed fuel. The Zhuzhou coal blending station, with a capacity of 1.8 Mtpa, is being redeveloped to produce washed coke, washed coal, blended coal, briquettes and coal-water mixtures (CWM) from low-sulphur coal. Such an approach offers economies of scale for the provision of processed fuels and feedstocks, although at this stage its economic viability has still to be proved. The World Bank is supporting technical assistance to assess, demonstrate and publicise the economic value of using various forms of processed coal in selected applications.

Market-related technology assessments

EU-China co-operation on cleaner fossil fuel-fired power plants

A significant study was undertaken by a joint EU-China team drawn from research institutes, industrial support organisations and consultancy groups (Minchener, 2005). The purpose was:

■ To determine and quantify the opportunities for EU industry to transfer to China those techniques and equipment that would allow sustainable improvements in efficiency and environmental performance through retrofit and upgrading of existing coal-fired power plants.

■ To establish the basis for future co-operation for such technology introductions from Europe by drawing together EU industry and their Chinese counterparts.

Cost-effective opportunities were identified for which the market potential was estimated at EUR 2.4 to 4.2 billion over the period 2004-20. This assumes that much of the equipment is made in China under some form of co-operative venture between EU and Chinese companies. The study was presented to the European power plant industry, on a pre-competitive basis, to assist in bringing together EU and Chinese industrialists and enhance the prospects for co-operation and adoption of EU technology and expertise in China. Prior to 2007, power demand in China was in excess of supply and the priority was to keep plants on line: utility companies could not contemplate taking units out of service for extensive upgrades. Units have only come out of service for prolonged periods to retrofit FGD, as required by government directives, or for short periods of scheduled and unscheduled maintenance. However, from 2008 onwards, it is possible that the power supply shortages in China will ease as significant new capacity is brought on line. At that time, the Chinese power companies may start to look for ways to improve the overall performance of existing power plants and build relationships with the EU equipment suppliers involved in the study.

Co-operation on co-firing power generation market opportunities

Currently, a complementary study is underway, led by IEA Environmental Projects Ltd. and involving Aston University, VTT Processes, Exergia, the European Biomass Association, Tsinghua University, the Energy Research Institute of NDRC and the China Electricity Council (Minchener, 2007b). The objective is to assess how EU industrial companies might enter the coal-biomass co-firing power generation market in China. Besides the opportunity for technology transfer, this approach reduces GHG emissions and projects can benefit from credits under the CDM. It is intended to determine the commercial attractiveness and market potential of co-firing in China and offer policy advice. Further initiatives will assist co-operation between stakeholders, prior to technology demonstrations and subsequent deployment in China.

OPET Network

Work undertaken by the OPET Network has included one work package within a broad-ranging project on cleaner fossil fuels, which, as well as China, also considered Russia, South Caucasus countries, the Balkans and India (CERTH/ISFTA, 2003). The co-ordinator was CERTH/ISFTA of Greece while the Chinese partner was the Guangzhou Institute of Energy Conversion. The aim was to promote advanced European technologies in promising new markets – technologies which could lead to increased efficiency and improved environmental performance in power generation. Tasks included the collection of analytical information for each country, evaluation of the existing situation and identification of retrofit options. Developments in CCTs and possible projects involving European technology transfer were disseminated via many routes including regional workshops and site visits. In overall terms, this action supported the enhancement of security of energy supply in partner countries through the use of advanced technologies for domestic fuel utilisation and fuel diversification. It was also seen as one possible means to preserve local coal mining, resulting in improved employment prospects in related sectors.

Strategic and commercial studies

ABARE studies

The first major study of this kind was undertaken by ABARE in 2003 and was driven by Australian concerns about the rapid rise of China as a major competitor in Australia's traditional steam coal markets of Japan, the Republic of Korea and Chinese Taipei (Ball *et al.*, 2003). The report provides a historical review of how China emerged in 2001 as the world's second largest exporter of coal after Australia (until 2004 when Indonesia took second place). It notes that there had been significant changes in China's domestic coal consumption, production and distribution, including some expansions in rail and port capacity. Many of these changes were the result of government reform policies aimed at increasing the energy efficiency and export focus of China's economy. At the same time, exports were a way to dispose of excess domestic coal supplies since internal supply and demand were not in balance. As such, there was also some focus on the role that government policies, such as assistance to coal producers, might have played in underpinning the expansion in China's coal exports. These developments raised a number of questions about the sustainability of China's export levels in the medium to longer term, with the ensuing volatility of world coal markets, concerns that have since been borne out. Various modelling scenarios considered the evolution of coal exports and the impact on world coal markets. The results demonstrated the potential impact that China could have in the medium term, particularly the challenges that any further export *expansion* would pose for competing exporters in the northeast Asian market.

Subsequently, in 2004, this was followed by a second ABARE report that took a broader look at energy issues within China (Schneider, 2004). It noted that under the pressure of strong economic growth, the energy supply-demand balance had become strained and the energy outlook was challenging. It stressed that, even if economic growth slowed to the government's target rate of 7-8% (which it did not), significant investment would be required to meet energy demand growth. At the same time, China was struggling to ensure that this could be achieved in an environmentally sustainable manner. The report also noted that the policy challenge in this context was magnified because of China's significant dependence on coal. With the impact on Australia in mind, the report noted that, for coal, the increasing emphasis on meeting domestic demand was likely to mean that China's presence on export markets, would be more volatile than in recent years and would create great uncertainty for other market participants. On the positive side, China's energy sector was gradually becoming more commercially oriented, with a greater proportion of coal supply priced through direct negotiation between buyers and sellers, and with the commencement of market-based reforms in the electricity sector. However, the rapid increase in coal exports had stalled and China was becoming a large importer of coal. This had the potential to make China a strategically important player in the global coal market and to provide increased export opportunities for Australia, although the lack of stable trends would make investments to boost Australian export capacity more risky.

The third and more recent review from ABARE, in late 2006, provided some valuable information on recent trends in Chinese coal production and supply, with emphasis on the impact on prices and possible future trends in coal demand (Mélanie and Austin, 2006). From 2000 onwards, the Chinese coal industry underwent a process of significant realignment in response to a general expansion of industrial capacity and increase in demand for energy. NDRC reduced government intervention in the market. In 2003 and 2004, an increase in domestic demand for coal, coupled with infrastructure constraints, led to a significant rise in coal prices. Demand for both steam and coking coal was largely met by domestic production, albeit with an adverse impact on exports. At the same time, policy measures were being implemented by the Chinese government to increase mine safety and to reduce infrastructure bottlenecks, while pricing reform was aimed at maintaining the domestic coal supply-demand balance. In the opinion of the authors, this suggested that the coal market in China had returned to a more sustainable position in 2006, compared with the supply-constrained market of 2003 and 2004. However, while they noted that there were some positive signs arising from earlier investments in coal production and transport links, which had reduced upward pressure on coal prices, they also suggested that the coal sector would continue to face major issues if long-term stability were to be achieved. These included mine safety, transport infrastructure constraints during periods of high demand and ineffective pricing mechanisms that distorted price signals for both producers and consumers.

Studies by the Institute of Energy Economics, Japan

There are four reports by IEEJ that refer to Chinese coal issues. A report issued in 2003 was concerned with the impact on the Asian coal market of China's coal imports (Sagawa *et al.*, 2003). The conclusion was that the regional coal market would be seriously affected, and Japan's stable supply of coal severely compromised, if China's surplus coal production over demand collapsed or narrowed. It highlighted that the greatest portion of China's coal reserves lie in the northern and western regions, while the areas of demand are primarily located in the eastern and southern regions. In particular, the region comprising Shanxi, Shaanxi and the western part of Inner Mongolia will be expected to become the major supply centre for coal in China, so the transport of coal from this region will be key to stable coal supply. Projections for future coal supply and demand in China to 2020 and an assessment of transport requirements were made. Although the demand projection turned out to be a significant underestimate of the outturn, the study correctly determined that coal transport bottlenecks would be a key issue and that the railway capacity for coal transport from interior regions would be insufficient. Consequently, the authors predicted that domestic coal supply would be constricted and the volume for export reduced. In addition, they suggested that the increase in demand for coal imports by coastal provinces, in particular Guangdong and Fujian, which are close to coal exporting countries such as Indonesia, would have a significant impact on the overall Asian coal market. While they also examined the possible impact of energy conservation and fuel switching to natural gas on coal demand, they concluded that the predicted increases in coal consumption and transport volumes would greatly reduce the potential for Chinese coal exports.

The second IEEJ report considered environmental issues associated with steadily increasing demand for coal in northeast Asia, which was defined as a vast region comprising eastern Russia, Mongolia, China, the Republic of Korea, North Korea and Japan – together accounting for around half of global coal consumption (Fukushima, 2004a). The focus was on SO_2, NOx and dust emissions, since two of these pollutants contribute to the acid rain problem affecting wide areas across the region. The report noted that China has serious environmental problems, with coal combustion being the source of about 75% of CO_2 emissions, 90% of SO_2 emissions and 75% of NOx emissions in 2001 when China emitted more SO_2 than any other country in the world, and more NOx and CO_2 than any other except the US. The trends in emissions suggested that unless steps were taken, northeast Asia would become the world's biggest source of all three emissions. The report included the modelling of several scenarios for control of coal-based emissions. However, although these assumed strong policies on energy conservation and fuel switching to natural gas, the report stressed that, in reality, there would be a limit to the impact that these could have on the reduction of emissions of the greenhouse gas CO_2 and of the acid rain precursors SO_2 and NOx. In the author's opinion, it would be impossible to limit emissions of pollutants causing acid rain to the then current level or below solely by broad policy measures – there would need to be stringent legislation introduced on total emissions, supported by the installation of emission control technologies, such as FGD plant and deNOx equipment. The results of this study indicate that such technology could be installed in China's power and industrial sectors at a cost of about 0.5% of the nation's GDP. The author suggests that such expenditure would be affordable and would also be an opportunity to actively promote growth of the environmental industry, although whether this refers to Japanese or Chinese industry is not clear. It was further proposed that the countries in northeast Asia should co-operate to assure a stable trade in energy and to improve regional environmental standards.

The third IEEJ report to some extent repeated, and then built on the previous one by the same author (Fukushima, 2004b). The starting point was that northeast Asia would maintain high economic growth with a corresponding increase in energy demand. Modelling results suggest that primary energy demand in the region would increase at an average growth rate of 3.2% over the period to 2020. Within the energy mix, coal demand would be expected to increase from 1 570 Mt in 2000 to 2 280 Mt in 2010 and to 3 270 Mt in 2020 at an average growth rate of 3.8%. However, problems were anticipated in finding suppliers to meet these levels of demand, which have turned out to be underestimates. In particular, the report questioned China's capability to continue raising coal output to meet internal demand. As an example, it stressed that there were many issues to consider, which would determine if Shanxi, Inner Mongolia and Shaanxi could produce in excess of 1 200 Mt, when their output in 2000 was 440 Mt. These include environmental protection, safety and putting an end to illegal mining. The author suggests that the northeast Asia region needs to take the lead in establishing a stable coal supply and demand relationship, especially given the scope for energy saving and environmental protection through technical co-operation and integrated energy policies, including "thoughtful" incentives for the introduction of modern technologies and facilities.

The fourth IEEJ report was a further modelling study, examining the possible regional variations in coal supply and demand through to 2030 (Komiyama *et al.*, 2006). This noted that China had maintained high economic growth rates of around 10%, with the expectation that this would be maintained, with a commensurate increase in energy demand. China's energy challenges and evolving policies would have a growing influence on international markets. This meant that Japan, which depends on imports for most of its energy supplies, must understand China's energy challenges as accurately as possible and determine how its own energy needs can be met. Since regions of China vary widely, in terms of economic growth and patterns of energy supply and demand, the authors conclude that region-by-region analyses, including the oil, gas, coal, electricity and other energy markets, would be important for forecasting China's energy future. Accordingly, economic and energy supply-demand statistics were developed for each of China's 31 provinces, and an econometric model was built to make a quantitative projection of the expected energy supply-demand patterns in China's coastal and inland areas to 2030. The report then presented scenarios for coal production and transport to meet regional demand. It suggested that the fast-growing coastal regions will shift primary energy consumption from coal to natural gas and nuclear energy – coal's share falling from 71% to 58%, natural gas rising from 1% to 9% and nuclear from 1% to 5%, with oil remaining at about 26%. In contrast, in the inland regions, coal's share is predicted to fall less, from 76% to 68%, while oil rises from 16% to 17%, natural gas from 4% to 7%, hydropower from 3% to 5%, and a very limited introduction of nuclear. In due course, as part of a further study, IEEJ intends to consider the region-by-region effects of China's energy-saving technology introduction, based on predictions in this study, from the viewpoints of international energy security and global warming.

Topical reviews

China power plant optimisation study

The IEA initiated this project to recommend cost-effective efficiency and emission performance improvements at coal-fired power plants in China (DTI, 2004). The specific objectives were:

■ to audit two coal-fired power stations in China and benchmark against current Western best practice;

■ to recommend cost-effective improvements to the operational and environmental performance of the plants;

■ to promote the international collaborative nature of the project in China by hosting a workshop to disseminate the results; and

■ to show how the results of such a study could be replicated across China at other coal-fired plant.

The work was undertaken by a group of international experts on a 200 MW unit at the Tongliao power plant in Inner Mongolia and a 300 MW unit at Tianjia'an power plant at Huainan, Anhui. The key findings were that there would be considerable scope to improve efficiencies and reduce environmental impacts while also reducing unplanned outages. These could be achieved by changing operating practices, introducing more advanced monitoring and control equipment, plus upgrading certain key components. For example, the plant efficiency could be improved through the integration of the major sub-systems into the overall control system. Better environmental performance could be achieved with the use of higher grade electrostatic precipitators, while SO_2 and NOx emissions could be reduced through the use of FGD and NOx reduction technologies. Tighter control of coal quality was also identified as a key means to improve performance. The experts felt that it would be very worthwhile to improve turbine availability and performance through retrofit of the pressure parts, as this would provide additional useful energy and, hence, reduce specific emissions. Finally, they suggested a better monitoring regime to limit component failures. In the boiler, this could be achieved through the installation of continuous steam and water sampling instrumentation, circulation checks and tuning the combustion to optimise the location of the fireball in the furnace. The replication potential of the recommendations was considered high. By following them, plant owners in China could improve efficiency, availability and operability, while reducing operating costs and environmental impacts.

IEA Clean Coal Centre reports

Two IEA CCC reports review coal-related issues in China for the Centre's members from national governments and industry. The first provided an overview of the use of coal in China, its position in the energy mix, the structure of the coal industry and related infrastructure, as well as the implications for the environment of the projected increase in coal production and use (Minchener, 2004). There is some emphasis on the power generation sector, with data presented on the introduction of clean-up systems for SO_2, NOx and particulates. At the same time, there is consideration of the large amount of coal used for non-power applications, where opportunities to introduce CCTs are often more limited due to the cost and availability of pollution control equipment, especially for smaller industrial plant. The report also examined issues such as the impact of government policies, institutional and regulatory drivers, including the need for strong environmental legislation that is properly enforced, and the means and incentives to introduce improved coal production and utilisation technologies, with implications for effective technology transfer.

The second IEA CCC report concentrated on the production, transport and subsequent use of coal in China, again with emphasis on the power sector (Minchener, 2007a).

It discusses why it has proved difficult to supply enough coal to end users due to production and transport infrastructure problems, with coal increasingly being mined in regions far away from the industrial centres of southern and eastern China. In particular, it notes that, although these problems have eased recently, the weaknesses inherent within China's macro-economic control system, which attempts to balance energy production and use, mean that fundamental problems remain. The potentially adverse, knock-on effects to other economies in the Asia-Pacific region were also explored. Finally, it reviews the series of policy, institutional and regulatory initiatives that the Chinese government has started to implement.

Capgemini investment study

The Capgemini survey provides an analysis of the investment opportunities and risks for the power generation sector in China (Capgemini, 2006). It considers many aspects of the power sector, with an emphasis on coal-fired generation. The additional capacity requirements to meet GDP targets, which Capgemini estimates could average around 50 GW per year for the next 15 years, would require very significant investment. With regard to global environmental concerns, these would quickly become an issue and, in the longer term, would be likely to have an impact on the future energy mix. In the nearer term, there could well be an internal realignment of the power sector, which is seen as too fragmented, and this could precipitate a wave of mergers and acquisitions to deliver economies of scale. Another driver would be the vertical integration of fuel suppliers to ensure security of supply. Many of the activities reviewed would best be described as short-term and this has resulted in significant limitations in achieving a stable electricity supply and demand relationship. From an investment perspective, potential investors in thermal power companies are advised to be cautious, because of price controls and lagging price reforms. Consequently, the report suggests that the most practical way to invest at present is via co-operation with domestic companies and technology transfer, thus reflecting state policy.

Study by the Peterson Institute for International Economics and CSIS

A recent joint study by the Peterson Institute for International Economics and the Center for Strategic International Studies (CSIS) offers an overview of the Chinese energy sector, with insights on the drivers behind energy demand and the significant global impacts (Rosen and Houser, 2007). China's share of global energy use has grown to over 15%, so the country has had to rely on international markets for more of its oil, natural gas and coal. Since 2000, China's energy demand growth surged to an annual 13% – well in excess of the average annual 9% growth in the economy – creating shortages at home, market volatility abroad, and questions about the sustainability of China's growth trajectory. The critical point made in the study is that China's energy profile is in constant flux, in an effort to keep up with the rest of the economy. As such, it is a fusion of planned and market forces, of formal regulation and short-term expediency, and of central intentions and local interests. The report further notes that China's energy challenge is rooted in systemic conditions that go beyond the energy sector, and therefore changes to energy policy alone would not provide the solution. As such, it suggests that co-ordinating energy analysis with the broader policy agenda on macro and external imbalances is essential. However, while structural adjustments are necessary to address the root causes of many problems, the study concludes that conflicting pressures make it unlikely that China's efforts can proceed in an optimised manner, at least not in the short term.

CRITIQUE OF RECOMMENDATIONS MADE IN STUDIES AND PROJECTS

The studies have been grouped here into categories, each with quite different driving forces. In considering their impact, emphasis is given to those studies that were implemented with committed input from Chinese stakeholders.

Coal industry restructuring studies and projects

The work undertaken by the ADB is linked very closely to the aims and objectives of the Chinese government's Five-Year Plans. Its operational strategy in China is to help the country achieve economic growth in an efficient, equitable and sustainable manner. For the energy sector, this strategy includes developing cleaner energy sources, renovating and retrofitting existing facilities to improve efficiency and to reduce emissions, promoting commercial energy utilities, and introducing pricing and tariff reforms. This is complemented by the ADB's environmental strategy, which supports economic, supply-side and other measures to ensure the sustainable use of natural resources through the promotion of market-based pricing, full cost recovery and transparent disclosure of information on environmental performance.

Given that context, the few coal-based studies undertaken by the ADB in recent years have focused on socio-economic and environmental topics, consistent with certain aims of the Five-Year Plans. These include coal sector restructuring as small mines are closed, improving coal mine safety and helping to establish a viable CBM industry, which itself can be linked to mine safety. That said, the scope and number of coal-related projects are both very small. This is consistent with the political view in many OECD countries in recent years that coal was an old-fashioned fuel and that the future energy mix needed to be dominated by renewables as the way to achieve low- or even zero-carbon emissions. However, there is an increasingly realistic, global recognition that coal will continue to be used extensively and that it should therefore be used as cleanly and efficiently as possible. Certainly, for China, a country whose energy sector will continue to be dominated by coal, it can be strongly argued that the focus of ADB support needs to be adjusted towards sustainable coal use and away from the preoccupation with renewable energy, which cannot meet China's energy needs to any meaningful extent in the foreseeable future. This adjustment of the ADB programme to reflect the importance of coal should include full consideration of all aspects of the coal production, supply and utilisation chain. This should include both shorter-term issues, such as deploying CCTs, and longer-term issues such as establishing zero-emission technologies, notably carbon dioxide capture and storage (CCS). It should be stressed that if a framework for the widespread deployment of CCTs can be established then this would offer a stable basis for the subsequent introduction of CCS.

The World Bank actively supports clean coal and low-carbon technology projects within China. Over the years, it has supported major improvements to the Chinese coal-fired power sector, including:

■ introduction of the first 600 MW subcritical and the first 900/1 000 MW supercritical units;

- introduction of new, large-scale ultra-supercritical units that further increase plant efficiency;

- continuous emission monitoring, electrostatic precipitator, FGD and low-NOx burner investment projects;

- pilot sulphur trading and the "sulphur bubble" concept for optimal use of FGD; and

- rehabilitation of medium-size (200-300 MW) power plants to improve efficiency and retrofit of pollution control systems.

The World Bank technical assistance studies, reported above, considered technical and non-technical issues that affect CCTs. As such, they have assisted the Bank to determine how it might best support improvements in the coal utilisation sectors, particularly power generation (World Bank, 2001). Many recommendations, such as those on the commercial introduction of FGD, high-efficiency pulverised coal-fired plants and CFBC plants, have been embraced by the Chinese government, who has ensured that most components can be manufactured domestically to reduce costs and hence aid significant market penetration. Alongside this, the government has strengthened the environmental drivers and begun to enforce environmental monitoring to ensure utilities meet tightened emissions standards, at least for new coal-fired power plants. For future coal technology options, such as IGCC, the Chinese government has not shared the World Bank's enthusiasm because of the high cost compared to conventional combustion plant. However, an ongoing dialogue with NDRC could see the technology introduced for polygeneration applications, which may have merit in China.

In the non-power sectors, there is less evidence of positive Chinese actions on the sustainable use of coal. For example, some suggestions have not been taken forward – such as those on establishing a system to direct high-sulphur coal to power plants with FGD, where it would have least environmental impact, while ensuring low-sulphur coal is used in industrial boilers (World Bank, 2003b).

The Bank, with the UNDP in the case of one joint study, has made a number of recommendations to rationalise the coal mining sector. This has begun under the 11th Five-Year Plan. However, the sheer magnitude of what needs to be done means that changes will take time to formulate and require determination to implement. As such, it is understandable that many of the changes now being implemented were recommended by the World Bank several years ago. For example, the recommendation to introduce market-based pricing for coal, has been partly adopted, although problems remain as there is no equivalent freedom of action within the power sector for electricity pricing. Hence, power companies cannot readily pass on coal price increases to their customers (NDRC, 2004).

One area where the World Bank continues to be active is in promoting a viable market for CMM and CBM. Currently, the Bank is supporting two major demonstration projects, which include significant infrastructure components (World Bank, 2004c and 2006b). CMM exploitation appears to be moving towards achieving a meaningful market penetration, especially with the benefit of CDM credits, although most applications are localised within the immediate vicinity of a mine. The position for CBM is less clear. The high costs for degasifying coal seams ahead of mining are often not covered by the sale price of CBM. At the same time, the concentration of major coal mining activities in the remote northwest of China is not conducive to establishing a strong market for CBM produced ahead of mining, unless the amounts available were sufficient to justify a local collection network and a major pipeline to the coastal regions. There has been enormous financial support over the past decade to produce CBM from coal reserves that will not be mined, with more than 2 000 wells sunk, 1 700 in the period 2005-07. However, these are mostly still at the prospecting and testing stage, and far from commercial operation. The state-owned China United Coalbed Methane Corp. Ltd. has had, until recently, a monopoly, with an exclusive right to co-operate with foreign counterparts in CBM exploration (China Daily, 2007). This did not encouraged Chinese companies to seek CBM opportunities. It remains to be seen if World Bank support will resolve any of these critical issues.

In conclusion, there is a near- to medium-term requirement to stabilise the entire coal production and supply chain by encouraging more efficient mining practices across the sector, more effective and reliable coal transport and more rational coal use. It is now evident that coal imports may well provide a valuable means to balance coal demand and supply since this cannot be achieved through domestic action alone. Looking to the future, there is a longer-term need to establish near-zero emission, large-scale coal-fired power and industrial plants, and possibly CTL plants. Before this, there will be a need to introduce the CO_2 capture-ready concept at the large number of new coal-fired power plants that continue to be built each year. These issues all require very careful consideration and it is evident that the current bilateral and multilateral collaborative arrangements could greatly assist in addressing many of these issues (Chapter 9). At the same time, there needs to be a greater dialogue on policy and regulatory reform, so that China can adopt measures that have been proven elsewhere.

Market-related technology assessments

EU industry has made good use of the market assessment studies, so they have been deemed successful by the European Commission. Not only has market information been provided in a timely manner to EU stakeholders, but the various capacity building initiatives have brought together EU and Chinese counterparts, leading to productive dialogue on technical and commercial co-operation. A spin-off from the prime purpose of such studies has been the links established between EU and Chinese research institutes, which have been further strengthened by joint proposals within the current EU Framework Programme for R&D.

Strategic and commercial studies

The studies undertaken to provide guidance to other countries on the likely strategic and commercial impacts of Chinese policies must be considered valuable since the sponsoring organisations continue to fund further work. The ABARE reviews provide a perceptive insight into the issues influencing coal supply and demand within northeast Asia, highlighting the instabilities within China's own coal supply and use activities. The extensive studies by IEEJ are more philosophical in outlook and suggest more regional technical and policy co-operation. However, there is little evidence that this has been taken forward in any significant way. The Japanese projections of Chinese coal supply and demand have consistently underestimated the massive growth that has occurred since 2000, as have official projections from within China and projections from elsewhere.

The IEA-led power plant refurbishment study is an example of a technical exercise with significant potential benefit to the Chinese power sector. However, although dissemination of the results was undertaken at various workshops, there is little evidence that Chinese power companies have taken forward the suggested retrofit options. In part, as with the EC-funded market assessments, this may be because power plants have needed to be kept online to meet soaring demand, leaving little scope for extended outages for retrofits. However, beyond that, it does not appear that the messages, as set out in the comprehensive report, have been absorbed within China. This certainly raises questions about how best to achieve effective promotion and dissemination of best practices and application of readily available, commercial technologies within China.

REFERENCES

ADB (Asian Development Bank) (1999), *Study on Clean Coal Integrated Gasification Combined Cycle Technology*, Technical Assistance Report TA 2792-PRC, ADB, Manila, Philippines, www.adb.org/projects/project.asp?id=30396.

ADB (2004a), *Technical Assistance to the People's Republic of China: Waste Coal Utilization Study*, Project No. 36471, Terms of Reference TAR: PRC 36471, ADB, Manila, Philippines, www.adb.org/projects/project.asp?id=36471.

ADB (2004b), *Report and Recommendation of the President to the Board of Directors on a Proposed Loan to the People's Republic of China for the Coal Mine Methane Demonstration Project*, Loan No. 2146-PRC, Report and Recommendation of the President RRP: PRC 30403, ADB, Manila, Philippines, www.adb.org/projects/project.asp?id=30403.

ADB (2005a), *Technical Assistance to the People's Republic of China for Poverty Reduction in Coal Mine Areas in Shanxi Province*, Project No. 37616, Terms of Reference TAR: PRC 37616, ADB, Manila, Philippines, www.adb.org/projects/project.asp?id=37616.

ADB (2005b), *People's Republic of China: Alternative Livelihood Options to Facilitate Coal Sector Restructuring (formerly Impact of Closure of Coal Mines in Poverty Areas and Options for Job Creation)*, Project No. 39020, Technical Assistance Report TA 4680, ADB, Manila, Philippines, www.adb.org/Documents/PRF/PRC/TA4680-prc.asp.

ADB (2006a), *People's Republic of China: Country Strategy and Program Update (2007-2008)*, ADB, Manila, Philippines, www.adb.org/Documents/CSPs/PRC/2006/default.asp.

ADB (2006b), *People's Republic of China: Poverty Reduction in Coal Mine Areas*, Project No. 37616, Technical Assistance Consultant's Report TA 4566-PRC, ESSA Technologies Ltd. (for Shanxi Provincial Development and Reform Commission), Vancouver, Canada, www.adb.org/projects/project.asp?id=37616.

ADB (2006c), *People's Republic of China: Coal Mine Safety Study*, Project No. 39657, Terms of Reference TAR: PRC 39657, ADB, Manila, Philippines, www.adb.org/projects/project.asp?id=39657.

ADB (2006d), *People's Republic of China: Coalbed Methane II*, Project No. 37613, outline description of proposed project, December, www.adb.org/Projects/pids.asp.

ADB (2007a), *Technical Assistance to the People's Republic of China: Waste Coal Utilization Study*, Project No. 36471, Technical Assistance Completion Report TA 4389-PRC, ADB, Manila, Philippines, www.adb.org/projects/project.asp?id=36471.

ADB (2007b), *People's Republic of China: Alternative Livelihood Options to Facilitate Coal Sector Restructuring*, Project No. 37618, Technical Assistance Consultant's Report 37618/ TA 4680-PRC, PricewaterhouseCoopers (for China National Coal Association), New Delhi, India, www.adb.org/projects/project.asp?id=37618.

ADB (2007c), *People's Republic of China: Alternative Livelihood Options to Facilitate Coal Sector Restructuring*, Project No. 37618, Technical Assistance Completion Report 37618/TA 4680-PRC, ADB, Manila, Philippines, www.adb.org/projects/project.asp?id=37618.

ADB (2008a), *Annual Report 2007*, ADB, Manila, Philippines.

ADB (2008b), *Tianjin Integrated Gasification Combined Cycle Power Plant Project*, Environmental Assessment Report, Project No. 42117, Huaneng Greengen Co. Ltd., Beijing, www.adb.org/Documents/Environment/PRC/42117/42117-PRC-SEIA.pdf.

ADB (2008c), *Poverty Reduction in Coal Mine Areas in Shanxi Province*, Project No. 37616, Technical Assistance Completion Report TA: 4566-PRC, ADB, Manila, Philippines, www.adb.org/projects/project.asp?id=37616.

Ball, A., A. Hanstead, R. Curtotti and K. Schneider (2003), *China's Changing Coal Industry: Implications and Outlook*, eReport 03.3, Australian Bureau of Agricultural and Resource Economics, Canberra, Australia, www.abareconomics.com/publications_html/energy/energy_03/er03_coal.pdf.

Beijing Municipal Science and Technology Commission (北京市科学技术委员会) (2005), *Study on the Application of Clean Coal Technology in Beijing (2003-04)*, Beijing Municipal Science and Technology Commission, Beijing (see also scoping document *Clean Energy Plan for Beijing*, Beijing Municipal Sustainable Development and Science and Technology Promotion Center, Beijing, 16 September 2002, www.cct.org.cn/cea_asptest/undp_trend/energy_design/eng/beijing.pdf).

BRICC (煤科总院北京煤化工研究分院 – Beijing Research Institute of Coal Chemistry) (2004), 国家"十五"攻关专项-清洁能源行动 "洁净煤技术相关政策研究" ("Policy Research on Clean Coal Technology", *National Key R&D Project of the 10th Five-Year Plan)*, BRICC, Beijing.

BRICC (2005), 国家"十五"科技攻关课题"全面推广应用动力洗选煤和水煤浆政策研究" ("Policy Study on Promotion of Steam Coal Washing and Coal-Water Mixtures in China", *National Key R&D Project of the 10th Five-Year Plan)*, BRICC, Beijing.

Capgemini (2006), *Investment in China's Demanding and Deregulating Power Market*, executive summary, Capgemini, Paris, 27 February, www.capgemini.com/resources/thought_leadership/investment_in_chinas_demanding_and_deregulating_power_market/.

Carpenter, A. (2008), *Polygeneration from Coal*, Report No. CCC/139, IEA Clean Coal Centre, London.

CERS (China Energy Research Society – 中国能源研究会) (2005), 煤炭工业"十一五"发展战略及2020年设想 (*Coal Industry Development Strategy During the 11th Five-Year Plan Period and Prospects to 2020*), CERS, Beijing.

CERTH/ISFTA (Centre for Research and Technology Hellas / Institute for Solid Fuels Technology and Applications) (2003), *Cleaner Fossil Fuel OPET: Promotion of CCT Implementation Options in Existing Coal-Fired Power Plants*, final report to the European Commission on Contract No. NNE5/2002/97, CERTH/ISFTA, Thessaloniki, Greece, www.lignite.gr/OPET/CFF/WP3.htm.

China Daily (2007), "Fueling Debate", *China Daily*, Beijing, 18 June, www.chinadaily.com.cn/bw/2007-06/18/content_896185.htm.

CIEDR (China Industry Research Network – 中国产业研究网) (2006), 中国煤炭行业发展分析及投资预测报告(2006-2010) (*Development and Investment Forecasts for the Chinese Coal Industry* [2006-2010]), CIEDR, Beijing Fengjieliren Investment Consulting Co. Ltd., Beijing, http://ciedr.com/pages/browsereport.php?id=2165.

DTI (Department of Trade and Industry) (2004), *IEA-China Power Plant Optimisation Study: Executive Summary*, Report No. COAL R258, DTI/Pub URN 04/1015, DTI (now Department of Energy and Climate Change), London, www.berr.gov.uk/publications/index.html.

Fritz, J. J. (2004), "Environmental Performance of Coal-Fired Power Plants Financed by the World Bank", paper presented at the symposium on Urbanization, Energy, and Air Pollution in China: The Challenges Ahead, National Academy of Engineering / National Research Council / Chinese Academy of Engineering / Chinese Academy of Sciences, National Academies Press, Washington, DC, http://books.nap.edu/openbook.php?record_id=11192&page=187.

Fukushima, A. (2004a), *Coal and Environmental Issues in Northeast Asia*, Institute of Energy Economics, Tokyo, Japan, March, http://eneken.ieej.or.jp/en/data/pdf/242.pdf.

Fukushima, A. (2004b), *Energy Outlook and the Role of Coal in Northeast Asia*, Institute of Energy Economics, Tokyo, Japan, October, http://eneken.ieej.or.jp/en/data/pdf/264.pdf.

Komiyama R., Zhang P., Lü Z., Li Z. D. and Ito K. (2006), *Long-Term Energy Demand and Supply Outlook for the 31 Provinces in China through 2030*, Institute of Energy Economics, Tokyo, Japan, September, http://eneken.ieej.or.jp/en/data/pdf/353.pdf.

Mélanie, J. and A. Austin (2006), "China's Coal Sector: Recent Developments and Implications for Prices", *Australian Commodities*, Vol. 13, No. 3, Australian Bureau of Agricultural and Resource Economics, Canberra, Australia, September.

Minchener, A. (2004), *Coal in China*, Report No. CCC/87, IEA Clean Coal Centre, London.

Minchener, A. (2005), "Coal-Fired Power Generation in China: Prospects and Challenges", paper based on final report to the European Commission on Contract No. NNE5/2001/552 and presented at the 2nd International Conference on Clean Coal Technologies for our Future, hosted by IEA Clean Coal Centre, Sardinia, Italy, 10-12 May.

Minchener, A. (2007a), *Coal Supply Challenges for China*, Report No. CCC/127, IEA Clean Coal Centre, London.

Minchener, A. (2007b), "Development of Co-firing Power Generation Market Opportunities to Enhance the EU Biomass Sector through International Co-operation with China", international public domain announcement for the European Commission Contract No. 19668/TREN/05/FP6EN/S07.60657/019668 CH-EU-BIO, Brussels, Belgium, January.

MOST (科学技术部 – Ministry of Science and Technology) (2004), 煤炭开采低浓度瓦斯排放及治理途径研究 (*Emissions and Control of Low-Concentration Coal Mine Methane*), MOST 863 Program project, MOST, Beijing.

MOST (2005), *Guidelines for Urban Clean Energy Action Plan*, MOST, Beijing.

MOST (2006), 西北煤炭开发引发的生态环境问题及对策 (*Study on Ecological and Environmental Issues and Measures for Coal Development in Northwest China*), 研究项目为科学技术部社会公益专项项目 (special public-benefit research project), MOST, Beijing.

National Academies (2008), *Energy Futures and Urban Air Pollution: Challenges for China and the United States*, Committee on Energy Futures and Air Pollution in Urban China and the United States, National Academy of Engineering and National Research Council of the National Academies in collaboration with the Chinese Academy of Engineering and the Chinese Academy of Sciences, The National Academies Press, Washington, DC, http://books.nap.edu/openbook.php?isbn=0309111404.

NDRC (国家发展改革委 – National Development and Reform Commission) (2004), 《关于建立煤电价格联动机制的意见的通知》 ("Notice on the Establishment of a Coal-Fired Electricity Price Linkage Mechanism"), 发改价格[2004]2909号 (Fa Gai Jiage [2004] No. 2909), NDRC, Beijing, 15 December, www.ndrc.gov.cn/zcfb/zcfbtz/zcfbtz2004/t20080710_223762.htm.

NDRC (2005), 关于煤基多联产技术的调研报告 (*Investigation of Coal-Based Multi-Product Technologies*), NDRC, Beijing.

NDRC (2006a), 煤层气(煤矿瓦斯)开发利用"十一五"规划 (*National 11th Five-Year Plan for CBM and CMM Development and Utilisation*), NDRC, Beijing, 2006年6月26日 (26 June 2006), www.ndrc.gov.cn/nyjt/nyzywx/t20060626_74591.htm.

NDRC (2006b), 煤化工"十一五"发展规划及2020年展望 (*National 11th Five-Year Plan on Coal Chemical Projects and Outlook to 2020*), NDRC, Beijing.

NDRC (2006c), 替代能源研究领导小组 (*Study of Leading Group on Alternative Energy*), press release, 国家发展改革委能源局 (Energy Bureau of NDRC), Beijing, 26 June, http://nyj.ndrc.gov.cn/nygz/t20060626_74411.htm.

NDRC (2006d), *11th Five-Year Plan on Coke and Coal Chemical Projects for Shanxi Province (2005-06)*, NDRC, Beijing.

NDRC (2007), 煤炭工业发展"十一五"规划 (*11th Five-Year Plan for Coal Industry Development*), press release, NDRC, Beijing, 2007年01月22日 (22 January 2007), www.ndrc.gov.cn/nyjt/zhdt/t20070122_112661.htm (full text at www.ndrc.gov.cn/nyjt/zhdt/W020070306520943990664.doc).

ONELG (国家能源领导小组办公室 – Office of the National Energy Leading Group) (2007), 洁净煤技术重点产业化发展及政策研究 (*Commercial Development and Policy Study of Major Clean Coal Technologies*), ONELG, Beijing.

Pu Hongjiu (濮洪九) (2005), 在中国能源战略高层论坛上的讲话 (speech at China Energy Strategy Forum), Beijing, 2005年5月 (May 2005).

Rosen, D. and T. Houser (2007), *China Energy: A Guide for the Perplexed*, Peterson Institute for International Economics and Centre for Strategic and International Studies, Washington, DC, www.iie.com/publications/papers/rosen0507.pdf.

Sagawa, A., A. Fukushima, Y. Mimuroto and C. S. Chew (2003), *Prospects for the Supply and Demand of Coal and Related Coal Transportation Issues in China*, Institute of Energy Economics, Tokyo, Japan, October, http://eneken.ieej.or.jp/en/data/pdf/220.pdf.

Schneider, K. (2004), "China's Energy Sector: Recent Developments and Outlook", *Australian Commodities*, Vol. 11, No. 2, Australian Bureau of Agricultural and Resource Economics, Canberra, Australia, June.

State Council (2004), *Medium- and Long-Term Energy Development Plan Outline* (2004-20), State Council, June, Beijing.

State Council (国务院) (2005), 《国务院关于促进煤炭工业健康发展的若干意见》 ("Opinions of the State Council on Promoting the Sound Development of the Coal Industry"), 国发[2005]18号 (Guo Fa [2005] No. 18), Office of the State Council, Beijing, 2005年6月7日 (7 June 2005), www.gov.cn/zwgk/2005-09/08/content_30251.htm.

Takahashi, M. (2006), "Clean Energy for Development: Support for Low-Carbon Technology Development", presentation at IGCC and Zero CO_2 Emissions workshop, hosted by World Bank, Beijing, China, 27/28 June.

World Bank (1985), *Changcun Coal Project*, Project ID P003434, World Bank, Washington, DC, http://web.worldbank.org/external/projects/main?pagePK=64312881&piPK=64302848&theSitePK=40941&Projectid=P003434.

World Bank (2001), *Policy Recommendations to SDPC Supporting the Deployment of Clean Coal Technologies in China*, report prepared for the State Development Planning Commission, World Bank, Washington, DC.

World Bank (2003a), *Options and Opportunities for Clean Coal Technologies in China's Non-Power Sectors*, World Bank, Washington, DC.

World Bank (2003b), proceedings of the Clean Coal Technology Workshop held as a public forum for the dissemination of results from the World Bank study *Clean Coal Technology for Non-Power Applications*, hosted by World Bank and Mitsubishi Research Institute, Beijing, 16 September.

World Bank (2004a), *Toward a Sustainable Coal Sector in China*, final report for the joint UNDP/World Bank Energy Sector Management Assistance Programme (ESMAP), World Bank, Washington, DC, www.esmap.org/regions/region.asp?id=5#China.

World Bank (2004b), *Hunan Urban Development Project*, Project ID P075730, World Bank projects and operations database, World Bank, Washington, DC, http://web.worldbank.org/external/projects/main?pagePK=64283627&piPK=64290415&theSitePK=40941&menuPK=228424&Projectid=P075730.

World Bank (2004c), *Jincheng Coal Bed Methane Project*, Project Information Document: Project ID P087291, World Bank, Washington, DC, http://web.worldbank.org/external/projects/main?pagePK=64283627&piPK=73230&theSitePK=40941&menuPK=228424&Projectid=P087291.

World Bank (2006a), *Country Partnership Strategy for the People's Republic of China for the Period 2006-2010*, World Bank, Washington, DC, http://go.worldbank.org/81COGHVX80.

World Bank (2006b), *Shanxi Qingshui Coal Mine Methane Project*, Project ID P095640, World Bank, Washington, DC, http://web.worldbank.org/external/projects/main?Projectid=P095640&Type=Overview&theSitePK=40941&pagePK=64283627&menuPK=64282134&piPK=64290415.

World Bank (2007), *China Thermal Power Efficiency Project*, Project ID P098654, Project Identification Form, Global Environment Facility (GEF) Trust Fund, World Bank, Washington, DC, 15 September, http://gefonline.org/projectDetailsSQL.cfm?projID=2952.

World Bank (2008a), *Development and Climate Change – A Strategic Framework for the World Bank Group*, consultation draft, World Bank, Washington, DC, 21 August, http://siteresources.worldbank.org/EXTCC/Resources/407863-1219339233881/DevelopmentandClimateChange.pdf.

World Bank (2008b), *GEF China Thermal Power Plant Efficiency Project*, Project Appraisal Document, World Bank, Washington, DC, 30 September.

World Bank (2008c), *Shandong Power Plant Flue Gas Desulfurization Project*, Project ID P093882, Project Appraisal Document, Report No. 38067-CN, World Bank, Washington, DC, 18 April, http://web.worldbank.org/external/projects/main?pagePK=64283627&piPK=73230&theSitePK=40941&menuPK=228424&Projectid=P093882.

IX. CHINA'S INTERNATIONAL COLLABORATION ON CLEANER COAL

This chapter considers the role of international collaboration to push forward the development and deployment of clean coal technologies in China. It comprises a review of existing bilateral and multilateral arrangements, followed by a commentary on how such international collaboration might further speed cleaner technology deployment. It also considers and suggests frameworks for future collaboration to help meet China's energy and environmental challenges.

CURRENT BILATERAL AND MULTILATERAL COLLABORATION

This section summarises the major international collaborative activities, involving governments and commercial entities, with the common aim of promoting the cleaner production and use of coal. It reviews the progress achieved through these activities in deploying clean coal technologies in China.

Introduction

Over the years, as its coal-based economy expanded, China has established various bilateral government-to-government collaborations on clean coal technology. The rationale for such agreements, at least up to 2001, was primarily to assist China explore how to accelerate the deployment of clean coal technologies for environmental benefit, but with the expectation by both sides that there would also be economic benefits. The IEA, in a review of the situation in China up to 2003, noted that economic gain for Western companies was often seen as a priority, either through trade or technology transfer, and that collaborating governments tended to promote their own industries (Philibert and Podkanski, 2005). The IEA concluded that the results of these collaborations were mixed.

For example, Japan, through its Green Aid Plan, supported and financed four flue gas desulphurisation (FGD) demonstrations in the 1990s for pulverised fuel power plant applications and four circulating fluidised bed combustion (CFBC) demonstrations where sulphur removal was a key interest (Evans, 1999). Most of these were technically successful, but there was negligible diffusion beyond the demonstration projects (Oshita and Ortolano, 2002 and 2003). Reasons for this included an apparent lack of understanding of the Chinese situation by the Japanese equipment suppliers. The Chinese power plant operators aimed to minimise capital and operating costs, particularly since emission control standards were not strictly enforced, and therefore had no interest in the sophisticated and expensive systems on offer. The Japanese suppliers, on the other hand, were reluctant to lower costs by manufacturing components in China. The Japanese companies were concerned about the commercial risks of sharing intellectual property with local companies and so proposed only to import equipment into China. This offered negligible economic benefit to the Chinese economy, was not supported by the Chinese government and thus there was no wider adoption of these FGD systems.

Since 2001, however, there have been major changes in China. There has been a surge in demand for new coal-fired power generation and a wider recognition that these need to be state-of-the-art supercritical and ultra-supercritical units of at least 600 MW to take advantage of scale economies. Initially, such units were not universally adopted, but China has since become the major world market for advanced coal-fired power plants with high-specification emission control systems. This market has been driven by the increasing demand for power, alongside the introduction and enforcement of strict emission control standards for SO_2, NOx and particulates for new power plants (Minchener, 2005 and 2007). This technology introduction is through commercial, industry-to-industry arrangements, overseen by the Chinese government which approves each proposed power plant. Direct government-to-government collaboration would not be appropriate or necessary to further develop such business.

Table 9.1 .... Bilateral and multilateral collaborations with China related to cleaner coal

Bilateral	• Japan-China Climate Change Dialogue
	• China-Japan co-operation on CCS and enhanced oil recovery
	• Australia-China Joint Coordination Group on Clean Coal Technology
	• Australia-China Climate Change Partnership
	• China-US Working Group on Climate Change
	• EU-China Partnership on Climate Change
	• China-UK Climate Change Working Group
	• UK-China Cleaner Coal Technology Transfer Programme (now Carbon Abatement Technology Programme)
	• Germany-China environmental co-operation
	• South Africa-China co-operation on coal-to-liquids
	• Canada-China Working Group on Climate Change
	• India-China co-operation on climate change
Multilateral	• Asia-Pacific Partnership on Clean Development and Climate
	• Carbon Sequestration Leadership Forum
	• Methane to Markets Partnership
	• Asia-Pacific Economic Cooperation
	• FutureGen initiative
	• GreenGen initiative
	• Asia-Pacific Network on Global Change Research
	• ASEAN +3
	• China Council for International Cooperation on Environment and Development

Note: those with less relevance to cleaner coal issues at present are in coloured text (see Annex IV).

Emerging government-level collaboration with a focus on climate change issues, however, is a welcome development and is where cleaner coal can be part of a longer-term solution. This responds to the growing awareness, both within and outside China, that coal will remain an integral and major energy source for the Chinese economy. From this perspective, the way forward is to ensure the best possible environmental performance, with the longer-term goal of near-zero emission power plants with carbon dioxide capture and storage (CCS), subject to the appropriate economic, regulatory and institutional frameworks being established. Such environmental awareness is now reflected in many of the more recent bilateral and multilateral collaborations.

A number of bilateral and multilateral collaborations either focus on cleaner coal or have cleaner coal as a subsidiary concern to climate change issues (Table 9.1). Those of greatest importance are outlined below.

Bilateral collaboration agreements

Japan-China Climate Change Dialogue

Japan and China have an ongoing dialogue on climate change and, at a meeting in April 2007, they agreed to bolster bilateral economic, policy and technical co-operation with an emphasis on energy and the environment (METI, 2007). This included agreement for continued co-operation in such areas as efficient coal production, safety in coal mines and clean coal technologies, especially in the coal-fired power generation sector. Japan and China also agreed to take steps that would ensure a stable coal trade between the two countries. Later, in May 2008, the two nations agreed a six-point Sino-Japanese joint statement in which they agreed to strengthen co-operation in the fields of energy and environmental protection. This included a pledge to continue research on CCS linked to enhanced oil recovery (see below), together with a focus on energy efficiency and environmental performance in the iron, steel and cement industries (Xinhua, 2008).

This new focus on climate change issues can be contrasted with earlier co-operation under the Green Aid Plan. This was established by Japan as a means to transfer Japanese pollution control technologies to those developing countries where the environmental consequences of industrialisation were most severe. Much of the activity under the plan focused on demonstrations of Japanese-produced equipment at Chinese power plants.

China-Japan co-operation on CCS and enhanced oil recovery

In the May 2008 statement, Japan and China announced their intention to jointly develop a CCS and enhanced oil recovery (EOR) project which aims to recover 3-4 Mt of carbon dioxide (CO_2) per annum from two coal-fired power plants in China. The Japanese partners, under the Ministry of Economy, Trade and Industry (METI), include: JGC Corporation (a partner in the Algerian In Salah CCS project), Japan Coal Energy Center, Toyota Motors, Mitsubishi and the Research Institute of Innovative Technology for the Earth (RITE – a sustainable energy research establishment under METI). For China, the National Development and Reform Commission (NDRC) is the lead government department with input from: CNPC-PetroChina, Daqing Oil Field Ltd. (local oil field partner), Harbin district government, Harbin Utilities Company and China Huadian Corporation (Webb, 2008).

The USD 300 million project will be located in Heilongjiang province in north-east China, 100 km from the Daqing oil field. Daqing Oil Field Ltd. and RITE are jointly engineering the project and schedule to commission it in 2011. The intention is to use two 600 MW coal-fired power plants, retrofitted for post-combustion CCS and linked by pipeline to a near-by mature oil field to enhance oil production by 30-40 thousand barrels per day. The reason for using two power plants is to spread the 10-15% energy penalty associated with CCS and so limit any disruption to local electricity supply. Based on initial tests in China, the partners believe that it will be possible to achieve a CO_2-to-oil recovery ratio of 2:1.

Australia-China Joint Coordination Group on Clean Coal Technology

Building on a longstanding co-operation on energy under the Australia-China Bilateral Dialogue Mechanism for Resources Cooperation agreed in 2000, the Australia-China Joint Coordination Group on Clean Coal Technology was launched in January 2007. The group aims to advance the development and demonstration of the next generation of clean coal technologies that will be needed to reduce greenhouse gas (GHG) emissions (PM&C, 2007). Specific activities will include the following:

■ Identifying and implementing joint clean coal technology projects, including projects under the Asia-Pacific Partnership on Clean Development and Climate and the Australia-China Climate Change Partnership, both described below.

■ Sharing knowledge gained through clean coal and other relevant projects in Australia and China.

■ Identifying areas where co-operation on the development, demonstration and deployment of clean coal technologies can be enhanced. This includes low-emission technologies incorporating CCS.

Table 9.2............Projects under the Australia-China Joint Coordination Group on Clean Coal Technology

Title	Partners	Location
Post-combustion capture pilot project	Commonwealth Scientific and Industrial Research Organisation (CSIRO), Australia; Thermal Power Research Institute (TPRI), China; Huaneng Group, China	China Huaneng Group's Gaobeidian coal-fired power station in Beijing, China
400 MW integrated drying gasification combined cycle (IDGCC) power station	HRL Ltd, Australia; Harbin Power Equipment Co Ltd, China	Latrobe Valley, Victoria, Australia
Strategic workshops on clean coal technology	Australia-China Joint Coordination Group on Clean Coal Technology (ACJCG)	Australia and China
Co-operative research arrangements	Australian and Chinese universities	Australia and China

Source: PM&C (2007).

At the first meeting of the Group in March 2007, several projects were announced for which the Australian government had earmarked funding. The first two listed in Table 9.2 are the most important. The first is evidence of an active interest by Chinese industry in the application of post-combustion capture at a pulverised coal-fired power plant (Figure 5.6). The Australian government is supporting this project with a AUD 12 million grant, AUD 4 million of which supports the pilot project in Beijing. Prior to this project, China's focus for clean coal technology research and development had been almost exclusively on integrated gasification combined cycle (IGCC) and polygeneration applications. The second project, a 400 MW integrated drying gasification combined cycle (IDGCC) demonstration, involves input from Harbin Power Equipment Company, a major Chinese boilermaker, in an overseas power plant venture. The AUD 750 million project, with AUD 100 million support from the Australian government, aims to improve the efficiency, environmental performance and economics of power generation from brown coal. In April 2008, further funds of AUD 20 million were announced for the Australia-China Joint Coordination Group on Clean Coal Technology, coming from Australia's AUD 500 million National Low Emissions Coal Fund.

Australia-China Climate Change Partnership

The Australia-China Climate Change Partnership was established in 2003 to implement co-operation in (AGO, 2006a):

- climate change policies;
- climate change impacts and adaptation;
- national communications (GHG inventories and projections);
- technology co-operation; and
- capacity building and public awareness.

One aspect of this early work was Australian assistance to identify China's CO_2 storage potential and to help match this with major sources of CO_2, especially coal-fired power plants. With industry backing and in-kind support from Australia's research community, this survey drew on Australia's expertise and experience in mapping Australia's CO_2 reservoir potential under the GEODISC programme.[1]

Later, in October 2006, Australia and China announced eleven joint projects to reduce GHG emissions, to assist in adaptation to climate change, to improve coal mine safety and to enhance climate change expertise in both countries (AGO, 2006b). Of particular relevance are the four methane-capture projects (Table 9.3). All four use a largely untapped energy source, while increasing mine safety. The projects will be implemented by Australian and Chinese partners at mine sites in China, with the Australian government contributing a grant to meet part of the project costs.

Table 9.3 .... Projects under the Australia-China Climate Change Partnership

Title	Partners	Location
Ventilation air methane catalytic combustion gas turbine (VAM-CAT)	Shanghai Jiaotong University, China; Huainan Mining Group, China; Commonwealth Scientific and Industrial Research Organisation (CSIRO), Australia	A mine site in China
Study on maximisation of coal mine methane capture	China Coal Information Institute (CCII); Huainan Coal Mining Group, China; CSIRO	A mine site in China
Study on coal mine methane resources and potential project development	CCII; CSIRO	Key coal mining areas in China
Exploration of options for the generation of electricity from coal methane in China	CCII; ComEnergy Pty Ltd, Australia	Key coal mining areas in China

Source: AGO (2006b)

---

1. See IEA Greenhouse Gas R&D Programme project database for details of the GEODISC programme (www.co2captureandstorage.info).

China-US Working Group on Climate Change

The United States (US) has a longstanding co-operation with China on energy issues and continues to engage China on matters of energy policy, energy security, fossil energy, energy efficiency and renewable energy (Austin, 2005). A Protocol for Cooperation in the Field of Fossil Energy Technology Development and Utilization was agreed in 2000, replacing a 1985 agreement. It has included joint R&D and information exchange on IGCC, FGD and CO_2 sequestration. In February 2002, the China-US Working Group on Climate Change was established to promote bilateral research on key areas of policy and science, including non-CO_2 gases, hydrogen and fuel cell technology, and CCS. Recently, bilateral co-operation has been limited to information exchange and joint studies, rather than large-scale projects. For example, the US-China Energy Policy Dialogue was signed by the US Department of Energy (DOE) and NDRC in May 2004 and covers collaboration in the field of fossil energy (DOE, 2007). However, most co-operation in this field is undertaken through the various multilateral agreements considered below. Nevertheless, at the second meeting of the US-China Strategic Economic Dialogue[2] in May 2007, both countries agreed (Treasury, 2007):

■ to develop up to 15 large-scale CMM capture and utilisation projects in China;
■ to provide policy incentives to promote the full commercialisation of advanced clean coal technologies; and
■ to advance the commercial use of CCS technologies.

This suggests a possible change in approach and, in June 2008, a ten-year Energy and Environment Cooperation Framework was agreed with five broad goals and task forces that may include clean fossil fuel power generation (Treasury, 2008).

EU-China Partnership on Climate Change

This partnership, agreed at the European Union-China summit in September 2005, provides a high-level political framework for co-operation and dialogue on climate change and energy (EC, 2005). It incorporates the China-EU Action Plan on Clean Coal and the China-EU Action Plan on Energy Efficiency and Renewable Energies, both dating from March 2005. A major objective is "by 2020 to develop and demonstrate in China and the EU advanced, near-zero emissions coal technology through carbon capture and storage" (MOFA, 2006). A second co-operation goal is to significantly reduce the cost of key technologies and promote their deployment. The roles of the private sector, joint ventures, public-private partnerships, bilateral and multilateral financing instruments and the potential roles of carbon finance and export credits will all be explored. The priority areas of co-operation include:

■ energy efficiency, energy conservation, and new and renewable energy;
■ clean coal technologies and CCS;
■ methane recovery and use;
■ hydrogen energy and fuel cells;
■ power generation, transmission and distribution;
■ CDM and other market-based instruments;
■ impacts of and adaptation to climate change; and
■ capacity building and raising public awareness.

In recognition of the importance of developing and demonstrating near-zero emissions coal technology, various Chinese research institutes, universities and industrial

---

2.   www.ustreas.gov/initiatives/us-china

organisations have been included in CCS R&D initiatives that were either underway or were initiated by the final call of the EU's 6th Framework Programme for R&D. The EU is fully funding the Chinese input to such work. The major project, Cooperation Action within CO_2 Capture and Storage China-EU (COACH), is examining various technical and non-technical aspects of large-scale, clean coal energy facilities based on coal gasification with CCS.[3]

These initiatives build on the six-year, EUR 42.9 million EU-China Energy and Environment Programme that has been running since 2003.[4] Jointly funded by the European Commission (EC) and the Chinese government, the programme's objectives are to contribute to sustainable energy production and consumption, to improve energy security and to improve health conditions by strengthening government capacity. The programme promotes EU-China industrial co-operation and supports EU-China policy dialogue on energy, environment and climate change. The four areas of activity focus on energy policy development, energy efficiency (including at coal-fired boilers), renewable energy and natural gas (including CBM).

China-UK Climate Change Working Group

In 2005, China's NDRC and the Department for Environment, Food and Rural Affairs (Defra) of the United Kingdom (UK) signed a memorandum of understanding (MoU) to establish a China-UK Climate Change Working Group (Defra, 2006). It includes co-operation on the UK-China elements of a first-phase assessment and actions to encourage the implementation of subsequent phases of the EU-China Co-operation on Near-Zero Emissions Power Generation Technology through CCS (Defra, 2007). Thus, the UK, through Defra, is leading and funding Phase I of this Near-Zero Emissions Coal (NZEC) project.[5] This comprises a three-year, GBP 3.5 million feasibility study, examining the viability of different options for the capture of CO_2 emissions from power generation and the potential for their geological storage in China. The aim is to bring forward the time when new Chinese coal-fired power plants are built with CCS. Defra organised a competitive bidding process from which a UK-China consortium was selected to consider the key technologies for near-zero emissions power generation, together with CO_2 storage in either coal seams, aquifers or depleted oil wells, including the possibility of enhanced oil recovery. Discussions are underway to establish the basis, within an overall EU-China framework, for funding Phases II and III of NZEC, which are intended to support a demonstration project starting up between 2010 and 2015.

Also under this Working Group, in July 2006, Defra and the Ministry of Science and Technology (MOST) organised a major symposium between EU industrialists and researchers, and their counterparts in China, to discuss the technical and non-technical issues associated with CCS.

UK-China Cleaner Coal Technology Transfer Programme

Since 1998, under a MoU between the UK Department of Trade and Industry (DTI) and the Ministry of Foreign Trade and Economic Cooperation, the UK and China have had an established agreement on clean coal technology transfer, with an emphasis on co-operation in the development and dissemination of technologies that are most

---

3.   www.co2-coach.com
4.   www.eep.org.cn
5.   www.nzec.info

appropriate to China's needs (DTI, 2002). Separately, a MoU on energy research was signed by MOST in November 2005, replacing an earlier MoU dating from October 2001 that itself amended the original Protocol on Scientific and Technological Co-operation signed in November 1978 (*ibid.*; MOST, 2006). A specific objective has been to enhance technology transfer, both to demonstrate UK expertise as a source of components and know-how, and to assist China in reducing the environmental impacts of coal use. The UK Department for Business, Enterprise and Regulatory Reform (BERR – formerly DTI and now the Department of Energy and Climate Change) has supported workshops and seminars in China aimed at encouraging information exchange on both conventional and advanced energy technologies. Export-related activities have been undertaken in collaboration with the China-Britain Business Council and Trade Partners UK, including missions to China involving large companies, small and medium enterprises, and financial institutions. Several UK-China collaborative R&D projects have also been initiated under the MoUs, involving UK industry and universities (DTI, 2000).

Alongside these collaborative projects, BERR has supported the dissemination of information on technology transfer and export promotion activities in China: the China Coal Information Institute has translated and distributed throughout China key information and publications arising from the BERR Cleaner Coal Technology Programme.[6]

Germany-China environmental co-operation

Germany has an ongoing strategic co-operation with China, covering financial and technical activities, with an environmental focus. The objectives and terms of the co-operation have been adjusted over the years both to reflect China's emergence as a major economy and growing concerns over environmental issues, such as acid rain and global CO_2 emissions (BMZ, 2007). In the field of coal, co-operation includes a Sino-German Working Group on Coal that has brought together officials, scientists and industrialists from both countries with the aim of promoting and strengthening technical and commercial co-operation, notably in coal mining and coal mine safety. Much of the co-operation is channelled through the German Technical Co-operation Agency (GTZ), an enterprise that has been active in the Chinese coal and power sectors for nearly twenty years.[7] It has supported a number of projects ranging from policy advice to specific implementation measures that have included some coal-related activities, mostly concerned with making improvements to the efficiency of existing power generation units (Moczadlo, 2008). A noteworthy project to monitor and extinguish coal-seam fires in China's northern provinces has made real progress in identifying underground fires and sharing knowledge on how to tackle this pernicious problem.[8] In addition to noxious pollution, land subsidence and an annual loss of perhaps 100-200 Mt of coal resources, uncontrolled coal-seam fires in China consume an estimated 10-20 Mt of coal each year and release large volumes of CO_2 (Bauer, 2006).

South Africa-China co-operation on CTL

In September 2002, the Chinese government agreed to guidelines for co-operation with Sasol Ltd., owners of proprietary Fischer-Tropsch technology used to synthesise liquid

---

6.   www.coalinfo.net.cn/cnuk/eindex.htm

7.   www.gtz.de

8.   www.coalfire.caf.dlr.de

fuels from coal (Strauss, 2006). Then, in January 2004, NDRC appointed Shenhua Group and Ningxia Coal Industry Group to co-operate with Sasol on two possible 80 000 bpd coal-to-liquids (CTL) projects: one in Shanxi and one in Ningxia. The timetable since then has been:

- signed Feasibility Study Stage I MoU (September 2004);
- initiated pre-feasibility study after kick-off meeting of the engineering contractors (February 2005);
- completed pre-feasibility study which confirmed that the key drivers were in place for a viable CTL business (November 2005); and
- agreed to proceed with Feasibility Study Stage II to examine the capital cost, feedstock cost, water supply, market conditions and commercial viability (June 2006).

Multilateral collaboration agreements

Asia-Pacific Partnership on Clean Development and Climate

The Asia-Pacific Partnership on Clean Development and Climate (APP) was established in January 2006, bringing together six countries – Australia, China, India, Japan, the Republic of Korea and the US – to co-operate on meeting their increased energy needs and associated challenges, including air pollution, energy security and GHG emissions.[9] All six have signed the APP Charter, a non-legally binding commitment to create a co-operative framework that accelerates the development, demonstration, transfer and deployment of technologies to meet "development, energy, environment and climate change objectives" (APP, 2007a). In October 2007, during the second APP ministerial meeting in New Delhi, Canada became a partner.

The seven partners are to pursue initiatives, involving the public and private sectors, that are intended to complement activities under the Kyoto Protocol. The aim is to develop sustainable solutions through bottom-up, practical actions that build capacity, expertise and the research base in partner countries (APP, 2008a). To this end, experts from the public and private sectors and the research community, alongside leaders from each of the partner countries, have joined task forces covering eight key sectors, among them cleaner fossil energy and coal mining.

The Cleaner Fossil Energy Task Force assumes that coal, oil and natural gas will remain key fuels for all seven partners, and that energy demand in the Asia-Pacific region will rise (APP, 2007b). The Task Force seeks to improve the efficiency and environmental performance of this continued use of fossil fuels, and has identified those technologies that could significantly reduce GHG emissions, air-borne pollutants and other environmental impacts (Table 9.4). These included IGCC, hydrogen production from coal, ultra-supercritical pulverised coal-fired power plants and CCS. The Task Force's primary focus is to share best practices, eliminate market barriers to the deployment of clean technologies, increase the utilisation of natural gas and raise the efficiency of cleaner fossil energy use in general.

---

9. www.asiapacificpartnership.org

Table 9.4 .... APP Cleaner Fossil Energy Task Force: proposed activities

Code	Project title
CFE-06-01	CO_2 Capture and Storage Program
CFE-06-02	Ultra-Supercritical Pulverized Coal with Carbon Capture and Storage, Near-Zero Emissions Workshop and Design Guides for APP Countries
CFE-06-03	Ultra Clean Coal (UCC) Project
CFE-06-04	Oxy-fired Combustion Program
CFE-06-05	Callide A Oxy-Fuel Demonstration Project
CFE-06-06	Assessing Post-Combustion Capture and Storage Technologies for Emissions from Coal-fired Power Stations
CFE-06-07	Integrated Gasification Combined Cycle with Carbon Capture and Storage Workshop and Design Information for APP Country Coals
CFE-06-08	Asia Pacific Gas Market Growth
CFE-06-09	Evaluating and Reducing Emissions in Producing, Processing and Transporting Natural Gas
CFE-06-10	Information Exchange on LNG Public Education Campaigns
CFE-06-11	Asia Pacific Gas Hydrate Co-operation
CFE-06-12	Costs and Diffusion Barriers to Deployment of Low-emission Technologies
CFE-06-13	APP Enhanced Coal Bed Methane (CSIRO-JCOAL)
CFE-06-14	Development of Advanced Adsorption Process Technologies for Pre-Combustion Capture of CO_2 in Coal Gasification Processes (IGCC)
CFE-06-15	Coal Gasification Performance Assessments for Low Emissions IGCC Systems
CFE-06-16	Cooperative R&D on Cleaner Fossil Energy

Source: APP (2008b).

The Coal Mining Task Force has noted that, while coal is the world's most abundant and widely distributed fossil fuel, over 58% of the world's recoverable reserves are located in four APP partner countries, namely the US (27%), China (13%), India (10%) and Australia (9%) (APP, 2007c). The Task Force is working to improve coal mining and beneficiation efficiency, reduce the environmental impacts of coal mining and improve coal mining safety. This includes promoting best available technologies and practices in coal preparation, methane capture and improved mine worker health and safety (Table 9.5).

Table 9.5 .... APP Coal Mining Task Force: proposed activities

Code	Project title
CLM-06-01	Information Sharing on Coal Processing Technologies
CLM-06-02	Coal Beneficiation: Economic Modelling, Analysis and Case Studies
CLM-06-03	Fine Coal Beneficiation – Joint Venture Project
CLM-06-04	Information Sharing on Coal Drying
CLM-06-05	Joint Venture Project on Waste Coal Management
CLM-06-06	Extraction of Steep Seam Coal
CLM-06-07	Leading Practice Sustainable Development Program for the Mining Industry
CLM-06-08	Overburden Slope Stability
CLM-06-09	Coal Mine Health and Safety
CLM-06-10	Reclamation of Legacy Coal Mines to Abate Hazards
CLM-06-11	Increasing Recovery and Use of Coal Mine Methane
CLM-06-12	Integrated Coal and Methane Extraction
CLM-06-13	Thick Coal Seam Extraction
CLM-06-14	Underground Coal Gasification in India
CLM-06-15	Workforce Assessment and Training Needs
CLM-06-16	Technical Improvement for Control of Coalfield Fires
CLM-06-17	Underground Coal Mine Fires Prevention Control
CLM-06-18	Acceleration of Underground Coal Gasification in India – Phase 2 UCG Demonstration

Source: APP (2008b).

In October 2006, the APP Policy and Implementation Committee endorsed an initial set of projects and activities in eight sector-based action plans. The initial portfolio of projects is weighted towards activities such as sectoral assessments, capacity building, identifying best practices and technology research and demonstration (APP, 2006). There are no quantitative funding obligations placed on any of the partners. Rather, each partner may contribute funds, personnel or other resources at its sole discretion. The US announced that it would allocate USD 52 million to fund its role in the first year and Australia offered USD 15 million. The other partners have yet to announce their intended levels of financial support, but it is likely that developing countries will limit their inputs to in-kind contributions.

All of the coal-related projects involve business participation. In this sense, APP builds on successful public-private partnerships in the partner countries. Further projects will be added to the portfolio by the Task Forces and by partner countries as experience with the APP and opportunities to leverage resources increase.

Carbon Sequestration Leadership Forum

The Carbon Sequestration Leadership Forum (CSLF) was formed in 2003.[10] Its charter signatories comprise the world's major users and producers of fossil energy, *i.e.* Australia, Brazil, Canada, China, Colombia, Denmark, the European Commission, France, Germany, Greece, India, Italy, Japan, Mexico, the Netherlands, Norway, the Russian Federation, Saudi Arabia, South Africa, the Republic of Korea, the UK and the US (CSLF, 2003). The CSLF aims to bring together "intellectual, technical and financial resources from all parts of the world to support the long-term goal of the United Nations Framework Convention on Climate Change", namely the stabilisation of atmospheric CO_2 concentrations during the 21st century (CSLF, 2005). It is focused on the development of improved, cost-effective CCS technologies and making these universally available (CSLF, 2006). The Forum identifies wider issues relating to CCS and hopes to promote appropriate technical, political and regulatory environments for CCS development through activities overseen by its Policy and Technical Groups. Accordingly, task forces have been established to examine: capacity building; financial issues; risk assessment; storage capacity estimation; CO_2 capture and transport; and monitoring and verification of geological CO_2 storage.

In July 2005, G8 leaders endorsed the CSLF in the Gleneagles Plan of Action on Climate Change, Clean Energy and Sustainable Development, and sought to accelerate the development and commercialisation of CCS in co-operation with key developing countries (G8, 2005). To encourage such co-operation, the CSLF has introduced a system whereby it "recognises" projects that are funded by other bodies. By April 2008, twenty projects had received such recognition (CSLF, 2008).

Methane to Markets Partnership

The Methane to Markets Partnership (M2M) was launched in November 2004.[11] Partners now include: Argentina, Australia, Brazil, Canada, China, Colombia, Ecuador, the European Commission, Finland, Germany, India, Italy, Japan, Mexico, Mongolia, Nigeria, the Philippines, Poland, the Russia Federation, the Republic of Korea, Ukraine, the UK and the US (M2M, 2008a). It is a voluntary, non-binding framework for international co-operation to reduce global methane emissions, to enhance economic

---

10. www.cslforum.org
11. www.methanetomarkets.org and www.coalinfo.net.cn/coalbed/meeting/m2mb/index.htm (Chinese)

growth, to promote energy security and to improve the global environment. Initially, three major methane emission sources were targeted, *i.e.* landfills, coal mines and natural gas and oil systems. Later, agriculture, in particular animal-waste management, was added.

The aim is to promote cost-effective, near-term methane recovery and use technologies through international partnerships, in co-ordination with the private sector, researchers, development banks, experts and relevant governmental and non-governmental organisations. Currently, more than 800 organisations worldwide have made commitments to the Partnership network. Funding is voluntary. Each partner contributes funds, personnel and other resources; costs are borne by the partner that incurs them, unless other arrangements have been made.

Organisation of M2M is based around a committee system. The Steering Committee governs the overall framework, policies and procedures, while subcommittees are responsible for guidance and assessment of specific activities. Each subcommittee has developed an action plan that identifies needs, opportunities and priorities for project development, alongside key barriers and actions that would assist in overcoming these barriers. China chairs the Coal Mines Technical Subcommittee, reflecting its top ranking among the world's coal mine methane (CMM) emitters and its support for CMM and coalbed methane (CBM) development. The coal mines action plan centres on the following short-term actions (M2M, 2006):

- review methane recovery and use opportunities;
- describe available technologies and best practices;
- identify key barriers to project development;
- identify possible co-operative activities to increase methane recovery and use in the coal mining sector;
- discuss country-specific needs, opportunities and priorities; and
- reach out to engage M2M network members.

Barriers that many countries face include: lack of clarity about ownership of CMM; lack of transparency in legal and regulatory regimes covering gas rights; lack of information on economic factors such as taxes or incentives; difficulties in raising finance; lack of technical standardisation or harmonisation; inadequate or inaccessible infrastructure; and poor technical knowledge. An example of an M2M partner's response to such barriers is an investment guide for CMM/CBM in China published by the China Coalbed Methane Clearinghouse (M2M, 2004). There is also an M2M factsheet on CMM in English and Chinese (M2M, 2008b).

Asia-Pacific Economic Cooperation

The Asia-Pacific Economic Cooperation (APEC) organisation was established in 1989 to enhance economic growth and prosperity in the region and to strengthen the Asia-Pacific community.[12] It is the key forum for facilitating economic growth, voluntary co-operation, free trade and investment in the Asia-Pacific region. APEC has 21 member economies, which account for approximately 41% of the world's population, 56% of world GDP and 49% of world trade. They are: Australia; Brunei Darussalam; Canada; Chile; the People's Republic of China; Hong Kong, China; Indonesia; Japan;

---

12. www.apecsec.org.sg

the Republic of Korea; Malaysia; Mexico; New Zealand; Papua New Guinea; Peru; the Philippines; the Russian Federation; Singapore; Chinese Taipei; Thailand; the US and Vietnam.

In September 2007, APEC leaders pledged support for cleaner fossil fuels, as well as for renewables to address the fundamental and interlinked challenges of energy security and climate change (APEC, 2007). They stressed that fossil fuels will continue to play a major role at the regional and global level, but that development, deployment and transfer of low- and zero-emission technologies for their cleaner use, particularly for coal, will be essential, while energy efficiency enhancement and energy diversification, including renewable energy, will also be important. APEC also announced a programme of co-operative actions and initiatives designed to support economic growth, coherent policy and the reduction of GHG emissions, including an aspirational goal for the APEC-wide region of a 25% reduction in energy intensity by 2030, compared to 2005. The Action Agenda includes establishment of an Asia-Pacific Network for Energy Technology (APNet) in 2008, "to strengthen collaboration on energy research in our region" (*ibid.*).

FutureGen initiative

The FutureGen Alliance was formed in 2005 as a US DOE led and US-based, public-private partnership to design, build and operate a near-zero emissions coal-fuelled power plant.[13] A proposed commercial-scale prototype R&D facility would have produced electricity and hydrogen, while simultaneously capturing and permanently storing CO_2 emissions. The goal of the USD 1 billion project was for the 275 MW FutureGen plant to lead on to the development of similar power plants worldwide. The project was expected to use US coal gasification and gas turbine technologies integrated with combined cycle electricity generation and geological storage of CO_2 at a site in the US (DOE, 2003).

The FutureGen project is supported by the DOE Advanced Coal Technology RD&D Program, with an initial ten-year lifespan. It was led by the FutureGen Industrial Alliance, a non-profit, industrial consortium representing the coal and power industries. Active participation by industry in this project is intended to ensure that the public and private sectors share the costs and risks of developing the technologies necessary to commercialise the FutureGen concept (FutureGen Alliance, 2007). The thirteen Alliance members are: American Electric Power, Anglo American, BHP Billiton, China Huaneng Group, CONSOL Energy, E.ON US, Foundation Coal, Luminant, Peabody Energy, PPL Corporation, Rio Tinto Energy America, Southern Company and Xstrata Coal. In addition to the commercial partners, the governments of India, Japan, the Republic of Korea, China and Australia have also joined the FutureGen International Partnership. In May 2007, the Partnership met to review the proposed text of an agreement designed to encourage subsequent transfers of FutureGen technology.

A major milestone was the selection of a site at Mattoon, Illinois – announced by the Alliance in December 2007. To assist development of the project, the Illinois state legislature had passed a package of tax concessions, site zoning changes and also accepted potential CO_2 storage liabilities. Then, in January 2008, the DOE announced a restructuring of the FutureGen project. Rather than focus on a single project at

---

13. www.FutureGenAlliance.org

Mattoon – where costs had risen to an estimated USD 1.5-1.8 billion – the plan now is to support the additional cost of CCS at multiple commercial-scale IGCC plants across the US, each capturing and storing around 1 MtCO$_2$ each year (DOE, 2008). A competitive solicitation was issued by DOE in June 2008. Progress towards this new objective depends on the availability of federal funding.

GreenGen initiative

The GreenGen project aims to research, develop and demonstrate in China a high-efficiency, coal-based power generation system with hydrogen production through coal gasification, power generation from a combined-cycle gas turbine and fuel cells, and efficient treatment of pollutants with near-zero emissions of CO$_2$. The GreenGen Company was formed in December 2005 to implement the project. The founding shareholders comprise China Huaneng Group, China Datang Corporation, China Huadian Corporation, China Guodian Corporation, China Power Investment Corporation, Shenhua Group, China National Coal Group and the State Development and Investment Corporation.[14] Huaneng is the largest shareholder with 51% of the total investment; the other seven companies each hold 7% equity. GreenGen has sought international co-operation to take forward this project and, in December 2007, Peabody Energy of the US took a 6% equity stake in the GreenGen Company.

The objective of the USD 1 billion GreenGen project is to construct a 250 MW IGCC power plant and ultimately to demonstrate polygeneration for Chinese application using Chinese technology, where practicable. The official plan is to design, build and operate the first IGCC power plant in China by the end of 2009, to be followed within five years by an expansion to 650 MW when CCS will be considered – it would be the first coal-based, near-zero emissions power plant in China. Progress is well advanced: GreenGen Company has completed a feasibility study, selected a reclaimed coastal site at Tianjin, secured all necessary permits from the authorities and applied to the Asian Development Bank for financing (Section 8.4). As of June 2008, the engineering design was under way, with the main equipment procurement expected to be completed before the end of the year (Su, 2008). It is intended that all key components will be manufactured domestically. As such, bids are being solicited from GE, Mitsubishi Heavy Industries and Siemens to link up with Chinese partners for the supply of gas turbine technology to use with the domestic gasifier. Given this progress, completion of Phase I could be achieved by 2011-12.

GreenGen is a high profile initiative for China, with strong government support from NDRC. Certain key technologies have been included in the National Medium- to Long-Term Program for Scientific and Technological Development, and the Tianjin IGCC project itself has received support from MOST as a major project under the National High-Tech R&D Program ("863" Program).

IMPROVING INTERNATIONAL COLLABORATION

This section reprises the coal-related challenges China faces, then summarises the key issues that must be resolved for more effective clean coal technology transfer. The strengths and weaknesses of the current approaches to international collaboration

14. http://greengen.com.cn/en/index.asp and
   www.chng.com.cn/minisite/greengen/greengen_jihua_index.html

are considered, leading on to proposals for future collaboration that could improve economic, social and environmental conditions in China's coal sector and see clean coal technologies more widely deployed.

Challenges

As detailed in earlier chapters, China's economic growth is founded on extensive industrialisation, fuelled by coal. At the same time, the government recognises that coal use needs to be more sustainable. The coal mining sector needs to meet numerous environmental challenges, to improve its organisation and management to meet growing domestic demand from indigenous resources, to improve safety for workers and to build substantial new infrastructure. While China is quite capable of meeting these challenges, it may do so even more quickly and effectively if it acts in co-operation with other countries.

In June 2007, China released its first national plan for climate change (NDRC, 2007), prepared by the State Council, NDRC and seventeen other departments, in co-ordination with the United Nations Framework Convention on Climate Change (UNFCCC) Secretariat. It sets out the nation's guidelines and goals, and related policies and measures. It reiterates China's aim to reduce energy intensity by 20% between 2005 and 2010 and to increase renewable energy production to 10% of the national energy mix. It stresses China's long-standing position that it will not tackle the problem of climate change at the expense of economic development and thus will not cap GHG emissions. The plan also stresses the role of technology transfer and international co-operation in helping China move towards a low-carbon economy.

Fundamentally altering the nation's coal dependence in the short term is not feasible. Even with continued rapid progress in economic restructuring, technological innovation and energy saving, demand for coal will keep growing. In particular, the electricity, steel and building materials sectors will drive coal consumption, with the expectation that coal-to-chemicals and CTL sectors will further increase demand. Coal production and supply difficulties persist: resolving these is crucial to long-term supply security (Minchener, 2007). While China has massive coal reserves, the level of proven, exploitable reserves is rather low. Coal resources continue to be wasted through inefficient extraction techniques, particularly at small mines, further exacerbating the problem. China is making significant investments in efficient coal production at large mines, mostly in mining areas that are distant from the major industrial centres, leading to an ever-growing need for infrastructure investment to transport coal. At the same time, coal use is often inefficient. Balancing short-term needs with long-term sustainability, while not unique to China, is now a major concern due to the sheer scale of the energy needs in a country with a large and rapidly growing economy.

Towards more effective transfer of clean coal technologies

Within the framework of international co-operation, there are both government-to-government and industry-to-industry interactions. For successful co-operation, all sides must understand the needs and requirements of each participant.

Previous analysis of how best to ensure successful technology transfer, cited below, has not resulted in a definitive approach, only broad guidelines. In China, many mechanisms have been successfully used to transfer clean coal technologies, including joint ventures, licensing agreements and training programmes (Watson, 2002). It is not so much the choice of mechanism that is important, but the need for both sides to develop a level of trust that will allow collaboration to succeed. Thus, technology-transfer processes need to be comprehensive and long-term in order to succeed. The transfer of broader "know-how" and "know-why" skills is just as important as hardware equipment transfer, since the recipient Chinese institutions can only then assimilate the new technology.

The challenge for industrial companies and government institutions, on both sides, is to design transfer programmes which meet their multiple and varied objectives, many of which are potentially conflicting. These programmes must also address some of the non-economic barriers to successful transfer. Thus, state institutions need to frame their programmes to ensure that they positively encourage the broad technology transfer required for Chinese companies to improve their capabilities in technology development and management. In this respect, the Japanese Green Aid Plan, which has failed to involve Chinese equipment manufacturing companies in its demonstration projects, is an example of a poorly framed programme. In part, this was due to intellectual property concerns but, as is evident elsewhere, such problems can be resolved. State institutions also need to understand the regulatory drivers in China as these will have a significant impact on whether foreign technologies can become established within the Chinese market. Again, the Green Aid programme is an example of a mismatch between Japan's advanced technologies and a market in which demand, at the time, was for low-cost, relatively low-performance, pollution control equipment. Finally, state institutions need to recognise that China will not become a testing ground for unproven technology. An example is the US DOE's efforts to promote clean coal power plant technologies in China, which failed because of the preoccupation in the US with gasification technologies that were uneconomic and unproven (DOE, 1996; Watson, 2005 and MIT, 2007).

Technology transfer covers equipment transfer as well as essential knowledge transfer for operations, maintenance and further technical evolution. For clean coal technology transfer, the focus used to be just on those technologies that could make an improvement to the environmental and economic performance of Chinese industry in the short to medium term (Watson, 2002). This situation is now changing. The power sector now dominates coal use in China and the introduction of advanced coal-fired power plants is well under way, such that the role of government in technology promotion is now limited, except in cases where demonstrations of new technologies may require support to mitigate risk. At the same time, technology transfer for CTL applications is progressing on a commercial footing, after links between international technology suppliers and technology users in China were brokered at government level. However, with climate change firmly on the agenda, the technology that cannot now be divorced from advanced power generation and CTL is CCS. This is a very high-profile, longer-term deployment issue that very much requires government-to-government dialogue to establish methods of co-operation, since CCS is not yet proven at large commercial scale anywhere, nor is there yet a regulatory framework for its deployment. As such, industry in both OECD member countries and China are requesting government support to take forward its development and potential deployment.

Critique of bilateral collaboration

The majority of bilateral collaborations that include actions on cleaner coal are linked to declarations on climate change. Nevertheless, their terms of reference are often broad enough to incorporate work on not only CCS, but also advanced clean coal technologies with higher efficiency and improved environmental performance, in terms of conventional pollutants such as SO_2, NOx and particulates.

The Australia-China Joint Coordination Group on Clean Coal Technology appears to be a classic, bilateral agreement where Australia sees China as a potentially enormous local market for clean coal technology goods and services. Some projects have been announced which will involve know-how transfer from Australian industry to Chinese industry, including a joint Australia-China cleaner coal gasification technology demonstration and an innovative pilot plant to remove CO_2 from coal combustion flue gas. Australia is providing financial support, but whether it will be enough to seriously engage China in areas that are not near-term, such as commercial-scale CCS, remains to be seen. Certainly, there is evidence to suggest that China will only participate in significant CCS activities if the other party provides funding to cover all direct and indirect costs in China (see EU-China details above). Alongside is the Australia-China Climate Change Partnership with a broad remit that includes some coal technology activities, specifically on novel technologies for CMM utilisation. Again, these will result in know-how transfer from Australia to China, with the prospect of equipment transfer to follow if the projects are successful. It is presumed here that, because of its near-term, mine safety benefits, China will make a balancing, in-kind contribution by providing access to Chinese mines for testing and will cover some of its own research costs.

The China-UK bilateral programme is a longer-standing agreement than the Australian initiatives, with a focus on clean coal technology transfer. It has covered all aspects of technology transfer, including know-how dissemination via many routes. Such dissemination supports one of the UK government's environmental objectives, namely to assist developing countries to deploy clean coal technologies. Hence, the programme has supported technologies and techniques that are seen by the Chinese partners as important, rather than those which the UK necessarily sees as commercially exploitable. The programme is well regarded in China and has been well supported by UK industry as a means to engage with Chinese industry and research institutes (MacKerron *et al.*, 2004). That said, the focus has been on medium- to longer-term opportunities and the benefits have been difficult to quantify. Now that the programme has been renamed the Carbon Abatement Technology Programme to reflect its new focus, benefits are likely to remain distant.

The EU-China Partnership on Climate Change is a significant collaboration, which includes the declared intention of building a large-scale CCS demonstration at a new, commercial-scale, coal-fired power plant in China. This is complemented by the

China-UK Working Group on Climate Change, which has identical wording within its MoU on CCS activities. However, while China has signed this agreement, it is not yet prepared to invest its own funds in the project. Thus, Phase I activities are fully funded by the EC and Defra, including the staff and associated costs of the Chinese partners. Currently, the EC and Defra have started EU-wide discussions on the future costs for Phases II and III, while also sounding out nation states and industry on options to secure their involvement and financial commitment.

Critique of multilateral collaboration

The three key agreements are the APP, the M2M Partnership and the recently announced APEC coal-based R&D initiative. As in most areas of multilateral activity, there seems to be an enormous amount of overlap, with either similar – and sometimes the very same – projects announced under different multilateral and even bilateral agreements (this seems to be especially true of the two Australia-China agreements). At present, funds committed to these multilateral agreements are limited. For example, the direct funding committed across the APP by the US and Australia seems wholly inadequate to achieve the aims of the initiative, while the lack of public commitments from the developing countries suggests their inputs will be solely in-kind contributions. Consequently, while an attractive portfolio of relevant coal-based projects has been proposed, it appears that only low-cost, capacity-building workshops and training schemes are actually proceeding. Such capacity-building activities are necessary steps before moving to large-scale R&D projects, but, unless large-scale commitments are made soon, these agreements risk losing credibility within the international community, to the detriment of all global climate change initiatives.

In principle, the APP seems to be a strong partnership with great potential. From a Chinese perspective, recommendations made by the Cleaner Fossil Energy Task Force offer a range of projects that would be of considerable benefit. These include projects to accelerate demonstration, deployment and transfer of key technologies that would improve the environmental and economic performance of fossil fuel use, such as near-term coal gasification, advanced coal combustion and longer-term CO_2 capture techniques and storage opportunities. Alongside these, the Coal Mining Task Force has proposed a similarly relevant series of projects. These include ways to accelerate the deployment of technologies and practices that could improve the economics and efficiencies of coal mining and processing, while continuing to improve mine safety and reduce environmental impacts. All proposals are compatible with the aims and objectives of China's 11th Five-Year Plan. However, this ambitious programme could be severely constrained by a lack of adequate funding.

The Methane to Markets Partnership is also of significant potential benefit to China. In principle, it will develop strategies for the recovery and use of methane from coal mines, through technology development, demonstration, deployment and diffusion, together with implementation of effective policy frameworks, identification of the means to support investment and removal of barriers to collaborative project development. This sounds admirable, but in practice it will require significant funding to achieve any lasting impact. As noted above (Chapter 8), there are a large number of unresolved technical and non-technical difficulties associated with establishing CMM operations and, even more so, CBM extraction on a commercially viable basis. While it is evident that the international financial institutions, such as the World Bank, and commercial companies are addressing many of these already in China, it is not clear where this partnership will add value.

The recent APEC declaration that coal will continue to play a major role in regional and global energy needs and calls for the development, deployment and transfer of low- and zero-emission technologies for its cleaner use are welcome recognitions of the reality of future fossil fuel use in the Asian region. The associated announcement on the establishment of an Asia-Pacific Network for Energy Technology (APNet) to strengthen collaboration on energy research in the APEC region, particularly on clean fossil energy and renewables, is also potentially significant. However, whether that potential will be realised will depend on exactly what is proposed and how projects are funded.

With regard to the CSLF, China should gain from exposure to global discussion on the many difficulties of establishing CCS in the market place and learn as these difficulties are overcome. Beyond that, it is understood that Chinese involvement on the various task forces is rather limited.

The other two agreements concern large-scale technology demonstrations. As originally conceived, the US-led FutureGen project was a single project based on coal gasification; its reincarnation as a number of projects has important implications since it could provide a template for demonstrations elsewhere of many CCS technologies. However, it has been a project designed and led by committees with heavy political influences. As a result, it takes time to reach consensus and there is every likelihood that other projects will move forward faster and so reduce the relevance of FutureGen within the global framework. That said, China can gain considerable benefit from participation, if only to progress its own GreenGen and other initiatives. GreenGen is a Chinese-based and -led project to demonstrate IGCC technology, possibly with CCS in a second stage, using as much domestic technology as possible. It is interesting to note that GreenGen is already ahead of FutureGen in terms of achievements to date, which reflects the need for decisive action if projects are to succeed.

Proposals for future collaboration

To achieve a sustainable coal supply chain, it is evident that the best way forward for China will be through international collaboration and technology transfer on a commercial basis in order to gain rapid access to know-how and equipment. There are many challenges, over various time horizons, but all must be tackled with some urgency. These include technical, policy, regulatory and institutional issues, all of which must be addressed in a coherent, consistent manner. It is the longer-term, climate issues that are highest on the international agenda, while China has to deal effectively, in the near term, with the rationalisation of coal production and use in order to establish a stable framework on which to then address the medium- to longer-term issues. Indeed, the international drive to engage with China on climate change issues could well count for nothing if China does not stabilise its domestic coal production and supply chain, and complete effectively the rationalisation of the major energy users within the power and industrial sectors. The Chinese government has begun reforms, but there is a long way to go. There is great scope for those OECD member countries that have previously undertaken such reforms to provide advice and assistance. Chapter 7 provides an insight into coal-sector reform issues in many countries and how they have been tackled. Regulation of the major coal-using sectors in IEA member countries is well advanced, in terms of achieving economic efficiency, fair competition, security of supply and acceptable environmental performance. In China, the international community needs to offer the benefit of its own past experiences. For example, there is a rich agenda for international support to the ongoing regulatory reform of China's electricity sector (Austin, 2005), with the following priorities (see also Box 10.1):

■ reform of the electricity pricing mechanism;

■ reform of new projects approval mechanism;

■ deepening the reform of regional grid companies and provincial power companies;

■ possible establishment of regional (sub-national) electricity markets and regulatory organisations; and

■ revision of the Electric Power Law 1995 and associated regulations.

The decentralisation of many of the decision-making functions within China's energy sector has resulted in an inconsistent approach when implementing national energy policy. The State Council has responded with a more robust approach, centralising the planning and approval processes for the coal and power sectors. It seems likely that this ongoing conflict between national and provincial decision-making bodies will pose additional challenges to effective GHG mitigation, with the central government seeking to deal with sustainability and the provincial governments continuing to focus on economic growth and local employment. Consequently, it is important for the international community to increase bilateral and multilateral collaboration with China, at all levels of government, to address shared energy and environmental concerns (Logan *et al.*, 2007).

China has ratified the primary international accords on climate change, *i.e.* the UNFCCC and the Kyoto Protocol. However, as a developing country, China has no binding emission limits under either accord, although it is an active participant in the Clean Development Mechanism (CDM) established under the Kyoto Protocol. The CDM grants emission credits for verified reductions in developing countries, which can be used toward meeting the Kyoto Protocol targets of those countries who have committed to absolute reductions (*i.e.* Annex I countries). This can provide lower-cost reductions and generates investment in clean energy in developing countries. China is by far the largest source of CDM credits, accounting for 52% of those expected from projects registered to date (UNFCCC, 2008). Nine CMM projects have been registered with the CDM Executive Board and six more are awaiting registration, together accounting for a total estimated emission reduction of 7 $MtCO_2e$ per year (*ibid.*). However, for CDM project activity to achieve any significant impact on China's emissions from coal combustion (4 200 $MtCO_2$ in 2005 [IEA, 2007]) would require a major transformation of the eligibility criteria – for example, allowing large-scale CCS projects. In any event, China will presumably wish to maintain this revenue source under the close supervision of NDRC which currently sets the base price for CDM credits.

It is essential to recognise that China, and other developing countries, are unlikely to accept absolute GHG emission targets as they see these as a constraint on their economic growth. Rather, China is likely to propose intensity-based targets as these provide a measure for the efficiency of economic growth without limiting absolute energy use. The 11th Five-Year Plan has a significant range of domestic policy initiatives aimed at reducing energy intensity. Consequently, any future international climate agreement would need an arrangement whereby emissions mitigation measures fit or advance the national priorities of developing countries, such as economic growth, energy security and public health (Lewis, 2007).

Given this, any collaborative agreements on climate change are likely to be closer to the APP and M2M models. In principle, these could help countries meet their broader economic development strategies and, if implemented effectively, would also serve to mitigate GHG emissions. However, such agreements only work if there is sufficient funding provided to allow the necessary suite of projects to be formulated and implemented. The APP and M2M lack the funding to achieve a material impact, yet they do bring together an important grouping of nations, and therefore have the potential to lay the groundwork for future actions (Logan *et al.*, 2007).

However, reducing energy intensity does not reduce absolute GHG emissions. Consequently, CCS will be an essential part of any CO_2 abatement strategy for China. China's own climate change plan has identified priority areas for international collaboration, which include co-operation on advanced coal technologies. Although CCS is not a priority, it is important to build on this starting position by designing a framework whereby China has an incentive to introduce CCS at power generation plants, at large industrial plants and, if eventually established, at plants in the CTL sector. In this regard, the EU-China Partnership on Climate Change could have an important role to play in establishing large-scale demonstrations of CCS in China and Europe before 2020. There are many important and difficult steps to be taken, but the EU and China have both made a commitment to achieving this target. At a technical level, the IEA Greenhouse Gas R&D Programme offers China a forum for exchange of information and ideas, should it choose to join.[15] With 17 member countries, the European Commission and 17 multi-national industrial sponsors, this collaborative research programme evaluates and promotes technologies to reduce greenhouse gas emissions, disseminates results and facilitates practical research, development and demonstration activities.

For clean coal technologies – including CCS, once demonstrated – to have a material impact on pollutant and GHG emissions in China and elsewhere demands their widespread adoption. Bilateral and multilateral co-operation between governments can never achieve this alone – they can only sow the seeds. Only national government diktat or commercialisation within regulated markets that value clean plants can achieve the penetration needed to make deep cuts in emissions. Few national governments have unilaterally imposed stringent pollution emission standards, and fewer still would

---

15. www.ieagreen.org.uk

unilaterally restrict CO_2 emissions. Countries have tended to move forward together, gradually strengthening environmental legislation, but not stepping too far out of line for fear of suffering economically. Thus, we should look to establish a global market for clean coal technologies. With many older coal-fired plants now scheduled for replacement in OECD member countries, and China now the world's largest manufacturer of coal-fired units, it seems that establishing a genuine market for low-emission coal plant in OECD member countries should be a priority. Chinese industry might be encouraged to form international partnerships to compete for business in this market – partners would benefit from a combination of China's low-cost base and the technological know-how found in more industrialised countries. The same, low-emission technologies would then be available to the Chinese market at a more affordable cost. Establishing a clean energy market in China is therefore also a priority. The CDM and any successors can be an important driver, if allowed to reach their full potential by including CCS. However, China will also need to learn from the experience in OECD member countries of creating clean energy markets. Their development is ongoing and no blueprint exists to copy. Hence, government-to-government co-operation will be needed to develop ideas and share experiences – precisely the role of the IEA. Technology transfer is a natural outcome of this suggested approach. By working together in commercial ventures, companies share and profit from application of their existing and new technology developments. The role of government here should be limited to support of early stage technology development. Today, that means supporting the demonstration of a broad range of CCS technologies. While such support is already seen in many of the bilateral and multilateral agreements reviewed above, the level of commitment and level of funding falls short of that needed for these technologies to be widely deployed within the timescales that many world leaders now believe is necessary to avoid the worst impacts of climate change.

REFERENCES

AGO (Australian Greenhouse Office) (2006a), *Australia-China Climate Change Cooperation – Progress and Achievements 2003-2005*, AGO, Department of the Environment and Heritage, Canberra, Australia and Office of the National Coordination Committee for Climate Change, National Development and Reform Commission, Beijing, 12 January,
www.climatechange.gov.au/international/publications/pubs/aust-china.pdf.

AGO (2006b), "Australia-China Climate Change Partnership: New Partnership Projects", press release, AGO, Canberra, Australia, 20 October,
www.greenhouse.gov.au/international/china/pubs/fs-projects.pdf.

APEC (Asia Pacific Economic Cooperation) (2007), *APEC Leaders' Declaration on Climate Change, Energy Security and Clean Development*, Sydney, Australia, 9 September,
www.apec2007.org/documents/Declaration%20Climate%20Change.pdf.

APP (Asia-Pacific Partnership on Clean Development and Climate) (2006), *Work Plan*, APP, Washington, DC, 11-12 January,
www.asiapacificpartnership.org/WorkPlan.pdf.

APP (2007a), *Charter*, adopted at inaugural ministerial meeting, Sydney, 11-13 January 2006 and amended at 2nd ministerial meeting, New Delhi, 14-15 October 2007, APP, Washington, DC, www.asiapacificpartnership.org/Charter.pdf.

APP (2007b), *Cleaner Fossil Energy Task Force Action Plan*, APP, Washington, DC, 3 May, www.asiapacificpartnership.org/APPProjects/CFETF/Cleaner%20Fossil%20 Energy%20Action%20Plan%20030507.pdf.

APP (2007c), *Coal Mining Task Force Action Plan*, APP, Washington, DC, 3 May, www.asiapacificpartnership.org/APPProjects/Coalmining/Coal%20Mining%20 Task%20Force%20Action%20Plan%20030507.pdf.

APP (2008a), *Asia-Pacific Partnership on Clean Development and Climate*, Department of State publication #11468, Washington, DC, August, www.asiapacificpartnership.org/brochure/APP_Booklet_Aug2008.pdf.

APP (2008b), "Project Roster", APP, Washington, DC, www.asiapacificpartnership.org/ProjectRoster.htm, accessed 18 October 2008.

Austin, A. (2005), *Energy and Power in China: Domestic Regulation and Foreign Policy*, Foreign Policy Centre, London.

Bauer, S. (2006), "Coal Seam Fires", Coal Fire Research Platform, Cologne, Germany, 11 September, www.coalfire.org/index.php?option=com_content&task=section&id=12&Itemi d=43, accessed 10 October 2008.

BMZ (Bundesministerium für wirtschaftliche Zusammenarbeit und Entwicklung – Federal Ministry for Economic Cooperation and Development) (2007), "Strategic Cooperation with China in Germany's Interest", press release, BMZ, Berlin, Germany, 27 July, www.bmz.de/en/press/pm/2007/july/pm_20070727_90.html, accessed 18 September 2008.

CSLF (Carbon Sequestration Leadership Forum) (2003), *Charter for the Carbon Sequestration Leadership Forum: A Carbon Capture and Storage Technology Initiative*, CSLF, Washington, DC, 25 June, www.cslforum.org/documents/CSLFcharter.pdf.

CSLF (2005), "CSLF Technology Aims at Largest Sources of CO_2 Emissions – Joins Developed and Developing Nations in Climate Activity", press release, CSLF, Berlin, Germany, 29 September, www.cslforum.org/documents/cslf_background.pdf.

CSLF (2006), *Strategic Plan*, CSLF, Washington, DC, www.cslforum.org/documents/CSLFStrategicPlan.pdf.

CSLF (2008), *CSLF Recognized Projects*, CSLF, Washington, DC, April, www.cslforum.org/documents/CSLFRecognizedProjectsApr2008.pdf.

Defra (Department for Environment, Food and Rural Affairs) (2006), *Memorandum of Understanding between the National Development and Reform Commission of the People's Republic of China and the Department for Environment, Food and Rural Affairs on Establishing a China-UK Climate Change Working Group*, Defra, London, 13 September, www.defra.gov.uk/environment/climatechange/internat/devcountry/pdf/china-memorandum.pdf.

Defra (2007), "UK-China Clean Coal Initiative Launched", press release, Defra, London, 20 November, www.defra.gov.uk/news/2007/071120b.htm, accessed 18 September 2008.

DOE (US Department of Energy) (1996), *The United States of America and the People's Republic of China Experts Report on Integrated Gasification Combined-Cycle Technology (IGCC)*, commissioned by the Office of Coal and Power Import and Export, sponsored by the Chinese Academy of Sciences (CAS) and the Federal Energy Technology Center, organised by the Institute of Engineering Thermophysics at CAS and Tulane University, DOE Report No. DOE/FE-0357, US DOE, Washington, DC, December.

DOE (2003), *FutureGen: A Sequestration and Hydrogen Research Initiative*, fact sheet, Office of Fossil Energy, DOE, Washington, DC, http://fossil.energy.gov/programs/powersystems/futuregen/futuregen_factsheet.pdf.

DOE (2007), "US-China Energy Cooperation", DOE, Washington, DC, www.energy.gov/news/5080.htm, accessed 18 October 2008.

DOE (2008), "DOE Announces Restructured FutureGen Approach to Demonstrate CCS Technology at Multiple Clean Coal Plants", press release, DOE, Washington, DC, 30 January, www.energy.gov/news/5912.htm, accessed 18 September 2008.

DTI (Department of Trade and Industry) (2000), *Transfer of Cleaner Coal Technologies to China: A UK Perspective*, Project Summary No. 274, DTI/Pub URN 00/1430, DTI, London.

DTI (2002), "Lord Sainsbury Signs MoU with MOST", *Clean Coal Research Matters*, newsletter No. 11, DTI, London, April, www.berr.gov.uk/files/file20990.pdf.

EC (European Commission) (2005), "EU and China Partnership on Climate Change", press release, MEMO/05/298, EC, Brussels, Belgium, 2 September, http://ec.europa.eu/external_relations/china/summit_0905/index.htm, accessed 18 September 2008.

Evans, P. (1999), "Japan's Green Aid Plan: The Limits of State-Led Technology Transfer", *Asian Survey*, Vol. 39, No. 6, University of California Press, Berkeley, CA, pp. 825-844.

FutureGen Alliance (2007), "FutureGen Alliance Project Update Presentation", FutureGen Alliance, Columbus, OH, 5 March, www.futuregenalliance.org/publications/fg_project_update_030507_v9.pdf.

G8 (2005), *Gleneagles Communiqué on Africa, Climate Change, Energy and Sustainable Development*, G8, Gleneagles, UK, 8 July, www.fco.gov.uk/Files/kfile/PostG8_Gleneagles_Communique.pdf.

IEA (2007), *CO_2 Emissions from Fossil Fuel Combustion*, OECD/IEA, Paris.

Lewis, J. I. (2007), "China's Climate Change Strategy", *China Brief*, Vol. 7, No. 13, The Jamestown Foundation, Washington, DC, www.jamestown.org/terrorism/news/uploads/cb_007_013.pdf.

Logan, J., J. Lewis and M. B. Cummings (2007), "For China, the Shift to Climate-Friendly Energy Depends on International Collaboration", *Boston Review*, Somerville, MA, http://bostonreview.net/BR32.1/loganlewiscummings.php.

M2M (Methane to Markets Partnership) (2004), *Investment Guide for China CMM/CBM*, China Coalbed Methane Clearinghouse, China Coal Information Institute, Beijing, April, www.coalinfo.net.cn/coalbed/report/200612.htm (full guide available from the US EPA Coalbed Methane Outreach Program at www.epa.gov/cmop/docs/guildline3.doc).

M2M (2006), *Coal Subcommittee Mines Action Plan*, M2M, US Environmental Protection Agency, Washington, DC, May, www.methanetomarkets.org/resources/coalmines/docs/coal_actionplan_5-06.pdf.

M2M (2008a), *Terms of Reference for the Methane to Markets Partnership*, M2M, US Environmental Protection Agency, Washington, DC, www.methanetomarkets.org/join/docs/termsofreference_signed.pdf.

M2M (2008b), *Underground Coal Mine Methane Recovery and Use Opportunities* (地下煤矿甲烷回收利用机会), M2M, US Environmental Protection Agency, Washington, DC, March, www.methanetomarkets.org/resources/factsheets/coalmine_eng.pdf and www.methanetomarkets.org/resources/factsheets/coalmine_chi.pdf.

MacKerron, G., E. Lieb-Doczy, J. Entress, R. Kleinsmann, and J. Watson (2004), *Evaluation of the Cleaner Coal Technologies Programme*, URN No: 04/2295, report prepared for the Department of Trade and Industry, NERA Economic Consulting, London.

METI (Ministry of Economy, Trade and Industry) (2007), *A Joint Statement by the Japanese Ministry of Economy, Trade and Industry and the People's Republic of China's National Development and Reform Commission on Enhancement of Cooperation between Japan and the People's Republic of China in the Energy Field*, METI, Tokyo, Japan, 11 April, www.enecho.meti.go.jp/english/data/070427jointstatement.pdf.

Minchener, A. (2005), "Coal-Fired Power Generation in China: Prospects and Challenges", paper based on final report to the European Commission on Contract No. NNE5/2001/552 and presented at the 2nd International Conference on Clean Coal Technologies for our Future, hosted by IEA Clean Coal Centre, Sardinia, Italy, 10-12 May.

Minchener, A. (2007), *Coal Supply Challenges for China*, Report No. CCC/127, IEA Clean Coal Centre, London.

MIT (Massachusetts Institute of Technology) (2007), *The Future of Coal: Options for a Carbon-Constrained World*, MIT, Cambridge, MA, http://web.mit.edu/coal/The_Future_of_Coal.pdf.

Moczadlo, J. (2008), "Sino-German Technical Cooperation in the Coal-Fired Power Sector: Activities and Achievements", presented at an IEA Working Party on Fossil Fuels workshop, International Energy Agency, Paris, 17-18 January, www.iea.org/textbase/work/2008/fossil_fuels/moczadlo.pdf.

MOFA (Ministry of Foreign Affairs) (2006), *China-EU Partnership on Climate Change: Rolling Work Plan*, MOFA, Beijing, 19 October, www.fmprc.gov.cn/eng/wjb/zzjg/tyfls/tfsxw/t283051.htm, accessed 18 September 2008.

MOST (Ministry of Science and Technology) (2006), "China-UK Seminar on Clean Energy", *China Science and Technology Newsletter*, No. 442, MOST, Beijing, 20 June, www.most.gov.cn/eng/newsletters/2006/200606/t20060621_34377.htm, accessed 10 October 2008.

NDRC (National Development and Reform Commission) (2007), *China's National Climate Change Programme*, NDRC, Beijing, June, http://en.ndrc.gov.cn/newsrelease/P020070604561191006823.pdf.

Oshita, S. B. and L. Ortolano (2002), "The Promise and Pitfalls of Japanese Cleaner Coal Technology Transfer to China", *International Journal of Technology Transfer and Commercialisation*, Vol. 1, No. 1/2, Inderscience, Geneva, Switzerland, pp. 56-81.

Oshita, S. B. and L. Ortolano (2003), "From Demonstration to Diffusion: The Gap in Japan's Environmental Technology Cooperation with China", *International Journal of Technology Transfer and Commercialisation*, Vol. 2, No. 4, Inderscience, Geneva, Switzerland, pp. 351-368.

Philibert, C. and J. Podkanski (2005), "International Energy Technology Collaboration and Climate Change Mitigation: Case Study 4 Clean Coal Technologies", Paper No. COM/ENV/EPOC/IEA/SLT(2005)4, Environment Directorate / International Energy Agency, OECD, Paris, www.iea.org/textbase/papers/2005/cp_clean_coal.pdf.

PM&C (Department of the Prime Minister and Cabinet) (2007), "Australia-China Joint Coordination Group on Clean Coal Technology", press release, PM&C, Canberra, Australia, 15 January, www.pmc.gov.au/climate_change/docs/au_china_clean_coal_group.rtf.

Strauss, L. (2006), "Sasol's Coal-to-Liquids Experience and its Potential in China", paper presented at the Deutsche Bank Access China Conference, Beijing, February.

Su Wenbin (2008), "GreenGen Project of China", paper presented at the Asia Clean Energy Forum 2008: Investing in Solutions that Address Climate Change and Energy Security, Asian Development Bank, Manila, Philippines, 2-6 June, www.adb.org/Documents/events/2008/ACEF/Session9-Wenbin.pdf.

Treasury (US Department of Treasury) (2007), "Second Meeting of the US-China Strategic Economic Dialogue", Fact Sheet No. HP-417, US Department of Treasury, Washington, DC, 23 May, www.ustreas.gov/press/releases/hp417.htm.

Treasury (2008), "US-China Ten-Year Energy and Environment Cooperation Framework", joint US-China fact sheet, US Treasury Department Office of Public Affairs, Washington, DC, 18 June, www.ustreas.gov/press/releases/reports/uschinased10yrfactsheet.pdf.

UNFCCC (United Nations Framework Convention on Climate Change) (2008), *Clean Development Mechanism (CDM): Project Activities and CDM Statistics*, Bonn, Germany, http://cdm.unfccc.int, accessed 19 September 2008.

Watson, J. (2002), "Cleaner Coal Technology Transfer to China: A Win-Win Opportunity for Sustainable Development?", *International Journal of Technology Transfer and Commercialisation*, Vol. 1, No. 4, Inderscience, Geneva, Switzerland, pp. 347-372.

Watson, J. (2005), "Advanced Cleaner Coal Technologies for Power Generation: Can They Deliver?", paper presented at the European Energy – Synergies and Conflicts Conference, British Institute of Energy Economics, St. John's College Oxford, 22-23 September.

Webb, M. (2008), personal communication, Foreign and Commonwealth Office, London, August.

Xinhua (2008), "China and Japan Agree to Strengthen Co-operation in Energy and Environmental Protection", Xinhua News Agency, Beijing, 8 May, www.gov.cn/misc/2008-05/08/content_965090.htm, accessed 10 October 2008.

X. CONCLUSIONS AND RECOMMENDATIONS

Earlier chapters have presented an overview of coal supply and utilisation in China. Coal will continue to be used in China for many years to come; all the analyses presented in reports from the Chinese government and others show rising consumption, to a greater or lesser extent. Protecting the well-being of individuals, society, the economy and the environment requires that China uses coal as efficiently and cleanly as possible. In this final chapter, conclusions and key recommendations are made that could move China towards such a more sustainable future by:

■ using the best-available, clean coal technologies for all applications, and investing time and money to create even cleaner ones;

■ fostering the uptake of these clean coal technologies through improved regulation that leads to clear, cost-reflective market signals, and through better reporting of environmental performance and implementation of environmental protection measures;

■ actively participating in international partnerships to speed research, development, demonstration and the deployment of cleaner coal technologies; and

■ working with other governments to create a global market for clean energy technologies and responding to this market with commercially relevant products, for local markets and for export.

Energy technologies – on the supply and the demand sides – do not exist outside their economic, physical, and social contexts. It is relatively simple to list "the best" clean coal technologies (Box 2.2), but this will only help if accompanied by an understanding of how to create the settings required for them to deliver real benefits. Just as power generation is the biggest use of coal globally, so it is in China. Therefore, identifying the best coal-fired power generation and emissions control technologies is important. Even then, a large, modern, coal-fired supercritical unit cannot be considered in isolation from the coal mines, transport infrastructure and coal markets that supply it. The setting that allows clean coal technologies to be deployed effectively at a power plant is complex and includes: the grids and power markets that receive its output; the regulatory apparatus that approves its construction, and oversees its operation and eventual decommissioning; the banks and investors that join the EPC (engineering-procurement-construction) contractor and utility company to build, manage and maintain it; the neighbouring residents who work at it and live with it; and the increasingly global technical community that designs, manufactures and services it.

China already has examples of most of the world's best clean coal technologies. In some areas, it is at the forefront, as in the installation of 1 000 MW ultra-supercritical power generation units at Huaneng's Yuhuan power plant, and has developed some unique technologies that other countries should sensibly adopt. While opportunities for technology transfer do exist, the rest of the world does not have a large reservoir of advanced technologies and techniques that are outside the experience of Chinese companies. The most urgent need is for China to create the conditions necessary for much wider adoption of technologies that are already well proven and commercially available. Recommendations made here focus on this need. In some cases, the recommendations may seem to bear only indirectly on cleaner coal utilisation, but each one of them concerns an important part of the technical, economic, regulatory and social infrastructure in which clean coal technologies can deliver their benefits.

Every country's circumstances are unique. It is up to its leaders and citizens to determine appropriate goals and the means to achieve them. The recommendations gathered together here, from the evidence presented in preceding chapters, are an attempt to apply in China the principles that have proven important in other countries to deploying cleaner coal technologies. The key recommendations were prepared in consultation with experts from inside and outside of China. Each is preceded by a brief summary of the situation in China, and how similar challenges have been dealt with in other countries. **Options that China may consider when putting the recommendations into action are printed in boldface type.** Few of the recommendations are entirely original; indeed, some have been discussed for years and may have been implemented already to some extent. The repetition of long-standing advice here should be seen as a measure of the continued need to act on it.

Recommendations cover four key areas. The first, on **coal resources and markets**, is essential since undistorted prices, free trade and secure investment frameworks enable energy markets to work efficiently and contribute to energy security. Tackling environmental, social and safety issues becomes much more manageable if the coal sector is economically efficient which, as experience in IEA member countries shows repeatedly, comes through competitive, well-regulated energy markets. Recommendations on the **contribution of cleaner coal technologies to a more secure environment** touch on institutional, technical, policy and regulatory matters, all of which must be addressed in a coherent and consistent manner. It is perhaps the longer-term environmental issues that are highest on the international agenda, yet China itself must act quickly to establish a stable framework within which to address its most pressing domestic problems. A high priority must be to improve **welfare and safety** in the coal sector, which is key to improving the sustainability of any energy system; recommendations are made based on the long history of improving mine worker safety in those IEA member countries where coal is mined. Finally, recommendations on **international collaboration** are presented – always with the aim of accelerating the deployment of cleaner coal technologies, and ultimately including carbon dioxide (CO_2) capture and storage (CCS) in response to the climate change challenge.

The international drive to engage with China on climate change issues could well count for nothing if China does not stabilise its own domestic coal production and supply chain, and complete the rationalisation of coal use, especially within the power sector. The Chinese government has already undertaken many reforms, but there is a long way to go. IEA member countries that have previously undertaken such reforms and reduced coal-related pollution, can provide advice and assistance, but ultimately China will need to decide for itself how to proceed. Its actions, more than those of any other country, will shape the global approach to cleaner coal utilisation and climate change. The chapter concludes by outlining China's potential role in such a global strategy.

COAL RESOURCES, MARKETS AND TRANSPORT

If China's economy and demand for energy continue to grow at current rates, there is a real and serious risk that its coal demand will exceed indigenous supply and strain global supply chains for all primary energy resources. In such circumstances, free markets can help set prices at which supply and demand balance, with some users reducing demand and other users exiting the market entirely. Current Chinese investment in offshore energy assets, including coal, will help expand global supplies, but should not be designated to sustain inefficient production in China, or elsewhere. While coal resources are abundant (WEC, 2007), higher exploitation and transport costs, and hence higher prices, should be allowed to dampen demand. However, more than resource economics, the real constraint to energy demand growth that has now clearly emerged is our ability to use coal without creating unacceptable environmental impacts.

The aims and objectives of China's Five-Year Plans are to achieve economic growth in an efficient, equitable and sustainable manner (NDRC, 2006) – the same aims and objectives as are found in OECD member countries. For the energy sector, these can be achieved by: improving efficiency and reducing emissions at existing facilities; continuing with sector reforms to establish commercial competition; accelerating pricing and tariff reforms; deploying alternatives to coal, such as nuclear and renewables; and developing new, cleaner energy technologies. A strategy for environmental protection is fundamental, ideally promoting the sustainable utilisation of natural resources through market-based pricing, full cost recovery, and reliable, transparent and timely disclosure of information on environmental performance. In other words, price levels for energy, water and other natural resources should reflect, as fully as possible, their scarcity value, including external costs. There is broad agreement that environmental externalities should be internalised into energy prices. It is an ideal that many IEA member countries are moving towards, but have not yet achieved. In co-operation with the European Environment Agency, the OECD has developed a database of instruments used for environmental policy and natural resource's management (Box 7.10). It is a rich source of information that is designed to assist anyone involved with drafting new laws and regulations.

Where higher prices have an adverse impact on poorer sections of a country's population and regions, government should target support mechanisms to shield sensitive groups, rather than employ price controls. For example, many OECD countries have programmes administered by electric utilities that provide low-cost supplies, energy audits and subsidised efficiency upgrades to customers on low incomes. In contrast, setting energy prices for social reasons is a blunt policy tool that is untargeted, inefficient and ineffective.

Recommendation 1

Market-based, energy and resource pricing should be used as the primary means of balancing supply and demand in China, so that resources are exploited, transported and used efficiently and effectively, including those that are imported and exported.

Coal resources

The challenges China faces to develop its coal resources were discussed in Chapter 3. While China has large coal reserves and a broad range of coal ranks, per-capita resources and reserves, particularly of coking coal, are below some of the world's other major coal-producing nations. There is a general trend towards deeper – and thus more expensive – mines in the west, distant from the key consuming regions. The Ministry of Land and Resources has identified a widening gap between the availability of prospected coal reserves and those needed to develop new mines (Section 3.1). Coal prospecting efforts in the past have not yielded the detailed data needed for mine planning. Some coalfields have been fragmented into mining blocks that are too small for efficient mine design. In the future, greater effort is needed to mitigate the environmental impacts of mining in regions suffering from a serious lack of water, desertification and vulnerable ecology. These are problems that can be and are being addressed to allow the continued growth of China's coal supply.

China has world-class expertise in high-capacity mining systems, but the current policy of restricting high-volume coal production to the state sector constrains the widest possible use of technical expertise, mining techniques, practical experience and management methods that have been proven elsewhere in the world. Greater international participation in coal prospecting, mine planning, equipment selection, coal-mining operations and supervision, and environmental management would speed the application of new knowledge and skills in the industry. While current Chinese legislation provides for private and foreign ownership of coal resources, the government does not encourage majority foreign investment in coal production. Asian American Coal Inc. has a majority stake in the Daning mine in Shanxi, but this is a unique example.

The first step would be for China to decide that the benefits accompanying foreign ownership outweigh the perceived security and other risks. Most developed nations have found that energy resource ownership does not equate to energy security, since many factors determine the destination of energy supplies. In any event, efficient domestic energy production and import diversity enhance energy security, regardless of ownership. The European Union, for example, no longer attempts to meet its coal demand solely from domestic sources. Under the same free-market policies, major coal users in the United Kingdom (UK), Spain, Netherlands, Belgium and Luxembourg, including the power generation and steel industries, have been privatised with many now foreign owned; although state-owned power companies remain the norm in other European countries.

Fair and transparent resource allocation, perhaps through auctioning, and non-discriminatory mine permitting would promote greater competition, including international participation, and consequently more rapid penetration of the most efficient mining practices from around the world. At the same time, strengthening the permitting system would aid in achieving national goals for pollution prevention and control since it is during permitting that authorities have most leverage over project developers, by imposing conditions on their future actions and performance.

The integration of environmental protection measures into land-use planning and regulation is common practice in IEA member countries and provides possible models to follow (Section 7.3). Conditions imposed during permitting of opencast mines in the UK, for example, define times when blasting is permitted, allowable truck movements, noise and dust limits, operating hours, standards and schedules of restoration, levels of compensation, project duration and many other details that must be adhered to before mining can commence and throughout the project. Effectiveness of this measure depends on there being independent local authorities who have the power and freedom to enforce regulations in the national and local interest and who are not influenced by narrow or short-term local pressures – a theme that recurs throughout these recommendations.

Limited control over coal mining in the past has allowed too much environmental damage in China. Investment needs to be made in landscape restoration, especially re-vegetation and slope stability to combat further erosion and loss of soil fertility. This legacy problem can only be resolved through government action. In future, mining companies must be held responsible for all these costs. To guard against a company's failure to perform, many IEA member countries have introduced a bond system whereby companies are obliged to establish independent funds or a charge is held over company assets sufficient to restore mining damage. The fund or charge is released back to the company once it has completed restoration, less any on-going maintenance costs. In the United States (US), the Surface Mining Control and Reclamation Act of 1977 includes such requirements (Box 7.11). It is administered by the Office of Surface Mining who also manage the Abandoned Mine Reclamation Fund. Similar provisions exist in Australia (Box 7.13). In the UK, the legacy of environmental damage from coal mining is now the responsibility of The Coal Authority which has been able to make good all valid claims for subsidence damage and is responsible for a programme to treat all water discharges from abandoned mines (Box 7.2).

Conditions imposed during permitting should set standards for land restoration and the treatment of subsidence damage. An industry-wide bond system could be considered to ensure the remediation of existing and new mines. A specialised agency could handle legacy claims and deal with any non-compliance at on-going operations in a consistent and transparent manner.

China's coal mine methane (CMM) potential is significant. CMM exploitation appears to be moving towards a meaningful market penetration, especially with the value of Clean Development Mechanism (CDM) credits that reflect the environmental benefits of reducing methane emissions under the Kyoto Protocol. Other countries, such as

Australia, the UK, Germany and the US, already have power-generation capacity based on this resource and there appears to be already a good transfer to China of knowledge through project activity and partnerships, often involving commercial finance houses seeking opportunities to invest in clean energy.

The position for coalbed methane (CBM) is less clear. There has been enormous financial support over the past decade, with over 2 000 CBM wells sunk, 1 700 in the period 2005-07. However, most remain far from commercial operation. Other countries have also faced difficulties in creating a business model for CBM exploitation; the CBM industry in the US developed under a generous subsidy programme. Today, CBM production is proceeding on a commercial basis in the US and elsewhere with some significant new projects announced in Australia and Indonesia. In China; the state-owned China United Coalbed Methane Corp. Ltd. has had, until recently, a monopoly, with exclusive rights to foreign co-operation. This attempt to create a national champion had drawbacks and the Chinese government now encourages other companies with CBM capabilities to work together with foreign companies to develop CBM opportunities using the most appropriate technologies.

China's CBM industry should be allowed to grow on a purely commercial footing, without subsidy – consistent with the approach for the coal industry as a whole.

Coal markets

Although some energy prices have been deregulated, the persistent under pricing of coal for power generation needs to be addressed. China is not unique in this respect; many other countries have a mixed record of energy price regulation. Some governments seek to keep energy prices low to stimulate economic growth, widen access to energy and maintain international competitiveness. But low energy prices have led to energy wastage and inefficiency in all countries where prices are held artificially low. Low electricity prices might be appealing in the short term, but they embed inefficiencies into a country's economy for decades to come. Freely determined prices, sometimes higher prices, mean that those who can make best use of resources are the ones who receive them, while inefficient users drop out of the market. China needs to consider its longer-term competitiveness, which will demand greater energy efficiency, and allow this to be driven by fundamental supply and demand factors. Otherwise, it may sell its natural resources too cheaply, embodied in products that benefit consumers in other countries, denying wealth to Chinese citizens.

China has an enormous influence on the international coal market. It could easily export more than 100 Mt per annum (equal to only 4% of current output); exports peaked at 94 Mt in 2003. At the same time, its import demand is growing significantly and the country is likely to become a net importer of coal. Given the sheer size of the country's domestic demand (2 543 Mt in 2007), predicting what its net import-export position will be over the coming years is fraught with error, and yet a 1% demand swing in China can translate into a 3-4% swing in international coal trade in what is, and is likely to remain, a tight market. In 2008 and at other times, the Chinese government has restricted coal exports by various means, including tax adjustments and quotas to ensure that domestic demand is met. In the absence of strategic coal stocks and free-market responsiveness in China, the recent coal price volatility – which affects all regions of the world – will remain a feature of international coal trade.

Recommendation 1, above, reflects the need to allow prices to be determined by market forces, and not be subject to government interference. This applies to all stages, upstream and downstream. In China, market-based coal pricing has been adopted to a large extent, but problems remain as there is no equivalent freedom of action within the power sector for electricity pricing (Box 10.1).

China should set out a timetable, so that all stakeholders can plan appropriately, for the incorporation of the full costs of fuel for power generation in wholesale and retail electricity rates. Alongside this, there will be a need to address energy poverty directly by establishing targeted support programmes based on the needs of low-income consumers.

Box 10.1 .... China's power sector reforms: where to next?

To take full advantage of opportunities to improve the efficiency and environmental performance of power generation requires continued progress in power sector reform. In this respect, there is great scope for international support and co-operation as China embarks on further regulatory reforms in its electricity sector, with the following recommended priorities:

■ Reaffirm China's strategy for power sector reform, and ensure that there are strong champions and mechanisms for its implementation.
■ Implement reforms for more cost-reflective electricity pricing and economically efficient investment that provide incentives for investments in grid expansion and cleaner, more efficient generation.
■ Establish fully competitive power markets overseen by strengthened regulatory authorities who are tasked with integrating environmental goals into the regulatory framework.
■ Introduce greater transparency in power sector data and analysis to improve the understanding of developments in supply and demand.

The OECD, in consultation with the IEA, is working with NDRC on regulatory reform, including the next stages in the evolution of China's power sector.

Source: IEA (2006).

Coal transport

Construction of the infrastructure to transport coal via the most efficient modes – rail and ship – has lagged behind production expansion and demand growth. This has forced too much of the incremental output to be moved by truck, clogging roads and adding to the demand for oil imports. This is an example of how subsidised transport fuels leads to inefficiency and, in this case, undesirable outcomes. **China will need to consider whether it makes more sense to allow a greater reliance on coal imports, or whether additional investment in rail and port loading facilities for internal coal movements can be justified. When new infrastructure is proposed, investment decisions should be made on the basis of its market-based value.**

In sum, there is a near- to medium-term requirement to stabilise the entire coal production and supply chain, including more competitive, better-regulated, market-based coal mine development, more efficient mining practices, and more effective and reliable coal transport. Coal imports and exports are an important component, as these can provide an effective means to balance coal demand and supply, and allow more transparent transmission of price signals that can, in turn, influence domestic consumption and investment patterns, thus strengthening energy security.

CLEAN COAL TECHNOLOGIES AND ENVIRONMENTAL SECURITY

China's economic growth, averaging 10% per year since economic reforms began in 1979, has been accompanied by rising pollution. Over the period 1990-2005, sulphur dioxide (SO_2) emissions grew by 37% and NOx emissions by 114%, although the economy quadrupled in size. Over the same period, particulate emissions grew only slightly and energy intensity has improved by about one half (Figure 6.2). Five-Year Plans have addressed priority environmental problems by establishing quantitative targets and framing investment programmes and budgets. The Chinese leadership has made environmental protection a national strategy and the 11th Five-Year Plan advocates a, "new economic model in which growth is guided by resource conservation" rather than by continued expansion of resource use. Reduced energy intensity and the adoption of a "circular economy" are recognised as keys to help reduce the pollution and resource intensity of the Chinese economy.[1] Measures have been taken to integrate environmental and economic decision making. The use of environment-related taxes has expanded, although these still account for only about 3% of total tax revenues (OECD, 2007), and are low compared with the 6-7% average across OECD countries.

Considerable progress has been made since the mid-1990s in the development of and access to environmental information, and wider participation on environmental issues. Each year, the Chinese government publishes environmental statistics and reports. The media and environmental non-governmental organisations (NGOs) reinforce the demand for environmental progress. However, environmental awareness needs to be widened, particularly in enterprises where environmental performance needs to be better linked to business goals. In IEA member countries, public awareness, public participation and public accountability have been the key drivers behind improvements in environmental performance, including the deployment of clean coal technologies. China will have to find a way forward that engages civil society in this area and suits its particular circumstances.

---

1. "Circular economy" (循环经济) is a key concept in current economic policy in China, similar to the ideas of industrial ecology, in which the waste streams from one process are used as material inputs to other processes.

One potential constraint to achieving the goals set out in the current Five-Year Plan is the relative complexity of the tiered administration systems (national, provincial and local), each of which sometimes appears to have divergent priorities. The earlier decentralisation of many of the decision-making functions within China's energy sector resulted in an inconsistent implementation of national energy policy. The government has responded by centralising the decision-making and implementation processes for the coal and power sectors, but it remains to be seen how successful these steps will be. The tensions between state and provincial decision-making bodies will pose challenges to effective environmental protection, with the central government seeking to deal with sustainability and some provincial governments continuing to focus on economic growth and local employment. To help avoid such conflicts of interest, the central government in many IEA member countries directly funds environmental regulators and their local branches. Fees and fines levied on polluters are typically put into general government funds to further separate environmental protection from revenue raising.

Recommendation 2

Greater accountability and transparency that allow reliable delegation to lower levels of Chinese government are prerequisites to the proper functioning of existing environmental laws and hence the successful deployment of clean coal technologies.

Institutional reform must be undertaken so that carefully designed strategic objectives are implemented effectively and are not treated as platitudes to be undermined by other interests. Severing budgetary ties between local branches of regulatory bodies and local governments would be one step towards ensuring independent environmental regulation. The international community can contribute through increased bilateral and multilateral collaboration with China, at all levels. For example, individuals with regulatory experience in other countries should share their first-hand experience on policy development, environmental impact assessment and cost-benefit analysis with Chinese counterparts who face similar challenges through official exchanges and secondments.

There are opportunities for China to clarify the various aspects of its legislative framework covering coal production, transport and utilisation, as well as environmental aspects, perhaps adopting a simpler structure that is easier to understand, interpret, administer and enforce. The UNEP Training Manual on Environmental Law provides a sound template to follow (Box 7.10). By doing this, opportunities for misinterpretation and non-compliance would be reduced significantly.

Coal mining

In the coal mining sector, the adoption of cleaner coal technologies and practices needs to be encouraged. It is usually not practical to re-use coal mine waste (*e.g.* spoil, overburden, discard, tailings); the "circular economy" concept does not apply:

large volumes of inert (non-combustible) waste must be inevitably dumped on land that should be subsequently restored to productive use or landscaped. Control of discharges of polluted mine water and efficiency in water use are also important. Pollution charges and water-use charges can ensure that treatment plants are properly managed and that water is used efficiently, especially when they are linked to the level of discharge and volume abstracted. China's existing environmental charges, however, are based on coal tonnage which, while convenient for purposes of collection, is less effective in encouraging mine operators to address environmental issues than linking environmental taxes directly to volumes of pollution, the extent of any damage caused or the volume of water used. In Australia, the creation of a nationally co-ordinated water market aims to address unsustainable water abstraction (Box 7.13). China and Australia should co-operate on developing workable solutions to water management in arid regions.

Recommendation 3

Environmental charges on coal mining have been introduced, but more should be done to directly link them to levels of pollution (i.e. the widely accepted "polluter-pays principle"). Funding for environmental protection agencies should be guaranteed separately and not be linked to revenues from environmental charges.

China may consider establishing an inter-ministerial group to examine how to restructure environment-related taxes to achieve environmental policy objectives. It could review existing pollution charges, user charges, emission trading and other market-based instruments to drive environmental improvements, and determine how to tie them more directly to pollution levels.

Revenues collected would not necessarily need to be directed to environmental improvements, and in any case should remain entirely separate from the funding of effective local environmental protection bureaus. One possibility would be to direct revenues from environmental fees, charges and taxes to general government revenues, and to fund provincial- and county-level regulatory agencies fully via the national budget. Any resulting burden of higher energy prices on poorer groups should be rectified through social policies and targeted support, not through broad-based energy subsidies or price controls, which inevitably compromise sound energy and environment policies.

Coal transport

Coal transport by truck in China has a huge impact on the countryside and communities bordering transport routes, in terms of both dust pollution and damage to the road infrastructure. Planning and investment should be made in alternatives to road transport for coal movements to major users, a task that should be made easier as the number of mines in operation is reduced.

Trucks will remain an important link in China's coal supply chain, so they must be better managed (*e.g.* wheel washes, dust sheeting, weight control and engine emission monitoring) and pay a greater share of their external costs through taxes, road-use charges and duties.

Coal-fired power generation

The Chinese government has strengthened environmental policy, regulations and monitoring to ensure power utilities meet tightened emission standards, especially for new coal-fired power plants. China already has experience in deploying a range of regulatory and economic instruments (*e.g.* pollution charges, user charges and pilot emission trading schemes). Campaigns and award schemes to support implementation at the local level have been organised; work with NGOs to develop procedures for public participation in environmental impact assessments is an important recent example (OECD, 2007). However, an implementation gap persists. Consistent implementation of environmental regulations at all coal-using plants should be given priority. The biggest obstacles to environmental policy implementation appear to be at the local level. China is taking an important step with the programme being rolled out in 2008 to link achievement of energy-intensity targets into the performance reviews of local officials. It should continue efforts to make local leaders more accountable for environmental performance objectives, and for better implementation of the pollution control regulations already adopted.

Looking further ahead, as China considers its actions to rein in emissions of greenhouse gases, CCS in the power sector stands out as an important option on the path towards a low-carbon economy. At present, commercial-scale CCS is neither a proven technology nor one that is easily affordable. It would be wrong and naive to think that any country would adopt this technology in isolation and without reward. Nevertheless, with China already consuming 46% of world hard coal production, a share that is rising, the country has a certain interest and responsibility to ensure that CCS does become a viable technology. China already participates in CCS R&D, information exchanges and pilot projects, providing a sound basis for future deployment activities. The country also has the opportunity to harness the evolving global carbon market to augment domestic financing. By establishing a compatible internal carbon market, it could pull affordable efficiency improvements into the energy supply system while establishing an income stream from emission credits to pay for more ambitious reductions through CCS deployment. Both clean coal technologies and non-carbon based energy systems would be encouraged by this means. Once technology developers in China and elsewhere can see this long-term market, they will respond with products. International collaboration will be important, and we consider this below.

Recommendation 4

Market-based mechanisms, such as sulphur and carbon trading, should be central to China's pollution abatement strategy and the key incentive to develop cleaner coal technologies for domestic and international markets.

In the near term, China needs to consider the "CO_2 capture-ready" concept for the unprecedented number of new coal-fired power plants that are built each year (IEA GHG, 2007). In the medium to longer term, establishing near-zero emission, large-scale, coal-fired power plants should be a strategic priority. These efforts can only happen if an attractive market for low-carbon technologies exists to make it more profitable to generate electricity at clean and efficient plants than at inefficient plants with poor pollution control.

Various means are available to signal the higher value of clean production: taxes on carbon and sulphur, feed-in tariffs, emissions trading and pollution charges. The alternative of micro-managing large and complex electricity supply systems is fraught with difficulty – not only at an intellectual level, but also because it can become subject to intense political interference that leads to sub-optimal outcomes.

Once these signals are properly in place, operators would find it more profitable to dispatch their cleanest and most efficient plants ahead of plants with poorer environmental performance. A plethora of rational decisions, made independently by market players, each optimising their own economic position, leads to a market outcome that benefits the environment and society at large. **Negotiations over the next couple of years are likely to shape a long-term international carbon market, so China needs to move swiftly and with determination so that its domestic actions are compatible with a global effort, and that international flows of funds can be harnessed to build CCS facilities in China.**

China is already a growing provider of equipment and know-how for power generation to other countries. The market for clean coal technologies is global – many countries must close old, inefficient coal-fired power plants over the coming decade and replace them with modern plant, some with CCS. China should view this as an opportunity to develop power plant equipment and systems that meet the needs of other countries, building on past achievements such as low-cost FGD systems that capture SO_2 and plasma-arc igniters that eliminate the use of fuel oil. Indeed, it is only with commercialisation that these technologies can be adopted on a wide enough scale to have a material impact. China already has positive experience in improving the competitiveness of its consumer appliances by harmonising its appliance efficiency standards with those in other countries. Establishing a global market for cleaner, more efficient coal technologies is crucial and common technical standards would allow this market to grow more quickly.

Recommendation 5

China should co-operate with other nations to establish common technical standards for coal-fired plants and their sub-systems, and so allow the wider deployment of more affordable clean coal technologies, both in China and elsewhere.

China's established plant-construction and technology industries provide the country with the capability to install and operate state-of-the-art combustion, gasification and other coal utilisation systems, thereby gaining major benefits through higher efficiencies and reduced environmental impacts. The share of supercritical (SC) and ultra-supercritical (USC) steam plants in new build is increasing. In 2006, around 20% of new build was either SC or USC, bringing the year-end total to over 30 GW or 8% of total coal-fired generation capacity. At the end of 2007, this rose to 12% including 10.7 GW of USC plant. By 2010, the IEA expects China to have 150 GW of SC/USC plant in operation or over 20% of total coal-fired generation capacity.

China can build on existing bilateral and multilateral energy dialogues to develop common or harmonised definitions, specifications and performance standards for key components of cleaner coal-fired power generation systems, in order to ensure that technologies brought into China meet national needs, and to facilitate exports of China's clean coal technologies to other countries.

Well-proven technologies and practices

Site visits made during this study revealed a number of examples where steps could be taken to improve the environment by using well-established and readily available technologies and techniques. These do not demand any technology transfer to China nor technology development, and they can be implemented immediately at low or even negative cost. For example, China did not meet its 10th Five-Year Plan target of washing 50% of the coal it burns by 2005 and while FGD has been installed at over 50% of coal-fired power plants (on a capacity basis), its use has been low to date. In some circumstances, simple administrative controls may be effective. For example, most IEA member countries prohibited the urban use of raw coal many decades ago, leading to dramatic improvements in air quality and human health. In the case of FGD, its effective use at power plants requires continuous emissions monitoring and inventory calculations to enable proper oversight of environmental performance. These are management issues, not technological failings.

Recommendation 6

Even as it pursues innovative new technical and policy solutions, China should quickly adopt well-proven technologies, management practices and policies that deliver immediate and sustainable improvements along the entire coal supply chain, from mine to end user.

Correct pricing signals, based on coal quality and emissions, and better implementation of existing environmental regulations would make a major contribution to pollution control by encouraging more coal washing and the greater use of FGD. This should see more of China's high-sulphur coal used only at power plants with FGD, while ensuring that only low-sulphur coal

is used in industrial boilers. In the residential sector, mandating the use of smokeless fuels manufactured from coal would see a major improvement in air quality in those towns and cities where coal continues to be used for cooking and heating.

COAL-INDUSTRY RESTRUCTURING, SOCIAL WELFARE AND MINE WORKER SAFETY

Coal industry restructuring

Rationalisation of the coal production and utilisation sectors is taking place today in China. New, integrated energy companies, encompassing mining, transport, power production and coal-to-liquids have international standing. Consolidation of mines improves economies of scale and allows the use of more appropriate mining techniques to improve both safety and resource recovery. Unfortunately, too little attention has been given to creating markets in which these large, new conglomerates can compete to deliver economic efficiency. They remain majority state-owned companies, even though many are now listed companies with private shareholders.

Past encouragement of township and village enterprise (TVE) coal mines by government succeeded both in increasing energy supplies and providing economic growth in rural regions, notwithstanding environmental and resource-recovery concerns. The recent turnaround in policy towards TVE mines has hit many local economies hard, particularly migrant contract miners, who have no job security once that work ends. China will have a continuing need for small coal mines to serve local markets. A policy that simply requires the closure of *existing* mines below 30 thousand tonnes per annum (ktpa) is difficult to implement in regions that offer few alternative employment opportunities. Moreover, small-scale mining need not be dangerous nor inefficient. Examples of viable small mines, properly regulated and inspected exist elsewhere – for example, in the Appalachian coalfields of the US and in the UK's Forest of Dean. Creating a workable framework for small-scale mining appears more tractable than attempts to eliminate this sector by banning any *new* mines with an annual capacity below 300 ktpa.

Moving to a market-based coal sector is not painless. For example, the German hard coal industry is in its twilight years and will cease production by 2018, but the cost to the state of maintaining uneconomic production has been enormous – around EUR 2.5 billion (USD 3.3 billion) in 2006 for a production of 24 million tonnes (USD 140 per tonne). The IEA estimates that the Chinese coal sector received implicit subsidies totalling around USD 3.7 billion in 2006 for a production of 2 320 million tonnes (USD 1.6 per tonne). The Chinese government must resist the temptation to support uneconomic coal mining activity because the costs, hidden in consumer tariffs or explicit in government budgets, could become crippling. Conversely, the current policy that demands the closure of small mines is indiscriminate. Just because a mine is small does not mean that it cannot be operated safely, efficiently and economically, especially where it meets local demand.

In order to be successful, coal-industry restructuring programmes must involve all stakeholders. More generally, strong community relations and civic pride are vital to the success of coal mine and power station projects during development, operation and closure. Greater community participation increases the awareness of the benefits of clean coal technologies and practices throughout the coal supply chain. Restructuring, without adequate consultation and investment in alternative employment opportunities, leads to high, long-term support costs for the affected communities or even community breakdown, migration and rural depopulation. Experience from IEA member countries in implementing well-planned, integrated coal-industry restructuring programmes can assist China as it goes through this process.

Recommendation 7

Coal-industry restructuring should be founded on a belief in the power of properly regulated markets to deliver economically efficient mines, operated by competing companies of varying sizes, from small to large.

It is important to engage stakeholders in plans to mitigate the impacts of mine closures and other restructuring activities through an open process, overseen by specialised agencies created for that purpose.

Social welfare

This study draws on experience in Europe and the US, where there are marked similarities to China in the problems faced, both during and after coal mine closures. Indeed, social impacts to communities, where the main source of income has been removed, are common worldwide. For example, small mines in the Appalachian region have been hardest hit by restructuring of the US coal industry since the early 1980s. Many were opened in response to rising coal prices during the 1970s, and in 1976 small mines produced nearly 34% of the US total. Subsequent refocusing on larger, more competitive production units operated by a few large companies, plus the impact of newly introduced health, safety and environmental legislation, left small-scale mining less viable, leading to widespread closures and job losses.

As in China, redundant miners often had little to offer in the way of transferable skills, and had few options other than to return to farming, seek low-wage labouring jobs in the locality, or leave to find work in other parts of the country. The latter option has had serious implications both for the communities that remain, and for the cities where, even if work were available, the social infrastructure may not be able to handle a large influx of low-skilled workers. In most IEA member countries where coal-industry restructuring has occurred, governments have tried to create the conditions for private-sector enterprises to drive regional regeneration. The Appalachian Regional Commission, created in the late 1960s in response to regional poverty, reduced unemployment and the numbers of people living in poverty.

China will need to increase investment in rural areas affected by coal-mine closures to fund training programmes that enhance workforce skills and to attract replacement job opportunities through programmes and organisations tailored to local conditions. Blanket programmes are unlikely to succeed.

Mine worker safety

Aside from humanitarian considerations, mining accidents result in productivity losses and economic costs associated with treating injuries and compensating dependents. A viable mining industry avoids these through improved safety. China does not lack legislation covering coal mine administration, health and safety, and environmental issues, but in many cases laws are ambiguous, having been developed and promoted by institutions with different agendas. Over the last eight years, since the State Administration of Work Safety was established, significant safety improvements have been made and the fatality rate is falling. However, there continue to be significant shortfalls in enforcement by the regulatory authorities for a variety of reasons, not least among which is the sheer number of mines that require inspection. Conflicts of interests have also been apparent, especially within the small-mine sector, since local governments derive income from this sector. This can lead to a largely uneducated, often migrant workforce open to exploitation in terms of workplace conditions and safety.

Other countries have faced similar challenges and devoted many years of effort to resolve them. Experience in IEA member countries over almost two centuries suggests that it is a combination of measures that create a safe working environment: legislation, inspection, penalties, training, culture, responsibility and empowerment. If any element is missing, safety is compromised. For example, a worker who is not aware of dangers, because he lacks knowledge, or is not empowered to question his management over safety failings, lacks the capacity to deal with dangerous situations and may suffer injury as a direct result. Providing workers themselves with the means to take responsibility for safety has proven to be a very successful approach in recent years.

Dealing successfully with current safety issues does not remove the need to look after those suffering from long-term injury. The costs of treating chronic illnesses associated with coal mining can become overwhelming. For instance, the UK government expects to pay a total of GBP 6.4 billion (approximately USD 13 billion) in compensation to mineworkers who continue to suffer health problems as a result of working underground. This is the biggest single personal injury compensation scheme in British legal history and possibly in the world.

Recommendation 8

A properly resourced, national mines inspectorate is central to ensuring mine worker safety. China needs to strengthen its own inspectorate, and complement this by training and empowering coal miners to take greater responsibility for their own safety.

Imaginative solutions are urgently needed to ensure that the thousands of small coal mines in China are regularly inspected by qualified, independent inspectors. While inspectors are independent, insofar as they are salaried from central government, they are too few in number. The Chinese government needs to consider alternatives and the IEA suggests that inspections by private-sector insurance companies might be a way forward – providing these are combined with a requirement for mine owners to carry employer liability insurance.

INTERNATIONAL COLLABORATION

There is a growing interest among other countries to work with China on improving its coal sector. A number of collaborations have been announced with that intent. Yet the scope and intensity of co-operative activity to date does not yet reflect that interest and intent, and above all the scale of need. The set of findings and recommendations summarised here is intended as a reminder of the need to boost co-operation and ensure it is focused on clear objectives.

The last decade has seen a dramatic increase in China's engagement with other countries to address environmental challenges (Box 10.2). This reflects a growing recognition in China of the important economic, social and ecological stakes linked to these challenges. China is now an active and constructive participant in a broad array of regional and global environmental conventions, institutions and programmes, and is drawing on international financial institutions and special mechanisms to pursue environment-related projects.

Box 10.2.............. Critique of multilateral and bilateral collaborations

Three key multilateral agreements are the Asia-Pacific Partnership for Clean Development and Climate (APP), the Methane to Markets Partnership (M2M) and the coal-based R&D initiative of the Asia-Pacific Economic Cooperation (APEC) organisation. In these and other multilateral groupings, there is frequently an overlap, with similar or even the same projects being announced under the various multilateral and bilateral agreements. At present, none of these three agreements is well funded. For example, the direct funding committed to the APP by the US and Australia seems inadequate to achieve the stated goals, while the lack of any public commitments from the developing country partners suggests their contributions will be chiefly in-kind. Consequently, while the APP proposes an attractive portfolio of coal-based projects, so far only low-cost (but still important) capacity-building and training activities are proceeding. Unless large-scale commitments to RD&D are made in the near future, these agreements risk losing credibility, harming the overall global climate change initiative.

Of interest to China should be the recommendations from the APP Clean Energy Task Force on projects that would accelerate demonstration, deployment and transfer of key technologies to improve the environmental and economic performance of fossil fuels. These include near-term coal gasification and advanced coal combustion projects, together with longer-term CO_2 capture techniques and CO_2 storage opportunities. The Coal Mining Task Force has proposed a series of projects, including some on mine safety and environmental impacts.

The M2M Partnership will develop strategies and markets for the recovery and use of methane from coal mines through technology development, demonstration, deployment and diffusion, together with policy frameworks, investment support and removal of barriers to collaborative project implementation. There are many unresolved technical and non-technical difficulties associated with CMM and even more with CBM. Multilateral donors such as the World Bank and the Asian Development Bank are addressing many of these problems already in China, so it is not yet clear what the M2M Partnership will add in China. In fact, the CDM credits available to CMM projects have generated considerable commercial project activity, leaving less need for government-supported activity.

The September 2007 APEC Declaration on Climate Change recognised the role of future coal use, and stated that R&D, deployment and transfer of low- and zero-emissions coal technologies will be essential. The establishment of an Asia-Pacific Network for Energy Technology (APNet) to strengthen collaboration on energy research in the APEC region, particularly in areas such as clean fossil energy and renewable energy sources, also holds promise. Positive outcomes will depend on what is proposed and how such proposals are funded.

Three other initiatives are important to mention. The Carbon Sequestration Leadership Forum counts China among its 22 members, but China may see few direct benefits beyond exposure to global views on the challenges of implementing CCS. The US-led FutureGen initiative is pursuing projects that could provide a template for subsequent demonstrations of CCS. With many partners and uncertain government funding, progress has been erratic. Nevertheless, China may draw benefit from its participation, if only to supplement the Chinese GreenGen initiative to demonstrate coal-based polygeneration technology with CCS in China, using as much domestic technology as possible. If successful, GreenGen would have far-reaching consequences on coal-fired power generation technology choice and equipment supply.

Bilateral collaborations with China have also been important, particularly those involving Australia, Germany, Japan, South Africa, the UK, the US and the European Commission. Where these have clear project objectives – for example, the post-combustion CO_2 capture pilot project at a power station in Beijing (Figure 5.6) – then good progress is seen. Progress is less visible when the objectives overlap with multilateral activities. The better co-ordination of multilateral collaboration to incorporate existing bilateral activity could be more productive and would certainly reduce the considerable administrative overhead associated with servicing multiple bilateral agreements. Bilateral agreements with ambitious objectives, such as the EU-China and China-UK agreements related to the Near-Zero Emissions Coal (NZEC) project, have a catalysing influence and should be allowed to gain wider support as they develop.

Source: Chapter 9.

In recent years, many studies have focused on socio-economic and environmental topics, consistent with the aims of China's Five-Year Plans (Box 10.3). These include coal sector restructuring, coal mine safety and coalbed methane utilisation. That said, the scope and number of international coal-related collaborative projects has been surprisingly small. This perhaps reflects the political view of coal in many OECD member countries as an old-fashioned fuel while renewables are seen as the way to reduce CO_2 emissions. However, there is a growing realisation that coal will continue to be used extensively in most of the world's large energy-consuming countries, so clean and efficient coal technologies must be an essential component of any clean energy strategy. In China, the focus of engagement needs to be adjusted towards sustainable coal use. *That is not to suggest that renewable energy should be neglected.* Rather, as renewables alone cannot meet China's rapidly growing energy needs in the foreseeable future, improving coal supply and the way it is used should be the focus of at least as much effort, if not more.

Box 10.3.............. Critique of international studies

The Asian Development Bank (ADB) has funded important technical, economic and social studies to inform its lending decisions, notably on CMM, mining safety, use of waste coal and prospects for IGCC. In the case of the latter, the ADB is considering an application to finance two-thirds of the first stage of the GreenGen project – an advanced IGCC with the possibility of incorporating CCS in a second stage. World Bank technical assistance studies have considered technical and non-technical issues that affect clean coal technologies, and projects have proceeded with the active involvement of the Chinese government which has ensured that most components can be manufactured domestically to reduce costs and aid significant market penetration. By contrast, the European Commission has funded market assessment studies on the assumption that China is a market for equipment exported from the EU. Any approach that assumes one-way trade is now dated given China's ability to compete as an equipment supplier in the global market.

An IEA-led power plant refurbishment study was a good example of a technical exercise with significant potential benefit to the Chinese power sector (DTI, 2004). However, although dissemination of the results was undertaken at various workshops, there is little evidence of Chinese power companies taking forward the suggested retrofit options. In part, this is because plants must be kept on line to meet demand, leaving little scope for extended outages for retrofits. The messages set out in the report remain relevant and the poor uptake begs the question about how best to achieve effective promotion and dissemination of plant improvements and better practices within China.

Some studies have provided guidance to other countries on the strategic and commercial impacts of Chinese policies, highlighting the internal dynamics of China's coal supply and utilisation activities. No domestic or international projections have come close to predicting the remarkable growth of the Chinese coal sector, so many earlier studies failed to grasp the magnitude of the issues or proposed solutions that were too narrowly focused.

Source: Chapter 8.

Examples of successful engagement exist in the coal sector. The World Bank has supported major improvements in the Chinese coal-fired power sector, including:

■ introduction of the first 600 MW subcritical and the first 900/1 000 MW supercritical units;

■ introduction of new large-scale ultra-supercritical units that further increase plant efficiency;

■ continuous emission monitoring, electrostatic precipitator, FGD and low-NOx burner investment projects;

■ pilot sulphur trading and the "sulphur bubble" concept for optimal use of FGD; and

■ rehabilitation of medium-size (200-300 MW) power plants to improve efficiency and retrofit of pollution control systems.

In moving forward, China can avoid the past mistakes of others. The development and commercialisation of clean coal technologies in IEA member countries has largely come about through multi-participant approaches. Governments have established long-term regulatory frameworks and, in some cases, carry out pre-commercial research that gives utilities and equipment suppliers sufficient confidence to demonstrate appropriate technologies. Support for major projects, such as development and demonstration of integrated gasification combined cycle (IGCC), comes from consortia in which all stakeholders are represented to spread risks and costs among participants – making these more manageable. More stringent pollution control legislation has led to the widespread installation of FGD, low-NOx burners and, in some cases, SCR and supercritical steam cycles. However, the higher costs of building, operating and maintaining plants that make use of state-of-the-art clean coal technologies, incorporating advanced pollution control systems and progressing from supercritical (SC) to ultra-supercritical (USC) steam conditions, have constrained their implementation. The same applies to IGCC technologies, with only demonstration plants built and operated to date. Hence, IEA member countries have not yet gained experience of the cost reductions that would come over time as commercial IGCC units are built.

Recommendation 9

International and national partnerships, supported by governments, industry and academia, can stimulate the development of new technologies before their commercialisation.

International collaboration and technology transfer in order to gain access to know-how and equipment is an important element of China's own strategy towards a sustainable coal supply chain. Beyond acquisition and learning, however, deployment of new technologies deserves high priority as well; market and regulatory conditions must combine to create an appetite among prospective users of new technologies. This is also an area where it is important to transfer know-how, and to use it to inform China's policy makers of regulatory systems that are practicable and applicable to its circumstances.

Protection of intellectual property (IP) in China remains of concern. Perceived shortcomings in terms of protection create a disincentive for foreign companies to commit their knowledge and experience to solving China's challenges in the coal sector and elsewhere. In consequence, Chinese companies still have considerable ground to make up in terms of achieving world-best technologies, while foreign companies have been more inclined to transfer out-of-date technologies. However, many commercial companies have already moved ahead in China, regardless of any outstanding IP concerns, to take advantage of China's low-cost manufacturing base. The IEA believes that companies will establish subsidiaries or joint ventures in China under normal commercial terms, and will protect their IP by a variety of practical means, not limited to the rights afforded by China's legal system. For example, where IP is associated with a manufacturing process, the IP owner may choose to maintain full ownership and control of key manufacturing facilities.

Technology transfer is not just about pieces of hardware and information held in engineering drawings and equipment manuals. To apply technology demands knowledgeable and skilled staff, engineering design rules, computer software and management systems. Foreign direct investment (FDI) and co-operative joint ventures are often an efficient means of transferring technology. Staff mobility between competing firms then allows the spread of knowledge and application of new technologies and best practices. Yet, joint ventures and FDI in the energy sector are viewed unfavourably by the Chinese government on security grounds, especially the security of scarce resources – a reaction seen among other governments that are keen to protect key strategic industries.

Recommendation 10

The government should further encourage joint ventures and foreign direct investments in the energy sector to promote technology transfer, both into and out of China.

Long-term, strategic partnerships with foreign enterprises can contribute to technological progress through the joint development of cleaner coal technologies that are appropriate for markets around the world. By reassessing the strategic value of foreign participation in key energy industries, the Chinese government might find that even majority foreign ownership can be encouraged to progress national goals.

China has ratified the primary international accords on climate change, *i.e.* the United Nations Framework Convention on Climate Change (UNFCCC) and the Kyoto Protocol. As a developing country, China has no binding emission limits under either accord, although it is a very active participant in the CDM established under the Kyoto Protocol. It is essential to recognise that China, and other developing countries, are unlikely to accept absolute greenhouse gas emission targets that they perceive as a constraint on their economic growth. Rather, China is likely to propose intensity-based targets as these provide a measure for the efficiency of economic growth without limiting absolute energy use. The 11th Five-Year Plan includes many domestic policy initiatives aimed at reducing energy intensity. Consequently, any future international climate agreement will need to develop and incorporate an arrangement whereby emissions-mitigation measures fit or advance the national priorities of developing countries, such as economic growth, energy security and public health. The current APP and M2M approaches cannot be considered as adequate, given their limited resources, yet undoubtedly they do bring together an important grouping of nations, and therefore have potential to lay the groundwork for future actions.

While such an approach would reduce energy intensity, it will not reduce absolute greenhouse gas emissions. Consequently, CCS appears to be an essential part of any CO_2 abatement strategy. China's National Action Plan on Climate Change, launched in 2007, identifies clean coal technologies and CCS demonstration as priorities, and sees international scientific and technological collaboration as key to technology transfer. This is a good foundation for designing national and international frameworks that encourage China to introduce CCS at its large-scale power generation plants and, if eventually established, its coal-to-liquids facilities. The EU-China Partnership on Climate Change could play an important role in establishing large-scale demonstrations of CCS, both in China and Europe, before 2020. There are many important and difficult steps to be taken, but the EU and China have both made a commitment to achieving this target. The early mapping of China's CO_2 storage potential is a clear priority.

China should continue its active international environmental co-operation, participating in international programmes that support deployment of cleaner coal technologies. To facilitate progress through competition, it should increase the opportunities for wider national and international input to the development of coal production, environmental protection and coal utilisation technologies. Competition would imply a further opening up of the Chinese economy, but the IEA firmly believes that a global market for cleaner coal technologies is the way forward. Hence, Recommendation 5 above envisages a global market in which China is both a user and a prominent supplier of cleaner coal technologies. China's opening-up policy has allowed the country to take a leading position in the manufacture of many goods. This should now extend to cleaner coal technologies, so that technologies are developed and deployed through normal commercial relationships within a competitive market. The role of policy makers in China and elsewhere is to create that competitive market and allow it to flourish.

A CLEANER COAL STRATEGY FOR CHINA

The preceding recommendations, like the report as a whole, cover a tremendous range of activities. How might the suggested options be brought together as a coherent, flexible strategy that supports achievement of China's own national development goals? Such a strategy would make China a world leader in the adoption of clean coal technologies. Doing so requires major amendments to the investment and regulatory structures governing the coal cycle, from production to end use. The licensing and administration systems for resource acquisition and tenure would need to be strengthened; mine safety monitoring and regulation would require unequivocal support from all stakeholders; environmental licensing, monitoring and rehabilitation would entail greater commitment from national government downwards; and the investment climate for constructing and operating state-of-the-art power plants, coking plants and other industrial facilities would need to be improved. A comprehensive strategy would address all aspects of the coal production, supply and utilisation chain. It would require commitment by the authorities responsible for government finances, taxes, investment, environment and technological innovation. It would cover both shorter-term needs, such as deploying more efficient power plants and retrofitting FGD, and longer-term needs such as establishing near-zero emissions technologies. If a framework for the widespread deployment of cleaner coal technologies can be established in China and elsewhere, this would offer a stable basis for the subsequent introduction of CCS.

With supercritical pulverised-coal power stations already in operation in China, the country has an opportunity to bypass existing subcritical technologies for new plants and to achieve higher efficiencies for all its new generating capacity. While it is still too early to comment on the impact of the National Action Plan on Climate Change, this is obviously a springboard for the more widespread use of existing cleaner coal technologies and the development of their next generation. The milestone to be reached is commercial CCS, but unless there is greater near-term commitment to developing and demonstrating CCS systems through the use of grants and investment incentives from national governments and international bodies alike, the demonstration of these technologies will not happen until there is a significant and universal cost attached to carbon emissions. Similarly, a longer-term perspective that values improved performance over some decades is often needed to justify the additional investment cost of supercritical, ultra-supercritical and other advanced technologies when subcritical power plants are readily available at low capital costs.

Among IEA member countries, there is considerable diversity in approaches to coal utilisation. In general, the alternatives, including natural gas, nuclear and renewables, all face constraints, so coal remains an important component of the energy mix. While existing emission trading schemes have not resulted in the deployment of cleaner coal technologies to cut carbon emissions, draft EU directives, published in January 2008, would create structures to encourage CCS and more efficient coal technologies (EC, 2008). The European Commission also recognises that large-scale CCS demonstration projects are needed to prove their viability in commercial operation and that this will require government support in one form or another. The UK government has launched a competition with the intention of funding the first such demonstration.

Globally, two market imperfections currently limit the uptake of cleaner coal technologies: it costs less to pollute than to control pollution and barriers, such as high development costs, slow technological change. Accelerating deployment will require changes at the national and international levels. Commercial deployment of cleaner coal technologies requires investment certainty through stable policies that recognise the costs and risks of long-term capital investment in pollution control, ultra-supercritical, IGCC and CCS technologies. Hence, the **three priorities for international engagement** with China are:

■ negotiations leading to successful international accords that create national, regional and global markets for clean, low-carbon technologies;

■ government-industry partnerships to develop and demonstrate low-carbon, cleaner coal technologies; and

■ technology transfer and deployment of cleaner coal technologies through commercial arrangements that respond to the market demand created in China and elsewhere.

Finally, the cleaner and safer exploitation of coal is needed to sustain economic well being during the transition to a global, low-carbon energy system. This entails more than just state-of-the-art mining and high-efficiency power generation technologies. Social and environmental aspects have to be taken into consideration as well, especially now that large sections of China's population are becoming more environmentally aware. China has the opportunity to demonstrate a holistic approach that takes these aspects into account, and to set an example for other nations.

REFERENCES

DTI (Department of Trade and Industry) (2004), *IEA-China Power Plant Optimisation Study: Executive Summary*, Report No. COAL R258, DTI/Pub URN 04/1015, DTI (now Department of Energy and Climate Change), London, ww.berr.gov.uk/publications/index.html.

EC (European Commission) (2008), "Climate Action and Renewable Energy Package", EC, Brussels, Belgium, 23 January, http://ec.europa.eu/environment/climat/climate_action.htm.

IEA (International Energy Agency) (2006), *China's Power Sector Reforms: Where to Next?*, OECD/IEA, Paris.

IEA GHG (IEA Greenhouse Gas R&D Programme) (2007), *CO_2 Capture Ready Plants*, Technical Study Report No. 2007/4, IEA GHG, Cheltenham, UK, May, www.iea.org/Textbase/Papers/2007/CO2_Capture_Ready_Plants.pdf.

NDRC (National Development and Reform Commission) (2006), *Outline of the Eleventh Five-Year Plan for National Economic and Social Development of the People's Republic of China*, NDRC, Beijing, China, 19 March, http://en.ndrc.gov.cn/hot/t20060529_71334.htm.

OECD (Organisation for Economic Co-operation and Development) (2007), *OECD Environmental Performance Reviews: China (2007)* (OECD 中国环境绩效评估报告), OECD, Paris.

WEC (World Energy Council) (2007), *2007 Survey of Energy Resources*, WEC, London.

ACRONYMS, ABBREVIATIONS AND UNITS[1]

ABARE	Australian Bureau of Agricultural and Resource Economics
ADB	Asian Development Bank
ADHS	Appalachian Development Highway System (US)
AGO	Australian Greenhouse Office
APEC	Asia-Pacific Economic Cooperation
APNet	Asia-Pacific Network for Energy Technology
APP	Asia-Pacific Partnership on Clean Development and Climate
ARC	Appalachian Regional Commission (US)
ARP	*Agencja Rozwoju Przemysłu SA* (Industrial Development Agency, Poland)
ASEAN	Association of Southeast Asian Nations
ASEAN+3	ASEAN plus China, Japan and the Republic of Korea
AUD	Australian dollar
BAT	best-available techniques (or technology)
BCE	British Coal Enterprise (UK)
bcm	billion cubic metres (10^9 cubic metres)
BERR	Department for Business, Enterprise and Regulatory Reform (UK – now Department of Energy and Climate Change)
BFBC	bubbling fluidised bed combustion (or combustor)
BGR	*Bundesanstalt für Geowissenschaften und Rohstoffe* (Federal Institute for Geosciences and Natural Resources, Germany)
BJ	Barlow Jonker (Australia)
BLM vzw	*Begeleidingsdienst Limburgs Mijngebied* (Limburg Mining Region Counselling Service, Belgium)
BMZ	*Bundesministerium für wirtschaftliche Zusammenarbeit und Entwicklung* (German Federal Ministry for Economic Cooperation and Development)

---

1. See unit converter at www.iea.org/Texbase/stats/index.asp.

BREF	Best Available Techniques Reference Document (EIPPCB)
BRICC	Beijing Research Institute of Coal Chemistry (China)
Btu	British thermal unit
°C	degree Celsius (or centigrade)
CAS	Chinese Academy of Sciences
CASS	Chinese Academy of Social Sciences
CBM	coalbed methane
CCC	IEA Clean Coal Centre, (IEA Coal Research Implementing Agreement)
CCERC	Coal Industry Clean Coal Engineering Research Center (China)
CCFC	*congé charbonnier de fin de carrière* (end-of-career "holiday" for coal miners, France)
CCICED	China Council for International Cooperation on Environment and Development
CCIDRC	China Coal Industry Development Research Center
CCII	China Coal Information Institute
CCRI	China Coal Research Institute
CCS	carbon dioxide capture and storage
CCT	clean coal technology
CCTM	clean coal technology model (software developed by CCRI)
CCTMA	China Coal Transport and Marketing Association
CdF	Charbonnages de France
CDM	Clean Development Mechanism (Kyoto Protocol)
CDQ	coke dry quenching
CEC	China Electricity Council
CEMAC	China Economic Monitoring and Analysis Center
CERS	China Energy Research Society
CERTH/ISFTA	Centre for Research and Technology Hellas / Institute for Solid Fuels Technology and Applications (Greece)
CFBC	circulating fluidised bed combustion (or combustor)
CFRR	catalytic flow-reversal reactor
CHP	combined heat and power
CIAB	IEA Coal Industry Advisory Board
CIEDR	China Industry Research Network
CIF	carriage-insurance-freight

CLG	Department for Communities and Local Government (UK)
CMM	coal mine methane
CNCA	China National Coal Association
CNNC	China National Nuclear Corporation
CO_2	carbon dioxide
COACH	Cooperation Action within CO_2 Capture and Storage China-EU
COAG	Council of Australian Governments
CPPCC	Chinese People's Political Consultative Conference
CPR1000	Chinese Pressurised Water Reactor (powering a c.1 000 MWe unit)
CRESR	Centre for Regional Economic and Social Research (UK)
CSIRO	Commonwealth Scientific and Industrial Research Organisation (Australia)
CSIS	Center for Strategic International Studies (US)
CSLF	Carbon Sequestration Leadership Forum
CTL	coal-to-liquids
CWM	coal-water mixture
CZT	Changsha-Zhuzhou-Xiangtan region (China)
DCL	direct coal liquefaction
Defra	Department for Environment, Food and Rural Affairs (UK)
DFAIT	Foreign Affairs and International Trade Canada
DFAT	Department of Foreign Affairs and Trade (Australia)
DME	dimethyl ether (CH_3OCH_3 – a colourless, clean-burning gas)
DOE	US Department of Energy
DRC	Development and Refom Commission (China)
DSM	demand-side management
DTI	Department of Trade and Industry (UK – now Department of Energy and Climate Change)
EC	European Commission or European Community
ECSC	European Coal and Steel Community
EEC	European Economic Community
EIA	US Energy Information Administration
EIPPCB	European Integrated Pollution Prevention and Control Bureau (Spain)
ELV	emission limit value

EOR	enhanced oil recovery
EPA	US Environmental Protection Agency
EPB	Environmental Protection Bureau (China)
EPC	engineering-procurement-construction
EPRI	Electric Power Research Institute (US)
ERI	Energy Research Institute (China)
ESP	electrostatic precipitator
est.	estimate
ETS	EU Emissions Trading Scheme
EU	European Union
EUR	euro
EUR-Lex	EC web portal to Official Journal of the EU and all EU legal documents
FDI	foreign direct investment
FGD	flue gas desulphurisation
FINORPA	*Société Financière du Nord-Pas de Calais* (Nord-Pas de Calais Investment Company, France)
FOB	free on board
G8	Group of Eight (Canada, France, Germany, Italy, Japan, Russia, UK, US)
GAP	*Górnicza Agencja Pracy* (Labour Mining Agency, Poland)
GBP	UK pound
gce	gramme of coal equivalent
GDP	gross domestic product
GEF	Global Environment Facility
GEODISC	Geological Disposal of Carbon Dioxide (Australian research programme)
GHG	greenhouse gas
GJ	gigajoule (10^9 joules)
GPS	*Górniczy Pakiet Socjalny* (Mining Social Package, Poland)
Gt	gigatonne (10^9 metric tonnes or billion tonnes)
Gtce	gigatonne of coal equivalent
GTZ	*Deutsche Gesellschaft für Technische Zusammenarbeit (GTZ) GmbH* (German Technical Co-operation Agency)
GVSt	*Gesamtverband Steinkohle* (German Hard Coal Association)

GW	gigawatt (10^9 watts)
GWe	gigawatt of electric power
ha	hectare (10 000 m^2)
HHV	higher heating value
H&S	health and safety
HUNOSA	Hulleras del Norte SA (Spain)
IDGCC	integrated drying gasification combined cycle
IEA	International Energy Agency (OECD)
IEEJ	Institute of Energy Economics, Japan
IEO	EIA International Energy Outlook
IFC	International Finance Corporation of the World Bank Group
IGCC	integrated (coal) gasification combined cycle
IP	intellectual property
IPPC	Integrated Pollution Prevention and Control [Directive] (EU)
J	joule
JACCS	Job and Career Change Scheme (UK)
JCOAL	Japan Coal Energy Center
JV	joint venture [company]
kcal	kilocalorie
kg	kilogramme
kgce	kilogramme of coal equivalent
kgU	kilogramme of uranium
km	kilometre
km^2	square kilometre
kt	kilotonne (10^3 metric tonnes)
ktpa	kilotonne per annum (year)
kW	kilowatt (10^3 watts)
kWh	kilowatt-hour (10^3 watt-hours)
LCPD	EC Large Combustion Plants Directive
LHS	left-hand scale
LHV	lower heating value
LNG	liquefied natural gas
LPG	liquefied petroleum gas

LRTAP	UNECE Convention on Long-Range Transboundary Air Pollution
m	metre
m²	square metre
M2M	Methane to Markets Partnership
m³	cubic metre
MASHAM	Management and Administration of Safety and Health at Mines Regulations (UK)
MCIS	McCloskey Coal Information Services (UK)
mcm	million cubic metres (10^6 cubic metres)
MEA	Ministry of External Affairs (India)
MEP	Ministry of Environmental Protection (China)
MESA	Mining Enforcement and Safety Administration (US)
METI	Ministry of Economy, Trade and Industry (Japan)
mg	milligramme
min	minute
MINER	Mine Improvement and New Emergency Response Act of 2006 (US)
MIT	Massachusetts Institute of Technology (US)
MLR	Ministry of Land and Resources (China)
mmBtu	million Btu
MOCI	Ministry of Coal Industry (China)
MOE	Ministry of Energy (China)
MOFA	Ministry of Foreign Affairs (China)
MOFCOM	Ministry of Commerce (China)
MOHURD	Ministry of Housing and Urban-Rural Development (China)
MOST	Ministry of Science and Technology (China)
MoU	memorandum of understanding
MSHA	Mine Safety and Health Administration (US)
MSW	municipal solid waste
Mt	megatonne (10^6 metric tonnes or million tonnes)
Mtce	million tonnes of coal equivalent
Mtoe	million tonnes of oil equivalent
Mtpa	million tonnes per annum (year)
MW	megawatt (10^6 watts)
MWe	megawatt of electric power

MWth	megawatt of thermal power
NBS	National Bureau of Statistics of China
NCB	National Coal Board (UK)
NDRC	National Development and Reform Commission (China)
NEA	National Energy Administration (China)
NEA	Nuclear Energy Agency (OECD)
NEDO	New Energy and Industrial Technology Development Organization (Japan)
NEPA	National Environmental Policy Act of 1969 (US)
NGO	non-governmental organisation
NIOSH	National Institute for Occupational Safety and Health (US)
Nm³	normal cubic metre [of a gas]
NMA	National Mining Association (US)
no.	number
NOx	nitrogen oxides
NPC	National People's Congress (China)
NWC	National Water Commission (Australia)
NZEC	Near-Zero Emissions Coal project (China-UK and EU-China)
OECD	Organisation for Economic Co-operation and Development
ONELG	Office of the National Energy Leading Group (China)
OPEC	Organization of the Petroleum Exporting Countries
OPET	Organisations for the Promotion of Energy Technologies (EC)
OSHA	Occupational Safety and Health Administration (US)
OSM	Office of Surface Mining, Reclamation and Enforcement (US)
PARP	*Polska Agencja Rozwoju Przedsiębiorczości* (Polish Agency for Enterprise Development)
PC	pulverised coal [combustion]
PCI	pulverised coal injection (for iron making)
PEM	proton exchange membrane or polymer electrolyte membrane (fuel cell)
PFBC	pressurised fluidised bed combustion (or combustor)
PLN	Polish zloty
$PM_{2.5}$	particulate matter of 2.5 micrometre (μm) diameter or less
PM_{10}	particulate matter of 10 micrometre (μm) diameter or less
PM&C	Department of the Prime Minister and Cabinet (Australia)
POEO	Protection of the Environment Operations Act 1997 (Australia)

POP	persistent organic pollutant
PV	photovoltaic [solar cell]
PWR	pressurised water reactor
RAG	Ruhrkohle AG (Germany)
R&D	research and development
RD&D	research, development and demonstration
RHS	right-hand scale
RITE	Research Institute of Innovative Technology for the Earth (Japan)
RMB	Chinese renminbi or yuan
RVR	*Regionalverband Ruhr* (Ruhr Regional Association, Germany)
SACMS	State Administration of Coal Mine Safety (China)
SAIC	State Administration for Industry and Commerce (China)
SAM	State Administration for Materials (China)
SASAC	State-Owned Assets Supervision and Administration Commission (China)
SAWS	State Administration of Work Safety (China)
SC	supercritical
SCIB	State Coal Industry Bureau (China)
SCIO	State Council Information Office (China)
SCR	selective catalytic reduction
SDPC	State Development Planning Commission (China)
SEPA	State Environmental Protection Administration (China)
SERC	State Electricity Regulatory Commission (China)
SETC	State Economic and Trade Commission (China)
SIC	State Information Center (China)
SinoU	China Nuclear International Uranium Corporation
SMCRA	Surface Mining Control and Reclamation Act of 1977 (US)
SME	small and medium enterprise
SNCR	selective non-catalytic reduction
SO_2	sulphur dioxide
SODECO	*Sociedad para el Desarrollo de las Comarcas Mineras SA* (Coalfields Development Agency, Spain)
SOFIREM	*Société Financière pour favoriser l'Industrialisation des Régions Minières et Régions en Mutations* (Industrial Development and Transformation Fund for the Coal Mining Regions, France)

SPC	State Planning Commission (China)
t	metric tonne (1 000 kg)
tce	tonne of coal equivalent
TFRR	thermal flow-reversal reactor
t-km	tonne-kilometre
tpa	tonne per annum (year)
tpd	tonne per day
tph	tonne per hour
TPRI	Thermal Power Research Institute (China)
tU	tonne of uranium
TVE	township and village enterprise
TWh	terawatt-hour (10^{12} watt-hours)
UCC	ultra clean coal
UCG	underground coal gasification
UK	United Kingdom
UMWA	United Mine Workers of America (US)
UMWACC	UMWA Career Center (US)
UNDP	United Nations Development Programme
UNECE	United Nations Economic Commission for Europe
UNEP	United Nations Environment Programme
UNESCO	United Nations Educational, Scientific and Cultural Organization
UNFCCC	United Nations Framework Convention on Climate Change
US	United States
USC	ultra-supercritical
USD	US dollar
VAM	ventilation air methane
VAM-CAT	ventilation air methane catalytic combustion gas turbine
VAT	value added tax
VM	volatile matter
VOC	volatile organic compound
WEC	World Energy Council
WEO	IEA World Energy Outlook
WNA	World Nuclear Association

BIBLIOGRAPHY

Listed here are works that have been consulted during the course of the project, though not referenced in the main report. While web links are provided here and in the reference lists at the end of each chapter, many will become "broken" over time. The IEA will endeavour to provide copies of any cited material, subject to copyright restrictions. Where only Chinese text is available, machine translation of web pages (using browser gadgets) can provide good visibility for non-Chinese readers, although such translations should be used with caution.[1] There are a number of government, commercial and not-for-profit services that provide on-line access to Chinese laws, regulations and other resources in Chinese or English.[2]

AAA Minerals (2003), "Categories and Terminologies of Coal Resources and Reserves in China and their International Correlation" and "Classification of Coal Ranks in China (1986 Standard)", AAA Minerals International Co., Beijing, www.aaamineral.com/web1/infoenshow.asp?id=113 and 114.

ADB (Asian Development Bank) (2008), *Country Partnership Strategy: People's Republic of China 2008-2010*, ADB, Manila, Philippines, February.

Andrews-Speed, P., Ma Guo, Shao Bingjia and Liao Chenglin (2005), "Economic Responses to the Closure of Small-Scale Coal Mines in Chongqing, China", *Resources Policy*, Vol. 30, No. 1, Elsevier, pp. 39-54.

Andrews-Speed, P., Yang Minying, Shen Lei and Shelley Cao (2003), "The Regulation of China's Township and Village Coal Mines: A Study of Complexity and Ineffectiveness, *Journal of Cleaner Production*, Vol. 11, No. 2, Elsevier, pp. 185-196.

APEC (Asia-Pacific Economic Cooperation) (2005), CO_2 *Storage Prospectivity of Selected Sedimentary Basins in the Region of China and South East Asia*, APEC Energy Working Group Project 06/2003, Report No. APEC#205-RE-01.6, prepared by Innovative Carbon Technologies Pty. Ltd. and Geoscience Australia for the Cooperative Research Centre for Greenhouse Gas Technologies (CO2CRC), APEC, Singapore, June, www.apec.org/apec/publications/all_publications/energy_working_group.html.

---

1. http://translate.google.com/translate_tools?hl=en

2. For example, the official web portal of the Central People's Government of the People's Republic of China (PRC) (www.gov.cn/zwgk/index.htm); the State Council Legislative Affairs Office (www.chinalaw.gov.cn); China Internet Information Center web portal authorised by the State Council Information Office (www.china.org.cn/english/government/205794.htm); ChinaCourt sponsored by the Supreme People's Court of the PRC (http://en.chinacourt.org); 西湖法律图书馆 (West Bookstore Law Library Network) (www.law-lib.com); iSinoLaw, an approved commercial service that provides English translations of laws, regulations and court judgements (www.isinolaw.com); LawInfoChina, a commercial service established by the Legal Information Center of Peking University with searchable databases of laws, regulations, cases, gazettes and law journals (www.lawinfochina.com); China Law and Practice, a Hong Kong, China-based commercial service (www.chinalawandpractice.com); databases maintained by the Asian Legal Information Institute (www.asianlii.org); the US Congressional-Executive Commission on China (www.cecc.gov) and other links to English-language web resources from the University of British Columbia (www.library.ubc.ca/law/chineselaw.html).

APEC (2007), *How Can Environmental Regulations Promote Clean Coal Technology Adoption in APEC Developing Economies?*, Expert Group on Clean Fossil Energy, APEC Energy Working Group Project 06/2006, Report No. APEC#207-RE-01.12, Science Applications International Corp. (SAIC) for APEC, Singapore, 30 November.

APERC (Asia Pacific Energy Research Centre) (2006), *APEC Energy Demand and Supply Outlook 2006*, APERC, Institute of Energy Economics, Tokyo, Japan, www.ieej.or.jp/aperc/outlook2006.html.

Arquit-Niederberger, A. and B. Finamore (2005), "Building an Efficiency Power Plant under the Clean Development Mechanism", *Sinosphere*, Vol. 8, No. 1, Professional Association for China's Environment, pp. 33-38.

Attwood, T., V. Fung and W. W. Clark (2003), "Market Opportunities for Coal Gasification in China", *Journal of Cleaner Production*, Vol. 11, No. 4, Elsevier, pp. 473-479.

Aunan, K., Fang Jinghua, H. Vennemo, K. Oye and H. M. Seip (2004), "Co-benefits of Climate Policy – Lessons Learned from a Study in Shanxi, China", *Energy Policy*, Vol. 32, No. 4, Elsevier, pp. 567-581.

Avato, P. and J. Coony (2008), *Accelerating Clean Energy Technology Research, Development, and Deployment: Lessons from Non-Energy Sectors*, Working Paper No. 138, World Bank, Washington, DC, May.

Bai Quan and Tong Qing (白泉，佟庆) (2005), 《美国的能源多样化战略及对我国的启示》 ("Lessons and Experience for China from the US Energy Diversification Strategy"), 能源研究所, 清华大学 (Energy Research Institute, Tsinghua University), 核心期刊宏观经济管理 (*Journal of Macroeconomic Management*), 2005年第04期 (2005, No. 4).

Bariş, E. and M. Ezzati (eds.) (2007), *Household Energy, Indoor Air Pollution and Health: A Multisectoral Intervention Program in Rural China*, Special Report 002/07, Energy Sector Management Assistance Program (ESMAP), World Bank, Washington, DC, June.

Barlow Jonker (2007), *China Coal Fourth Edition*, Barlow Jonker, Sydney, Australia.

Berrah, N., Fei Feng, R. Priddle, Wang Leiping (2007), *Sustainable Energy in China: The Closing Window of Opportunity*, World Bank, Washington, DC,

www.esmap.org/filez/pubs/517200740503_SustainableEnergyinChina.pdf.

Bhattasali, D., Li Shantong and W. Martin (eds.) (2004), *China and the WTO: Accession, Policy Reform, and Poverty Reduction Strategies*, World Bank and Oxford University Press.

Blaschke, W. and L. Gawlik (1999), "Coal Mining Industry Restructuring in Poland: Implications for the Domestic and International Coal Markets", *Applied Energy*, Vol. 64, No. 4, Elsevier, pp. 453-456.

Burnard, K. (2007), "Carbon Capture and Storage and ZETs in China", presented at a meeting of the IEA Working Party on Fossil Fuels, Brasilia, 28-29 June.

Cao Jing, R. Garbaccio and Ho Mun (2008), "Benefits and Costs of SO_2 Abatement Policies in China", paper presented at the 11th Annual Conference on Global Economic Analysis, Helsinki, Finland, 12-14 June, https://www.gtap.agecon.purdue.edu/resources/download/3858.pdf.

CASS (Chinese Academy of Social Sciences) (2006), *Understanding China's Energy Policy: Economic Growth and Energy Use, Fuel Diversity, Energy/Carbon Intensity, and International Cooperation*, background paper prepared for Stern Review on the Economics of Climate Change, Research Centre for Sustainable Development, CASS, Beijing, www.hm-treasury.gov.uk/d/Climate_Change_CASS_final_report.pdf.

CCICED (China Council for International Cooperation on Environment and Development) (2003), "Transforming Coal for Sustainability: A Strategy for China", report by the CCICED Task Force on Energy Strategies and Technologies, *Energy for Sustainable Development*, Vol. 7, No. 4, Elsevier for International Energy Initiative, pp. 5-14.

Chen Guifeng, Yu Zhufeng and Wu Lixin (2005), "Present Status and Future Prospect of Clean Coal Technology in China", *International Journal of Global Energy Issues*, Vol. 24, Nos. 3-4, Inderscience, pp. 228-240.

China-Britain Business Council (2007), *China Business Guide*, 2nd edition, China Markets Unit, UK Trade and Investment, London.

Cole, B. D. (2003), *"Oil for the Lamps of China" – Beijing's 21st-Century Search for Energy*, McNair Paper 67, Institute for National Strategic Studies, National Defense University, Washington, DC, October.

Constantin, C. (2008), *China's Energy Policy and Energy Cooperation: Opportunities and Challenges for Canada*, preliminary paper, Canadian International Council (*Conseil International du Canada*), Toronto, Canada, July.

Cornelius, P. and J. Story (2007), "China and Global Energy Markets", *Orbis*, Vol. 51, No. 1, Elsevier for Foreign Policy Research Institute, pp. 5-20.

Creedy, D., Wang Lijie, Zhou Xinquan, Liu Haibin and G. Campbell (2006), "Transforming China's Coal Mines: A Case History of the Shuangliu Mine", *Natural Resources Forum*, Vol. 30, No. 1, Blackwell, pp. 15-26.

DOE (US Department of Energy) (2006), *Country Analysis Brief – China*, Energy Information Administration, DOE, Washington, DC, August, www.eia.doe.gov/emeu/cabs/China/pdf.pdf.

Drysdale, P., Jiang Kejun and D. Meagher (eds.) (2007), *China and East Asian Energy: Prospects and Issues*, Vol. I, Proceedings of Conference held on 10-11 October 2005, Xindadu Hotel, Beijing, Asia Pacific Economic Paper No. 361, Australia-Japan Research Centre, Australian National University, Canberra.

DTI (Department of Trade and Industry) (2001), "UK-China Coalbed Methane Technology Transfer", Report No. COAL R207, DTI/Pub URN 01/584, Wardell Armstrong for DTI (now Department of Energy and Climate Change), London, February, www.berr.gov.uk/files/file18629.pdf (Project Summary No. 260 available in Chinese from www.coalinfo.net.cn/cnuk/eml1.htm).

DTI (2002), "Review of the Coal Preparation Sector in China", Project Summary No. 324, DTI/Pub URN 02/786, DTI, London, March, www.berr.gov.uk/files/file20079.pdf.

DTI (2002), *Coalbed Methane and its Commercialisation in China*, Case Study No. 009, DTI/Pub URN 02/1095, DTI, London, August, www.berr.gov.uk/files/file20815.pdf.

DTI (2004), "Clean Energy from Underground Coal Gasification: Promoting Commercially Viable UCG in China", Project Summary No. 290, DTI/Pub URN 03/1610, DTI, London, February, www.berr.gov.uk/files/file20060.pdf.

DTI (2005), *Enhancing Coal Mine Methane Utilisation in China*, Report No. COAL R298, DTI/Pub URN 05/1816, IT Power and Wardell Armstrong for DTI, London, December, www.berr.gov.uk/files/file29223.pdf.

Ekawan, R., M. Duchêne and D. Goetz (2006), "The Evolution of Hard Coal Trade in the Pacific Market", *Energy Policy*, Vol. 34, No. 14, Elsevier, pp. 1853-1866.

ERI (能源研究所 – Energy Research Institute) (2006), 中国天然气发电政策研究 (*Policy Study: Gas-Fired Power Generation in China*), Synthesis Report, ERI, National Development and Reform Commission, Beijing, www.efchina.org/csepupfiles/report/2006102695218105.98387588404353.pdf/NG_Power_Generation_Rept_ERI_EN_060302.pdf.

ESMAP (Energy Sector Management Assistance Program) (2006), *Policy Advice on Implementing the Clean Coal Technology Project*, ESMAP Technical Paper 104/06, World Bank, Washington, DC, September, www.esmap.org/filez/pubs/88200714458_China_Clean_Coal.zip.

Fairley, P. (2007), "China's Coal Future", *Technology Review*, Vol. 110, No. 1, Massachusetts Institute of Technology, pp. 56-61, www.technologyreview.com/energy/18069/.

Fang Yiping, Zeng Yong and Li Shiming (2008), "Technological Influences and Abatement Strategies for Industrial Sulphur Dioxide in China", *International Journal of Sustainable Development and World Ecology*, Vol. 15, No. 2, Sapiens Publishing, pp. 122-131.

Feickert, D. (2004), "Coal Mine Safety in China: Can the Accident Rate Be Reduced?", statement to the Congressional-Executive Commission on China, Rayburn House, Washington, DC, 10 December, www.cecc.gov/pages/roundtables/121004/Feickert.php.

Fu Ping (2007), "Status of and Perspectives on CCS in China", presented at Research Institute of Innovative Technology for the Earth (RITE) Carbon Capture and Storage Workshop, 15 February, Kyoto, Japan, www.rite.or.jp/English/lab/geological/ccsws2007/3_ping.pdf.

Glomsrød, S. and Wei Taoyuan (2005), "Coal Cleaning: A Viable Strategy for Reduced Carbon Emissions and Improved Environment in China?", *Energy Policy*, Vol. 33, No. 4, Elsevier, pp. 525-542.

Gnansounou, E., Jun Dong and D. Bedniaguine (2004), "The Strategic Technology Options for Mitigating CO_2 Emissions in Power Sector: Assessment of Shanghai Electricity-Generating System", *Ecological Economics*, Vol. 50, No. 2, Elsevier, pp. 117-133.

Goodell, J. (2006), *Big Coal: The Dirty Secret Behind America's Energy Future*, Houghton Mifflin Company, Boston and New York.

Graus, W. H. J., M. Voogt and E. Worrell (2007), "International Comparison of Energy Efficiency of Fossil Power Generation", *Energy Policy*, Vol. 35, No. 7, Elsevier, pp. 3936-3951.

Griffiths, C. (2002), "Restructuring of the Coal Industries in the Economies in Transition: An Overview of the Last Decade", *Minerals and Energy – Raw Materials Report*, Vol. 17, No. 2, Routledge, pp. 3-14.

Gunson, A. J. and Yue Jian (2001), *Artisanal Mining in the People's Republic of China*, Mining, Minerals and Sustainable Development Project Report No. 74, International Institute of Environment and Development / World Business Council for Sustainable Development, September,

www.iied.org/pubs/pdfs/G00719.pdf.

Gupta, J., J. Vlasblom and C. Kroeze (2001), *An Asian Dilemma: Modernising the Electricity Sector in China and India in the Context of Rapid Economic Growth and the Concern for Climate Change*, IVM Report No. E-01/04, Institute for Environmental Studies (IVM), Free University, Amsterdam, The Netherlands, June, http://dare.ubvu.vu.nl/retrieve/1745/ivmvu0754.pdf.

Hang Leiming and Tu Meizeng (2007), "The Impacts of Energy Prices on Energy Intensity: Evidence from China", *Energy Policy*, Vol. 35, No. 5, Elsevier, pp. 2978-2988.

Hayes, D. (2002), "Catching the Wind: Clean and Sustainable Solutions to China's Energy Shortfall", *Refocus*, Vol. 3, No. 6, Elsevier, pp. 18+20-21.

He Youguo (2003), "China's Coal Demand Outlook for 2020 and Analysis of Coal Supply Capacity", paper presented at China-IEA Seminar on Energy Modelling and Statistics, Beijing, 20-21 October, www.iea.org/Textbase/work/2003/beijing/4Youg.pdf.

Holdren, J. P., K. Gallagher, P. Mouthino, Zou Ji, R. Banerjee *et al.* (2007), *Linking Climate Policy with Development Strategy in Brazil, China, and India*, Final Report to The William and Flora Hewlett Foundation, Woods Hole Research Center, Falmouth, MA, 15 November.

Hu, Albert G. Z., G. H. Jefferson and Qian Jinchang (2005), "R&D and Technology Transfer: Firm-Level Evidence from Chinese Industry", *Review of Economics and Statistics*, Vol. 87, No. 4, MIT Press, pp. 780-786.

Huang Shengchu *et al.* (2007), *China Coal Outlook 2006*, China Coal Information Institute, Beijing, 18 March.

IEA GHG (IEA Greenhouse Gas R&D Programme) (2002), "Opportunities for Early Applications of CO_2 Sequestration Technology", Report Number PH4/10, IEA GHG, Cheltenham, UK, September, (summarised in file note by J. J. Gale, 20 May 2003, www.cslforum.org/documents/EarlyOppsFile.pdf).

Jiang Kejun (2007), "China's Options after 2012 on Climate Change", paper presented at Shaping China's Energy Security: The Environmental Challenge (打造中国能源安全：环境挑战) seminar, hosted by the Asia Centre (*Centre études Asie à Sciences Po*, Paris), Kempinski Hotel, Beijing, 1 December.

Justus, D. and C. Philibert (2005), *International Energy Technology Collaboration and Climate Change Mitigation: Synthesis Report*, COM/ENV/EPOC/IEA/SLT (2005)11,

Organisation for Economic Co-operation and Development / International Energy Agency, Paris, 18 November.

Kahrl, F. and D. Roland-Holst (2006), *China's Carbon Challenge: Insights from the Electric Power Sector*, Research Paper No. 110106, Center for Energy, Resources, and Economic Sustainability, University of California, Berkeley, CA, November.

Karplus, V. J. (2007), *Innovation in China's Energy Sector*, Working Paper No. 61, Program on Energy and Sustainable Development, Center for Environmental Science and Policy, Stanford University, CA, March, http://iis-db.stanford.edu/pubs/21519/WP61__Karplus_China__Innovations.pdf.

Kroeze, C., J. Vlasblom, J. Gupta, C. Boudri and K. Blok (2004), "The Power Sector in China and India: Greenhouse Gas Emissions Reduction Potential and Scenarios for 1990-2020", *Energy Policy*, Vol. 32, No. 1, Elsevier, pp. 55-76.

Lako, P. (2002), *Options for CO_2 Sequestration and Enhanced Fuel Supply*, monograph in the framework of the VLEEM (Very Long-Term Energy-Environment Model) project, ECN-C-01-113, Energy Research Centre of the Netherlands (ECN), 1 April, www.ecn.nl/docs/library/report/2001/c01113.pdf.

Lanhe Yang, Jie Liang and Li Yu (2003), "Clean Coal Technology – Study on the Pilot Project Experiment of Underground Coal Gasification", *Energy*, Vol. 28, No. 14, Pergamon, pp. 1445-1460.

Larson, E. D., Wu Zongxin, P. DeLaquil, Chen Wenying and Gao Pengfei (2003), "Future Implications of China's Energy-Technology Choices", *Energy Policy*, Vol. 31, No. 12, Elsevier, pp. 1189-1204.

Laslett, J. H. M. (ed.) (1996), *The United Mine Workers of America: A Model of Industrial Solidarity?*, Pennsylvania State University Press, PA.

Lester, R. K. and E. S. Steinfeld (2007), *The Coal Industry in China (and Secondarily India)*, Working Paper No. MIT-IPC-07-001, Industrial Performance Center, Massachusetts Institute of Technology, Cambridge, MA, January.

Li Gao (2006), "CCS in China: Background, Activities and Perspectives", presented at IEA/CSLF Workshop on Near-Term Opportunities for Carbon Capture and Storage, San Francisco, International Energy Agency / Carbon Sequestration Leadership Forum, August, www.cslforum.org/documents/iea_cslf_Gao.pdf.

Li Xiao-Chun, Liu Yan-Feng, Bai Bing and Fang Zhi-Ming (李小春, 刘延锋, 白冰, 方志明) (2005), 《中国深部咸水含水层CO_2储存优先区域选择》 ("Ranking and Screening of CO_2 Saline Aquifer Storage Zones in China"), 岩石力学与工程学报 (Chinese Journal of Rock Mechanics and Engineering), Vol. 25, No. 5, Science Press, Hubei, pp. 963-968.

Liu Chenglin, Zhu Jie, Che Changbo and Liu Guangdi (2008), "Potential Recoverable Natural Gas Resources in China", *Petroleum Science*, Vol. 5, No. 1, China University of Petroleum, Dongying City, Shandong, pp. 83-86.

Liu Hengwei, Ni Weidou, Li Zheng and Ma Linwei (2008), "Strategic Thinking on IGCC Development in China", *Energy Policy*, Vol. 36, No. 1, Elsevier, pp. 1-11.

Liu Hongjie and Li Weizhe (刘宏杰， 李维哲) (2006),《中国能源消费状况和能源消费结构分析》 ("*Analysis of China's Energy Consumption and Structure*"), 国土资源情报 (Land and Resources Information), No. 12, 国土资源部信息中心 (Ministry of Land and Resources Information Center), Beijing, www.lrn.cn/bookscollection/magazines/maginfo/2006maginfo/2006_12/2007 01/t20070131_28631.htm.

Louche, C., A. Lambkin and P. Oliver (2007), *Study on the Future Opportunities and Challenges of EU-China Trade and Investment Relations – Study 11: Sustainable Technologies and Services*, Emerging Markets Group and Development Solutions for Directorate General Trade, European Commission, Brussels, 15 February, http://ec.europa.eu/trade/issues/bilateral/countries/china/legis/index_en.htm.

Lu Hong (2005), *A Study of the Environmental Regulatory System for China's Power Industry: The Case of Jiangsu Province*, School of Public Policy and Management, Tsinghua University, Beijing, June.

Lu Xin, Yu Zhufeng, Wu Lixin, Yu Jie, Chen Guifeng and Fan Maohong (2008), "Policy Study on Development and Utilization of Clean Coal Technology in China", *Fuel Processing Technology*, Vol. 89, No. 4, Elsevier, pp. 475-484.

Lu Xuedu (2006), "Experiences and Opportunities for CCS in China", presented at the UNFCC Workshop on Carbon Dioxide Capture and Storage, Bonn, Germany, 20 May, http://unfccc.int/files/meetings/sb24/in-session/application/pdf/experience_ and_opportunity_in_china_by_lu_xuedu.pdf.

Luo Wei and Joan Liu (2003), *A Complete Research Guide to the Laws of the People's Republic of China (PRC)*, Law Library Resource Xchange, Silver Spring, MD, 15 January, www.llrx.com/features/prc.htm.

Mao Xianqiang, Guo Xiurui, Chang Yongguan and Peng Yingdeng (2005), "Improving Air Quality in Large Cities by Substituting Natural Gas for Coal in China: Changing Idea and Incentive Policy Implications", *Energy Policy*, Vol. 33, No. 3, Elsevier, pp. 307-318.

Mao Xianqiang, Peng Yingdeng and Guo Xiurui (2002), "Cost-Benefit Analysis to Substituting Natural Gas for Coal Project in Large Chinese Cites", *Environmental Sciences*, Vol. 23, No. 5, Institute of Environmental Sciences, Beijing Normal University, pp. 121-125.

Martinot, E. (2001), "World Bank Energy Projects in China: Influences on Environmental Protection", *Energy Policy*, Vol. 29, No. 8, Elsevier, pp. 581-594.

Martinot, E. and Li Junfeng (2007), *Powering China's Development: The Role of Renewable Energy*, Worldwatch Institute, Washington, DC, November.

Masaki, T. (2003), "The World Bank Group's Perspectives and Cases of Cleaner Coal Technology Projects", *Energy and Environment*, Vol. 14, No. 1, Multi-Science Publishing Co. Ltd., pp. 51-57.

Mei, M. (2005), *Coal Mining Equipment Market in China*, US Commercial Service, US Department of Commerce, www.buyusainfo.net/docs/x_2606339.pdf.

Meng, K. C., R. H. Williams and M. A. Celia (2007), "Opportunities for Low-Cost CO_2 Storage Demonstration Projects in China", *Energy Policy*, Vol. 35, No. 4, Elsevier, pp. 2368-2378.

Morrison, K. (2008), *Living in a Material World: The Commodity Connection*, Wiley.

MOST (Ministry of Science and Technology) (2007), 《863计划资源环境技术领域2007年度专题课题申请指南》 ("863 Program –Application Guidelines for Proposals in the Fields of Resource Exploitation and Environmental Technologies"), 27 March, MOST, Beijing, www.most.gov.cn/tztg/200703/t20070327_42369.htm.

MOST (2007), *China's Scientific and Technological Actions on Climate Change*, MOST / National Development and Reform Commission / Ministry of Foreign Affairs / Ministry of Education / Ministry of Finance / Ministry of Water Resources / Ministry of Agriculture / State Environmental Protection Administration / State Forestry Administration / Chinese Academy of Sciences / China Meteorology Administration / National Natural Science Foundation / State Oceanic Administration / China Association for Science and Technology, Beijing, June.

NDRC (National Development and Reform Commission) (2006), 《火电厂烟气脱硫工程后评估费用计算标准(试行)》 ("Standard for the Assessment of FGD Project Costs" [trial implementation]), NDRC, Beijing, 21 March, www.cec.org.cn/news/showc.asp?id=28783.

NDRC (2006), 《我国火电厂烟气脱硫产业化现状及有关建议》 ("Current Status and Future Prospects of Flue Gas Desulphurisation Equipment Suppliers to Thermal Power Plants in China"), NDRC, 4 September, www.ndrc.gov.cn/hjbh/huanjing/t20060904_82826.htm.

Neil, C., M. Tykkyläinen and J. Bradbury (eds.) (1992), *Coping with Closure: An International Comparison of Mine Town Experiences*, Routledge, London.

Nelson, R. H. (1983), *The Making of Federal Coal Policy*, Duke University Press, Durham, NC.

Ni Weidou (2007), "China's Energy – Challenges and Strategies", *Frontiers of Energy and Power Engineering in China*, Vol. 1, No. 1, Higher Education Press and Springer-Verlag, pp. 1-8.

Ni Weidou and T. B. Johansson (2004), "Energy for Sustainable Development in China" *Energy Policy*, Vol. 32, No. 10, Elsevier, pp. 1225-1229.

Nolan, P. and Rui Huaichuan (2004), "Industrial Policy and Global Big Business Revolution: the Case of the Chinese Coal Industry", *Journal of Chinese Economic and Business Studies*, Vol. 2, No. 2, Routledge, pp. 97-113.

OECD (Organisation for Economic Co-operation and Development) (2005), *OECD Economic Surveys: China*, OECD, Paris.

ONELG (国家能源领导小组办公室 – Office of the National Energy Leading Group) (2007), 《浪费岂能持久—我国煤炭资源回采率问题调查》 ("How Much Waste – Survey of Coal Resource Recovery Rates in China"), ONELG, Beijing, 6 November, www.chinaenergy.gov.cn/news_21058.html.

Oskarsson, K., A. Berglund, R. Deling, U. Snellman, O. Stenbäck and J. Fritz (1997), *A Planner's Guide for Selecting Clean-Coal Technologies for Power Plants*, World Bank Technical Paper No. 387, World Bank, Washington, DC, November.

Pan Kexi (2005), "The Depth Distribution of Chinese Coal Resources", presented at Global Climate and Energy Project (GCEP) International Workshop on Exploring the Opportunities for Research to Integrate Advanced Coal Technologies with CO_2 Capture and Storage in China, Tsinghua University, Beijing, 22-23 August, http://gcep.stanford.edu/pdfs/wR5MezrJ2SJ6NfFl5sb5Jg/10_china_pankexi.pdf.

Pittman, R. W. and Vanessa Yanhua Zhang (2008), *Electricity Restructuring in China: The Elusive Quest for Competition*, discussion paper, Economic Analysis Group, US Department of Justice, Washington, DC, April, www.usdoj.gov/atr/public/eag/232668.pdf.

Qian Jinging and Fan Yaqing (2006), "Summary of Carbon Storage Potential and Activities in China" (《中国碳封存潜力和有关活动概要》), presented at US-China Clean Coal Forum, Taiyuan, Shanxi, 12-13 September, www.chinacleanenergy.org/docs/general/CarbonStoragePotential.pdf.

Rabanal, N. G. (2003), "Distinctive Features of Coal Reconversion in the European Union", *Applied Energy*, Vol. 74, No. 3, Elsevier, pp. 281-287.

Rui Huaichuan (2004), *Globalization, Transition and Development in China: The Case of the Coal Industry*, Routledge.

Sagawa, A. and K. Koizumi (2007), *Present State and Outlook of China's Coal Industry*, Institute of Energy Economics, Tokyo, Japan, December.

Senior, B. (2006), "Addressing the Challenge of Coal Use in China through Carbon Capture and Storage", presented at Italy-UK Clean Coal/CCS Workshop, London, 3 February, www.ukerc.ac.uk/Downloads/PDF/C/clean_coal__senior.pdf.

SEPA (国家环境保护总局 – State Environmental Protection Agency) (2003), 火电厂大气污染物排放标准 *(Emission Standard for Air Pollutants from Thermal Power Plants)*, GB 13223-2003 代替 GB 13223-1996, 2004-1-1 实施 (GB 13223-2003 replacing GB 13223-1996, effective 1 January 2004), SEPA, Beijing, 23 December, www.mep.gov.cn/image20010518/5297.pdf.

SEPA (2008), 清洁生产标准煤炭采选业 *(Cleaner Production Standard: Coal Mining and Processing Industry)*, State Environmental Standard No. HJ 446-2008, SEPA, Beijing, 21 November, www.sepa.gov.cn/info/bgw/bgg/200811/W020081125540373667202.pdf.

SERC (State Electricity Regulatory Commission) (2007), 2007 年电力行业节能减排情况报告（摘要）*(Energy Saving and Emission Reductions in the Electric Power Industry in 2007 [Summary])*, SERC Research Group, Beijing, 15 April, www.serc.gov.cn/jgyj/ztbg/200804/t20080415_8883.htm.

Shen Lei and P. Andrews-Speed (2001), "Economic Analysis of Reform Policies for Small Coal Mines in China", *Resources Policy*, Vol. 27, No. 4, Pergamon, pp. 247–254.

Sinton, J. E., R. E. Stern, N. T. Aden and M. D. Levine (2005), *Evaluation of China's Energy Strategy Options*, Paper No. LBNL-56609, Lawrence Berkeley National Laboratory, University of California, Berkeley, CA, 16 May.

Skeer, J. and Wang Yanjia (2006), "Carbon Charges and Natural Gas use in China", *Energy Policy*, Vol. 34, No. 15, Elsevier, pp. 2251-2262.

SPERI (国网北京经济技术研究院 – State Power Economic Research Institute) (2004), 促进中国提高能效的电力定价研究 (*Study on Promoting Energy Efficiency in China through the Electricity Pricing Mechanism*), SPERI (原：国电动力经济研究中心 – formerly State Power Economic Research Center), 国家电网公司 (State Grid Corporation), Beijing, www.chinasperi.com.cn/upload/attachment/2007829144814185.pdf.

State Council (2007), 《节能减排综合性工作方案的通知》 ("Notice on a Comprehensive Energy-Saving Emission-Reduction Programme"), 国发[2007]15号 (Guo Fa [2007] No. 15), Office of the State Council, Beijing, 2007年6月3日 (3 June 2007), www.gov.cn/zwgk/2007-06/03/content_634545.htm.

State Council (2007), *China's Energy Conditions and Policies*, White Paper, Information Office of the State Council, December, http://en.ndrc.gov.cn/policyrelease/P020071227502260511798.pdf.

Steenblik, R. P. and P. Coroyannakis (1995), "Reform of Coal Policies in Western and Central Europe", *Energy Policy*, Vol. 23, No. 6, Butterworth Heinemann, pp. 537-553.

Steenhof, P. A. and W. Fulton (2007), "Factors Affecting Electricity Generation in China: Current Situation and Prospects", *Technological Forecasting and Social Change*, Vol. 74, No. 5, Elsevier, pp. 663-681.

Steenhof, P. A. and W. Fulton (2007), "Scenario Development in China's Electricity Sector", *Technological Forecasting and Social Change*, Vol. 74, No. 6, Elsevier, pp. 779-797.

Steinfeld, E. S., R. K. Lester and E. A. Cunningham (2008), *Greener Plants, Grayer Skies? A Report from the Front Lines of China's Energy Sector*, China Energy Group, MIT Industrial Performance Center, Massachusetts Institute of Technology, Cambridge, MA, August, http://web.mit.edu/ipc/publications/pdf/08-003.pdf.

Stewart, T. A. (2005), *China's Coal Industry: Evolution and Opportunities*, Battelle Memorial Institute for the McCloskey Group, Columbus, OH, 1 June.

Stracher, G. B. and T. P. Taylor (2004), "Coal Fires Burning Out of Control Around the World: Thermodynamic Recipe for Environmental Catastrophe", *International Journal of Coal Geology*, Vol. 59, No. 1-2, Elsevier, pp. 7-17.

Suding, P. H. (2005), *China's Energy Supply: Many Paths – One Goal*, German Member Committee (DNK), World Energy Council, Berlin, Germany, May.

Sun Guodong (2005), "Advanced Coal Technologies in a Sustainable Energy System: Preparing and Preserving the Technological Options in China", Belfer Center for Science and International Affairs, Harvard University, Cambridge, MA, http://belfercenter.ksg.harvard.edu/files/Harvardmostactworkshop.pdf.

Suwala, W. and W. C. Labys (2002), "Market Transition and Regional Adjustments in the Polish Coal Industry", *Energy Economics*, Vol. 24, No. 3, Elsevier, pp. 285-303.

Tao Zaipu and Li Mingyu (2007), "What is the Limit of Chinese Coal Supplies – A STELLA Model of Hubbert Peak", *Energy Policy*, Vol. 35, No. 6, Elsevier, pp. 3145-3154.

Thomson, E. (2003), *The Chinese Coal Industry: An Economic History*, Routledge.

Torrens, I. M. (2007), *National and International Activities Related to CCS and ZETs in China*, working paper presented at a meeting of the IEA Working Party on Fossil Fuels, Brasilia, 28-29 June.

Tsinghua University and CCAP (Center for Clean Air Policy) (2006), *Greenhouse Gas Mitigation in China: Scenarios and Opportunities through 2030*, Tsinghua University, Beijing and CCAP, Washington, DC, November, www.ccap.org/docs/resources/61/Final_China_Report_(Nov_2006).pdf.

Tu JianJun (2007), "Safety Challenges in China's Coal Mining Industry", *China Brief*, Vol. 7, No. 1, Jamestown Foundation, pp. 6-8.

Tu JianJun (2007), "Coal Mining Safety: China's Achilles' Heel", *China Security*, Vol. 3, No. 2, World Security Institute, pp. 36-53.

USAID (United States Agency for International Development) (2007), *From Ideas to Action: Clean Energy Solutions for Asia to Address Climate Change – Annex 1: China Country Report*, USAID, Bangkok, Thailand, 22 June, http://usaid.eco-asia.org/programs/cdcp/reports/Ideas-to-Action/annexes/Annex%201_China.pdf.

USAID (2007), *Designing a Cleaner Future for Coal: Solutions for Asia that Address Climate Change*, USAID, Bangkok, Thailand, October, www.cleanenergyasia.net/upload/resources/file/file_93.pdf.

US-China Joint Economic Research Group (2007), *US-China Joint Economic Study: Economic Analyses of Energy Saving and Pollution Abatement Policies for the Electric Power Sectors of China and the United States*, Summary for Policy Makers, US-China Strategic Economic Dialogue, US Environmental Protection Agency, Washington, DC, November, www.epa.gov/airmarkets/international/china/JES_Summary.pdf.

Vennemo, Haakon (2008), *International Experiences in Air Pollution Control* (有关空气污染控制的国际经验), Centre for Law and Economics for Environment and Development, University of Cambridge, UK, www.landecon.cam.ac.uk/research/eeprg/cleed/researchprojects/china_environmentstrategy.htm (summary available in Chinese).

Wang, A. L., B. Finamore and C. Williams (2007), *Environmental Governance in China: Recommendations for Reform from International Experience*, Natural Resources Defense Council, New York, June.

Wang Bing (2007), "An Imbalanced Development of Coal and Electricity Industries in China", *Energy Policy*, Vol. 35, No. 10, Elsevier, pp. 4959-4968.

Wang Hao and T. Nakata (2009), "Analysis of the Market Penetration of Clean Coal Technologies and its Impacts in China's Electricity Sector", *Energy Policy*, Vol. 37, No. 1, Elsevier, pp. 338-351.

Wang Mingyuan (王明远) (2007), 中国清洁发展机制（CDM）监管的法律分析：虚假的法治主义与真实的政府干预主义 ("A Legal Analysis of China's Clean Development Mechanism Supervision: A Loosely Defined Law and the Reality of Government Intervention"), paper presented at Shaping China's Energy Security: The Environmental Challenge (打造中国能源安全：环境挑战) seminar, hosted by the Asia Centre (*Centre études Asie à Sciences Po*, Paris), Kempinski Hotel, Beijing, 1 December.

Wang Qingyi (2000), *Coal Industry in China: Evolvement and Prospects*, Nautilus Institute for Security and Sustainability, Berkeley, CA, www.nautilus.org/archives/energy/eaef/C5_final.PDF.

Wang Yanjia (2006), *Energy Efficiency Policy and CO$_2$ in China's Industry: Tapping the Potential*, paper presented at Working Together to Respond to Climate Change: Annex I Expert Group Seminar in conjunction with the OECD Global Forum on Sustainable Development, Paris, 27-28 March, www.oecd.org/dataoecd/58/28/36321399.pdf.

Wang Yi, Xiao Yunhan, Zhang Shijie, Zhou Honghchun and Qian Jingjing (2004), *Strategic Direction in Clean Utilization of Coal: Advantages of Coal Gasification-Based Polygeneration and the Barriers to and Options for its Development in China*, Chinese Academy of Sciences, Beijing for Natural Resources Defense Council, New York, September.

Watson, J. (2005), "Rising Sun: Technology Transfer in China", *Harvard International Review*, Vol. 26, No. 4, Winter, http://hir.harvard.edu/articles/1295/.

Watson, J., Liu Xue, G. Oldham, G. MacKerron and S. Thomas (2000), *International Perspectives on Clean Coal Technology Transfer to China*, Final Report to the Working Group on Trade and Environment, China Council for International Cooperation on Environment and Development (CCICED), Beijing, August.

World Bank (2001), *China: Air, Land, and Water – Environmental Priorities for a New Millennium*, World Bank, Washington, DC, August.

World Bank (2007), *Catalyzing Private Investment for a Low-Carbon Economy: World Bank Group Progress on Renewable Energy and Energy Efficiency in Fiscal 2007*, World Bank, Washington, DC, November, http://siteresources.worldbank.org/INTENERGY/Resources renewableenergy12407SCREEN.pdf.

Wright, T. (2004), "The Political Economy of Coal Mine Disasters in China: Your Rice Bowl or Your Life", *The China Quarterly*, No. 179, Cambridge University Press, pp. 629-646.

Xinhua Info Link and The McCloskey Group (2007), *China's Coal Industry 2007: Production, Consumption and Outlook*, The McCloskey Group, Petersfield, UK.

Xu Yi-chong (2002), *Powering China: Reforming the Electric Power Industry in China*, Ashgate Dartmouth, Farnham, UK, June.

Yan X. and R. J. Crookes (2007), "Study on Energy Use in China", *Journal of the Energy Institute*, Vol. 80, No. 2, Maney Publishing, pp. 110-115.

Ye Jianping, Feng Sanli, Fan Zhiqiang, B. Gunter, S. Wong and D. Law (2005), "CO_2 Sequestration Potential in Coal Seams of China", presented at Global Climate and Energy Project (GCEP) International Workshop on Exploring the Opportunities for Research to Integrate Advanced Coal Technologies with CO_2 Capture and Storage in China, Tsinghua University, Beijing, 22-23 August, http://gcep.stanford.edu/pdfs/wR5MezrJ2SJ6NfFl5sb5Jg/14_china_jiangping.pdf.

Yu Hongguan, Zhou Guangzhu, Fan Wietan and Ye Jianping (2007), "Predicted CO_2 Enhanced Coalbed Methane Recovery and CO_2 Sequestration in China", *International Journal of Coal Geology*, Vol. 71, No. 2-3, Elsevier, pp. 345-357.

Yu Huanzhang, Jiang Renzhong and Xu Yani (2007), *China's Coal-Fired Power Plants and Environment*, Department of Environment, Technology and Social Studies (TekSam), Roskilde University, Denmark.

Yu Zhufeng and Yu Jie (2001), "Policy Study in the Development of Clean Coal Technology", paper presented at International Conference on Cleaner Production in China, Beijing, 4-5 September, www.chinacp.org.cn/eng/cpconfer/iccp01/iccp30.html.

Yu Zhufeng and Zheng Xingzhou (2005), "Coal Market Outlook in China", *International Journal of Global Energy Issues*, Vol. 24, No. 3-4, Inderscience, pp. 211-227.

Zhang Chi, T. C. Heller and M. M. May (2005), "Carbon Intensity of Electricity Generation and CDM Baseline: Case Studies of Three Chinese Provinces", *Energy Policy*, Vol. 33, No. 4, Elsevier, pp. 451-465, (drafted as "Electricity Industry Development and Global Warming Impact: Case Studies of Three Chinese Provinces", Institute for International Studies, Stanford University for EPRI (under Contract 192E032) and Bechtel Initiative on Global Growth and Change, January 2003, http://powermin.nic.in/research/pdf/chinese_provinces.pdf).

Zhang Liang and Huang Zhen (2007), "Life Cycle Study of Coal-Based Dimethyl Ether as Vehicle Fuel for Urban Bus in China", *Energy*, Vol. 32, No. 10, Elsevier, pp. 1896–1904.

Zhang Ruihe (2007), "Should China Develop Coal to Oil?", *China Chemical Reporter*, Vol. 18, No. 34, China National Chemical Information Center, pp. 20-21.

Zhang Yue (2006), *China's "11th Five-Year Guidelines" with a Focus on Energy Policy*, Institute of Energy Economics, Tokyo, Japan, April, http://eneken.ieej.or.jp/en/data/pdf/327.pdf.

Zhao Lifeng and K. S. Gallagher (2007), "Research, Development, Demonstration, and Early Deployment Policies for Advanced-Coal Technology in China", *Energy Policy*, Vol. 35, No. 12, Elsevier, pp. 6467-6477.

Zhao Lifeng, Xiao Yunhan, K. S. Gallagher, Wang Bo and Xu Xiang (2008), "Technical, Environmental, and Economic Assessment of Deploying Advanced Coal Power Technologies in the Chinese Context", *Energy Policy*, Vol. 36, No. 7, Elsevier, pp. 2709-2718.

Zhou Nan, M. A. McNeil, D. Fridley, Jiang Lin, L. Price, S. de la Rue du Can, J. Sathaye and M. Levine (2007), *Energy Use in China: Sectoral Trends and Future Outlook*, Paper No. LBNL-61904, Lawrence Berkeley National Laboratory, University of California, Berkeley, CA, January, http://repositories.cdlib.org/cgi/viewcontent.cgi?article=6369&context=lbnl.

Zhu Fahua, Zheng Youfei, Guo Xulin and Wang Sheng (2005), "Environmental Impacts and Benefits of Regional Power Grid Interconnections for China", *Energy Policy*, Vol. 33, No. 14, Elsevier, pp. 1797-1805.

The Online Bookshop

International Energy Agency

/books

IEA B

Tel: +33 (0)
Fax: +33 (o)
E-mail: bo

ergy Agency
ration
ex 15, France

IEA PUBLICATIONS, 9, rue de la Fédération, 75739 PARIS CEDEX 15

PRINTED IN FRANCE BY STEDI MEDIA, April 2009

(61 2008 24 1P1) ISBN: 978-92-64-04814-0